DEGREES AND RADIANS

π rad $= 180°$
1 rad $\approx 57.29578°$
    $1° \approx 0.01745329$ rad
    $\pi \approx 3.14159\ 2654$

SPECIAL ANGLES

$\theta°$	$0°$	$15°$	$30°$	$45°$	$60°$	$75°$	$90°$
$\sin \theta°$	0	$\tfrac{1}{4}(\sqrt{6} - \sqrt{2})$	$\tfrac{1}{2}$	$\tfrac{1}{2}\sqrt{2}$	$\tfrac{1}{2}\sqrt{3}$	$\tfrac{1}{4}(\sqrt{6} + \sqrt{2})$	1
$\cos \theta°$	1	$\tfrac{1}{4}(\sqrt{6} + \sqrt{2})$	$\tfrac{1}{2}\sqrt{3}$	$\tfrac{1}{2}\sqrt{2}$	$\tfrac{1}{2}$	$\tfrac{1}{4}(\sqrt{6} - \sqrt{2})$	0
$\tan \theta°$	0	$2 - \sqrt{3}$	$\tfrac{1}{3}\sqrt{3}$	1	$\sqrt{3}$	$2 + \sqrt{3}$	undef.

BASIC IDENTITIES

$$\tan \theta = \frac{\sin \theta}{\cos \theta}$$

$$\cot \theta = \frac{1}{\tan \theta} = \frac{\cos \theta}{\sin \theta}$$

$$\sec \theta = \frac{1}{\cos \theta}$$

$$\csc \theta = \frac{1}{\sin \theta}$$

$$\sin^2 \theta + \cos^2 \theta = 1 \qquad 1 + \tan^2 \theta = \sec^2 \theta \qquad 1 + \cot^2 \theta = \csc^2 \theta$$

$$\sin(\alpha + \beta) = \sin \alpha \cos \beta + \cos \alpha \sin \beta$$
$$\cos(\alpha + \beta) = \cos \alpha \cos \beta - \sin \alpha \sin \beta$$

$$\tan(\alpha + \beta) = \frac{\tan \alpha + \tan \beta}{1 - \tan \alpha \tan \beta}$$

$$\sin 2\theta = 2 \sin \theta \cos \theta$$
$$\cos 2\theta = \cos^2 \theta - \sin^2 \theta$$

$$\tan 2\theta = \frac{2 \tan \theta}{1 - \tan^2 \theta}$$

SOLVING TRIANGLES

Right triangle

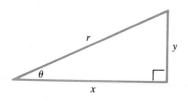

Oblique triangle

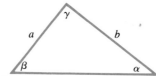

Law of sines

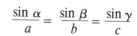

$$\frac{\sin \alpha}{a} = \frac{\sin \beta}{b} = \frac{\sin \gamma}{c}$$

Law of cosines

$$c^2 = a^2 + b^2 - 2ab \cos \gamma$$

$\cos \theta = x/r \qquad \sec \theta = r/x$
$\sin \theta = y/r \qquad \csc \theta = r/y$
$\tan \theta = y/x \qquad \cot \theta = x/y$

TRIGONOMETRY

TRIGONOMETRY

**Harley
Flanders**
Florida Atlantic University

and

**Justin J.
Price**
Purdue University

Flanders/Price Series 2nd Edition

SAUNDERS COLLEGE PUBLISHING

Philadelphia New York Chicago
San Francisco Montreal Toronto
London Sydney Tokyo Mexico City
Rio de Janeiro Madrid

Address orders to:
383 Madison Avenue
New York, NY 10017

Address editorial correspondence to:
West Washington Square
Philadelphia, PA 19105

This book was set in Times Roman by Progressive Typographers, Inc.
The editors were Jim Porterfield and Janis Moore.
The art director was Richard L. Moore.
The cover design was done by Adrianne Onderdonk Dudden.
The art was drawn by Linda Savalli.
The production manager was Tom O'Connor.
Fairfield Graphics was the printer.

TRIGONOMETRY 2/e   ISBN   0-03-057802-7

2345   144   987654321

CBS COLLEGE PUBLISHING
Saunders College Publishing
Holt, Rinehart and Winston
The Dryden Press

PREFACE

There are different ways to cover this text. The chart (see next page) shows the dependence of each chapter on previous ones. Chapters 1 to 8 include the central topics of a trigonometry course, plus a few optional sections. Chapters 9 to 12 provide a selection of optional material to be used as time allows or for special assignments.

A course that stresses the trigonometric functions might follow the chapter order

     1 2 3 4 7 8 5 6.

A course that requires applications early might use the order

     1 2 3 5 6 4 7 8,

while a reasonable compromise is

     1 2 3 5 4 6 7 8.

OBJECTIVES

The main objectives of this course are to learn

(1) the trigonometric functions: their graphs, their properties, their inverse functions,
(2) how to solve right and oblique triangles,
(3) how to do scientific calculation,
(4) how to model a geometric or real-life problem: set-up, solution, numerical answer.

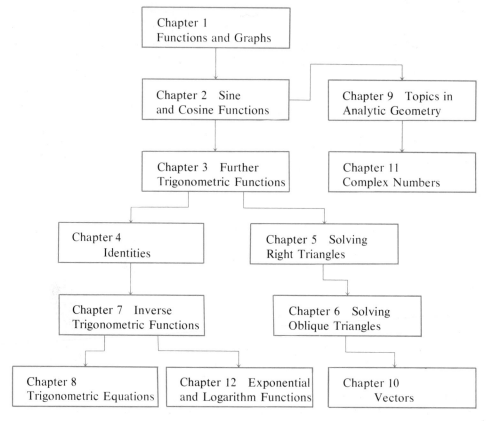

Notes
1. Chapter 5 requires only Section 1 from Chapter 3.
2. Chapter 10 requires only Section 1 from Chapter 6.
3. Chapter 11 requires only Section 1 from Chapter 9, but it also requires the addition laws from Chapter 4, Section 3.
4. Chapter 12 requires only Sections 1 and 2 of Chapter 7, and these can be taken directly after Chapter 1.

    All four are essential components of the course, but they are not equal in time requirements. Scientific calculation can be taught quickly and painlessly these days. It is correct for a trigonometry course because there is plenty of calculation to be done; it is generally not taught in other mathematics courses. Triangle solving is the next easiest and briefest to teach, because it is fairly mechanical. Setting up applied problems is hard for students because of little previous experience and a cultural fear of "story problems." Yet it is a most important skill, of immense value in subsequent calculus, physics, and engineering courses. Finally, learning the functions is hard because of many new concepts and because considerable agility with algebra is required for proving identities.

    The prerequisites are a year of geometry and $1\frac{1}{2}$ to 2 years of high school algebra or one quarter or semester of college-level algebra.

FEATURES OF THIS EDITION

We have rewritten about 75% of the text, mostly to improve its teachability, and have added many realistic applications.

Examples

The examples form the most important part of the text. Through them the student learns how to do mathematics himself rather than be told about it. There are many more informal examples (set off with the heading *Example*) than in the first edition, and the number of formal (numbered) worked examples has been increased 201%, from 76 to 153. A particular effort has been made to illustrate every type of exercise.

Exercises and Answers

The number of exercises has been increased 69%, from 1214 to 2048. Almost all of this increase has been at the easy and middle levels of difficulty. As always, we try to make the exercises interesting and to make the applications realistic and relevant.

While the production of the book was in the galley proof stage, we ourselves wrote the Solutions Manual (available to instructors), which includes a solution for every exercise in the text. In this way the exercise sections themselves were thoroughly shaken out, and should be error-free. An accuracy reviewer independently checked all solutions, so the *Solutions Manual* is guaranteed reliable, as is the *Answer Section* for odd-numbered exercises (printed in the back of the book and including all graphs).

Each chapter of the book closes with a section of about 10 review exercises.

Figures

We have taken great pains to have the art work simple, uncluttered, and clear. Our goal was figures that students can read easily and reproduce themselves, and that illustrate what they are supposed to. In the text alone there are 48% more illustrations than in the first edition.

Computations

Up to about 10 years ago, hand computation using log tables and log-trig tables was part of almost every trig course, generally hated by students. As hand-held scientific calculators became available, first at high prices, gradually at reasonable prices, the texts have changed their approach to tables and to instruction in calculation. Degrees and minutes have gradually shifted to decimal degrees, and five-place tables have given way to four. (One edition a few years ago gave six-place tables, as if *that* would answer calculators!) Today almost every text gives some calculator material. A few continue the tradition of teaching triangle solving by hand calculation using tables, and include the necessary log, trig, and log-trig tables. This uses up two to three weeks of class time, but we can find very few instructors who actually cover the topic.

These days the typical approach to computations in trig texts is this. Calculators are mentioned with little work on how to use them. Some tables are included, perhaps out of fear that traditionalists will resent their absence. Little or no instruction is given on how to do the computation of triangle solving. Thus, one will have a text with a table of trig functions, but no log-trig tables. The student is not *taught* how to carry out the computation for something like

$$53.68 \, \frac{\sin 21.49°}{\sin 48.83°}$$

but is merely told that from the tables $\sin 21.49° \approx 0.3663$, $\sin 48.83° \approx 0.7528$, and the answer is 26.12. Is the student supposed to know how to do this from a previous course? We think he or she should be taught how to use a scientific calculator in *this* course. This takes little time and effort; in fact, calculator techniques can be slipped in as needed without fuss. At the end of the course the student will finally know how to exploit the calculator and what a powerful tool it is.

Our 1975 algebra and trigonometry series were the very first precalculus books to introduce calculators. In our opinion, hand calculations using log and trig tables are obsolete; it is a waste of students' time to drill them on skills that they will never use in life. We think the time has come to declare the use of tables as a computation tool dead and buried. The calculations required in courses at this level are done on a hand-held scientific calculator. Calculations are now so easy that we can use concrete numerical evidence and experimentation freely in a way no one imagined possible 15 years ago. We believe numerical evidence should be used to reinforce theoretical discussions when possible. For instance, the theoretical discussion on the existence of a positive root of $\sin \theta = \theta^2$ is greatly helped by a few keystrokes leading to 0.876726 as its six-place approximation.

As an innovation, our appendix looks to the present and the near future rather than to the past: some computer programs for solving triangles, rather than numerical tables. It will not be long until almost all students have access to a programmable calculator, a microcomputer, or a terminal, and will use such programs routinely.

Design

The design of this book, evolved from four previous designs of our textbooks, has two main purposes. First, the student always knows precisely where he is. For instance, important statements are boxed. Worked examples begin with a clear statement of the *problem*. Then the *solution* is set off. Finally, the *answer* is stated. The student knows the example has ended because there is a line down the left side from its beginning to its end. It is important for the student to do his own work this way: state the problem clearly, solve it, state the answer that is called for in the problem.

In the text, *italics* are used for stress. Defined terms are in **boldface,** so students can spot these terms easily. In general, the design is open while not wasteful of space.

The other main purpose of the design is to make mathematical expressions and formulas easy to read. This involves much detail on the horizontal and vertical spacing of the math. We have replaced most of the English punctuation marks (which often obscure math) by large spaces. Whenever in the body of a paragraph a math symbol is next to a word, the space separation is 50% more than the separation between words. We hope this quality feature (done over the printer's dead body!) will set a new standard. Careful scrutiny will show other details, such as extra spaces between math symbols and punctuation marks.

CONTENTS

The book splits naturally into three main parts.

Functions

This is the core of the volume. The main tool for understanding a function is its graph, and emphasis is placed on graphing.

Chapter 1 reviews cartesian coordinates, real functions and their graphs, and constructing new functions from old, particularly by composition.

Chapter 2 introduces the trigonometric functions. We begin with angles in the domain $0° \leq \theta \leq 360°$ in standard position and define $\sin \theta = y/r$, etc. As quickly as possible we compute the functions of $45°$, $30°$, etc., and plot rough graphs. This occupies the first three sections; in the fourth we introduce radian measure and $\sin \theta$, etc. for any real θ, essentially by periodicity. Thus quickly, without fuss and philosophy, we have the functions and are able to work with them.

We shun universal covering spaces (wrapping function), which we consider far too theoretical for this level.

Chapter 3 covers the other four trig functions, their graphs, and periodic functions in general, including non-trigonometric examples. Approximations like $\cos \theta \approx 1 - \frac{1}{2}\theta^2$ near 0 are discussed, with calculator experiments.

Chapter 4 covers the standard trigonometric identity material—it is certainly hard to say anything new here. We do claim some restraint—there are hundreds, not thousands, of exercises—consistent with the relative importance of the topic. (It *is* possible to go overboard.) We include an optional, purely descriptive, section on how trigonometric identities are important in studying vibrations. Beats and audio modulation are discussed.

In Chapter 7, the inverse trigonometric functions are introduced and studied via their graphs.

Practical Trigonometry

Chapters 5 and 6 are on triangle solving, the first on right triangles and the second on oblique triangles. While degrees, minutes, and seconds are introduced, most numerical work is in decimal degrees. There are numerous interesting and practical applications, far more variety than usual, and far more realism. We include flowcharts for the law of sines and a neat proof of this law via area. There is an optional section on the tools of practical distance and angle measurement and their accuracy, including problems on sensitivity.

Chapter 8 covers trigonometric equations. Besides the standard material on periodic equations, there is some calculator work on root finding with applications to equations involving periodic and non-periodic elements.

Chapter 9 covers polar coordinates, polar graphs, and ellipses and hyperbolas, whose equations are derived first in polar form.

Chapter 10 covers vectors, inner products, and normal form.

Chapter 11 is the standard material on complex numbers through DeMoivre's theorem.

Chapter 12 on exponential and logarithm functions is included because this other family of elementary transcendental functions is sometimes covered in trig courses. It seems unrelated to the family of trig functions—until you get to the complex variable level, another story. Traditionally, logs had to be covered in trig because of their use in computation. Today there is no such obligation.

ACKNOWLEDGMENTS

We are grateful to Rudy Svoboda (Indiana University–Purdue University at Fort Wayne), Paul R. Swanson (Bismark Junior College), and Wesley W. Tom (Chaffey College), who reviewed the manuscript of this edition and supplied much useful criticism.

The staff at Saunders College Publishing has been most helpful and professional, particularly editors William Karjane, James Porterfield, Jay Freedman, and Janis Moore, and artist Linda Savalli.

Harley Flanders
Justin J. Price

CONTENTS OVERVIEW

TABLE OF CONTENTS

1 FUNCTIONS AND GRAPHS

1 COORDINATES IN THE PLANE

A large part of this course is devoted to the family of trigonometric functions. Much of this study uses the geometric tool of the graphs of these functions. So we shall begin our work with a review of coordinates and functions and their graphs.

When the points of a line are specified by real numbers, we say that the line is **coordinatized:** each point has a label or **coordinate.** It is possible also to label, or coordinatize, the points of the plane. Let us describe the most common way of doing this.

We start by drawing two perpendicular lines in the plane. We mark their intersection 0 and coordinatize each line as shown in Fig. 1a. We draw one line horizontal and name it the x-axis, the other line vertical and name it the y-axis.

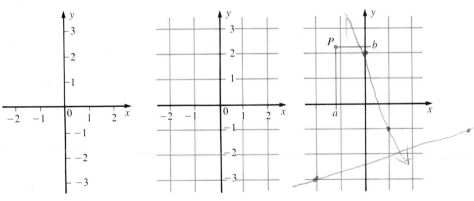

(a) coordinate axes in the plane   (b) rectangular grid   (c) coordinates of a point

Fig. 1

1

Consider all lines parallel to the *x*-axis and all lines parallel to the *y*-axis (Fig. 1b). These two systems of parallel lines impose a rectangular grid on the whole plane. We use this grid to coordinatize the points of the plane.

Take any point *P* of the plane. Through *P* pass one vertical line and one horizontal line (Fig. 1c). They meet the axes in points *a* and *b* respectively. We associate with *P* the ordered pair (a, b); it completely describes the location of *P*. The number *a* is the **x-coordinate** of *P*, and the number *b* is the **y-coordinate** of *P*.

Conversely, take any ordered pair (a, b) of real numbers. The vertical line through *a* on the *x*-axis and the horizontal line through *b* on the *y*-axis meet in a point *P* whose coordinates are precisely (a, b). Thus there is a one-to-one correspondence

$$P \longleftrightarrow (a, b)$$

between the set of points of the plane and the set of all ordered pairs of real numbers.

A coordinate system in the plane such as we have just described is called a **rectangular** or **cartesian** coordinate system. The point $(0, 0)$ is called the **origin.**

Remark 1   The pair (a, b) is also sometimes called (ungrammatically) the **coordinates** of *P*. Some writers refer to the *x*-coordinate of a point as its **abscissa** and the *y*-coordinate as its **ordinate.**

Remark 2   It is common practice to speak of "the point (a, b)". Logically, this is incorrect since an ordered pair of real numbers is not a point. Still, this slight inaccuracy rarely causes any confusion, but often saves some fuss.

The coordinate axes divide the plane into four **quadrants,** which are numbered as in Fig. 2a. Note that the signs of its coordinates determine the quadrant in which a point lies. Points of the form $(a, 0)$ lie on the *x*-axis; those of the form $(0, b)$ lie on the *y*-axis. Figure 2b shows a few examples.

It is not compulsory to use the same scales on both axes. For example, in plotting U.S. population for the last 20 years, you would probably choose a reasonable length (say 1 cm) to represent one year on the horizontal axis. But you would use the same length to represent, say, 5,000,000 people on the vertical axis.

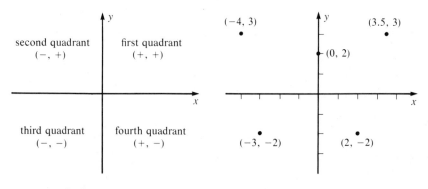

(a) the four quadrants   (b) some points plotted

Fig. 2

Subsets of the Plane

The part of the plane above the x-axis is the set of all points (x, y) for which $y > 0$. It can be described briefly by the simple inequality $y > 0$. Similarly, the part of the plane on and to the right of the y-axis is described by the inequality $x \geq 0$. Often a subset of the plane is conveniently described by an inequality or by a system of inequalities.

EXAMPLE 1   Sketch the set of all points (x, y) for which $y \leq 0$ and $|x| \geq 1$.

Solution   First we shade (Fig. 3) the points where $y \leq 0$, that is, the region on and *below* the x-axis. Then we shade the region where $|x| \geq 1$, which consists of two parts, the part on and to the right of the vertical line $x = 1$ (where $x \geq 1$), and the part on and to the left of the vertical line $x = -1$ (where $x \leq -1$). The region where both inequalities are satisfied is shaded twice.

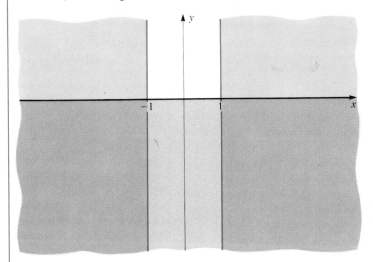

Fig. 3   The set where $y \leq 0$ *and* $|x| \geq 1$ is shaded dark.

The Distance Formula

Given two points (x_1, y_1) and (x_2, y_2) in the coordinate plane, what is the distance between them? We can answer this question easily by the Pythagorean theorem. See Fig. 4, next page.

No matter where the two points are located, we introduce the auxiliary point (x_2, y_1) forming a right triangle. Its legs have length $|x_2 - x_1|$ and $|y_2 - y_1|$, so by the Pythagorean theorem,

$$d^2 = |x_2 - x_1|^2 + |y_2 - y_1|^2 = (x_2 - x_1)^2 + (y_2 - y_1)^2.$$

Distance Formula   The distance between two points (x_1, y_1) and (x_2, y_2) is

$$\sqrt{(x_2 - x_1)^2 + (y_2 - y_1)^2}.$$

(The same unit of length on *both* axes is assumed.)

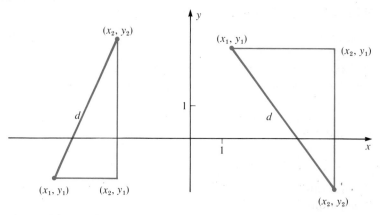

Fig. 4   Distance between two points (equal units on the two axes assumed)

If the points (x_1, y_1) and (x_2, y_2) lie on the same horizontal line or the same vertical line, their distance is $|x_1 - x_2|$ or $|y_1 - y_2|$; there is no need to introduce an auxiliary point. Nevertheless, the distance formula still yields the correct answer.

Examples

(a) distance from $(3, -1)$ to $(7, 2)$:
$$d^2 = (7 - 3)^2 + [2 - (-1)]^2 = 4^2 + 3^2 = 25 \qquad d = 5$$

(b) distance from $(2, 4)$ to $(-3, 1)$:
$$d^2 = (-3 - 2)^2 + (1 - 4)^2 = (-5)^2 + (-3)^2 = 34 \qquad d = \sqrt{34}$$

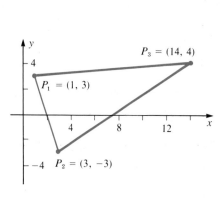

Fig. 5

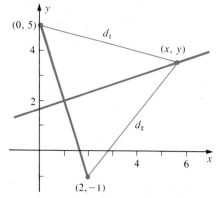

Fig. 6

EXAMPLE 2   Show that the three points $P_1 = (1, 3)$, $P_2 = (3, -3)$, and $P_3 = (14, 4)$ are the vertices of an isosceles triangle.

Solution   Plot the points (Fig. 5). If the triangle is actually isosceles, then the figure suggests that the equal sides are $\overline{P_1 P_3}$ and $\overline{P_2 P_3}$. Just check by the distance formula:

$$\overline{P_1P_3} = \sqrt{(14-1)^2 + (4-3)^2} = \sqrt{169+1} = \sqrt{170}$$

$$\overline{P_2P_3} = \sqrt{(14-3)^2 + (4+3)^2} = \sqrt{121+49} = \sqrt{170}.$$

Therefore $\overline{P_1P_3} = \overline{P_2P_3}$, so the triangle is isosceles.

Remark The solution of Example 2 shows how useful an accurate drawing can be.

EXAMPLE 3 Let (x,y) be any point on the perpendicular bisector of the segment joining $(0,5)$ and $(2,-1)$. Show that $y = \frac{1}{3}x + \frac{5}{3}$.

Solution Points on the perpendicular bisector are equidistant from $(0,5)$ and $(2,-1)$. Thus $d_1 = d_2$ in Fig. 6. Express d_1 and d_2 using the distance formula, and substitute into this equation. Better yet, to avoid radicals, substitute into the equivalent equation $d_1^2 = d_2^2$:

$$d_1^2 = d_2^2$$

$$(x-0)^2 + (y-5)^2 = (x-2)^2 + (y+1)^2$$

$$x^2 + y^2 - 10y + 25 = x^2 - 4x + 4 + y^2 + 2y + 1$$

$$-10y + 25 = -4x + 4 + 2y + 1$$

$$4x + 20 = 12y.$$

Solve for y:

$$y = \tfrac{1}{3}x + \tfrac{5}{3}.$$

Exercises

Plot and label the points on one graph

1 $(-4,1),(3,2),(5,-3),(1,4)$ **2** $(0,-2),(3,0),(-2,2),(1,-3)$

3 $(0.2,-0.5),(-0.3,0),(-1.0,-0.1)$ **4** $(75,-10),(-15,60),(95,40)$.

Choose suitable scales on the axes and label the points

5 $(150,0.3),(50,0.6)$ **6** $(-0.02,5),(0.03,8)$

7 $(0.1,-0.003),(-0.3,0.007)$ **8** $(-0.02,35),(0.00,-60)$.

Indicate on a suitable diagram all points (x,y) in the plane for which

9 $x = -3$ **10** $y = 2$

11 x and y are positive   **12** either x or y (or both) is zero

13 $1 \le x \le 3$ **14** $-1 \le y \le 2$

15 $-2 \le x \le 2$ and $-2 \le y \le 2$ **16** $x > 2$ and $y < 3$

17 both x and y are integers   **18** $x^2 > 4$

19 $|x| \ge 1$ and $|y| \le 2$ **20** $|x| \ge 2$ and $|y| \ge 2$

21 $xy > 0$ and $|x| \le 3$ **22** $|x| + |y| > 0$

23 $|x - 3| < 1$ and $y > 0$ **24** $|x - 1| < 2$ and $|y| < 1$

25 $x \ge 0$ and $|y - 2| = 1$ **26** $|x + 1| \le \frac{1}{2}$ and $|y - 1| \le 2$.

Write the coordinates (x,y) of the

27 vertices of a square centered at $(0,0)$, sides of length 2 and parallel to the axes

28 vertices of a square centered at $(1,3)$, sides of length 2, at 45° angles with the axes

29   vertices of a 3-4-5 right triangle in the first quadrant, right angle at $(0,0)$, hypotenuse of length 15, longer leg along the x-axis.

30   vertices of an equilateral triangle, sides of length 2, base on the x-axis, vertex on the positive y-axis.

Compute the distance between the points

31 $(4,1)$ and $(12,7)$ **32** $(-9,6)$ and $(7,-6)$
33 $(0,3)$ and $(8,5)$ **34** $(0,3)$ and $(0,11)$
35 $(-4,-1)$ and $(3,-3)$ **36** $(\frac{1}{2},\frac{3}{2})$ and $(2,5)$.

Show that the points are vertices of a right triangle

37 $(2,-3)\ (-2,1)\ (9,4)$ **38** $(-6,1)\ (-2,9)\ (4,-4)$.

Show that the points are vertices of a square

39 $(0,0)\ (3,4)\ (-1,7)\ (-4,3)$
40 $(-5,-6)\ (-12,18)\ (12,25)\ (19,1)$.

41   Recall that a **hexagon** is a six-sided polygon, and that a **regular** hexagon is one with all side lengths equal and all (internal) vertex angles equal. Show that the following points are the vertices of a regular hexagon with center at the origin:
$$(2,0)\ \ (1,\sqrt{3})\ \ (-1,\sqrt{3})\ \ (-2,0)\ \ (-1,-\sqrt{3})\ \ (1,-\sqrt{3}).$$

42   Verify that the distance between $(3x_1,3y_1)$ and $(3x_2,3y_2)$ is three times the distance between (x_1,y_1) and (x_2,y_2).

43   A circle has center $(1,-6)$ and radius 10. Another circle has center $(8,18)$ and radius 14. Do they intersect?

44   Show that the circle with center $(8,1)$ and radius 3 is contained in the circle with center $(3,2)$ and radius 9.

45   Find an equation that a point (x,y) must satisfy if it lies on the perpendicular bisector of the segment connecting $(0,0)$ and $(4,6)$.

46   Find all points on the x-axis whose distance from the point $(2,7)$ is 10.

47   Find all points on the line $y=2$ that are twice as far from $(3,0)$ as from the origin.

48   (cont.) Do the same for the line $y=4$.

49   Find all points at a distance 2 from the origin and having 1 as one of their coordinates.

50   (**Midpoint formula**) Using the distance formula, verify that the midpoint of the segment connecting (x_1,y_1) and (x_2,y_2) is $(\frac{1}{2}(x_1+x_2),\frac{1}{2}(y_1+y_2))$.

51   Show that the point $(6,9)$ is the same distance from the point $(0,1)$ as from the line $y=-1$.

52   (cont.) Show that in general a point (a,b) has the geometric property described in Ex. 51 provided $a^2=4b$.

2   FUNCTIONS

One of the fundamental concepts in mathematics is that of a function. Let us consider some examples.

1.   To each city in the U.S. (except Washington, D.C.) corresponds the state in which it is located. This correspondence is called a function from the set of U.S. cities to the set of U.S. states.

2. To each living person is associated his or her height in cm today. This association is a function from the set of living people to the set of positive real numbers.

3. To each non-negative number x is assigned the number $\sqrt{x}$. This assignment is a function from the set of non-negative real numbers to itself.

In general, a **function** f is a rule that assigns to each element of a set **D** an element of a set **Y**. We say that f is a function from **D** to **Y**. (The sets **D** and **Y** can be the same set, as in Example 3 above.) We call **D** the **domain** of the function f.

We emphasize that a function assigns to each x in its domain **D** a *unique* element y in **Y**. For instance, the function in Example 1 assigns to each city the *unique* state that city belongs to. Notice that it assigns both to Los Angeles and to San Francisco the same state, California. That's okay because each city individually gets a single state. What is not okay is assigning to California *both* Los Angeles and San Francisco. That is not a function. However, assigning to each state its unique capital is a legitimate function (from the set of states to the set of cities).

Remark Let us confess that our definition of function is not 100 percent rigorous because we have used the words ''rule,'' ''assignment,'' etc., which are not precise mathematical terms. Mathematical logicians and others who worry about the *foundations* of mathematics (as distinct from mathematics and its applicability) like to define a function as ''a set **S** of ordered pairs (x, y) in which each first element x occurs in at most one pair.'' This indeed bypasses the need for undefined words like ''rule'' and ''assignment'' since each ordered pair (x, y) of **S** ''assigns'' y to x. However, it passes the buck to equally vague words like ''collection'' and ''common property'' that go into the ''definition'' of a *set*.

Although the ''set of ordered pairs'' approach pleases logicians (for valid reasons), it has nothing to do with real life use of mathematics. In applications, the fruitful concept of functions is as *assignments* or *correspondences*. For example, to each $t > 0$ *corresponds* the distance traveled by a falling body during the first t seconds of fall; to each $r > 0$ *corresponds* the area of a circle of radius r; to each state is *assigned* its population, etc.

Real Functions

In this book we shall deal almost exclusively with functions for which both **D** and **Y** are the real numbers or subsets of the real numbers. So let us agree that unless stated otherwise, a function will mean an assignment of a real number y to each real number x in a set **D** of real numbers. The symbol x, representing the typical number of the domain **D**, is often called the **independent variable.** The symbol y, denoting the number assigned to x, is called the **dependent variable.** (The value of y depends on the value of x). The set of all assigned numbers y is the **range** of the function. Some people like to picture a function f as a machine or black box that converts inputs into outputs (Fig. 1). (Think of a TV receiver, an amplifier, or

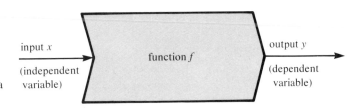

input x

(independent
variable)

function f

output y

(dependent
variable)

Fig. 1 A function as a
''black box''

a synthesizer as a function, or black box, that processes an input into an output.) The set of all acceptable inputs is the domain of f; the set of all possible outputs is the range.

Functional Notation

The number assigned to x by a function f is often denoted by $f(x)$, read "f of x" or "f at x". If you like inputs and outputs, think of $f(x)$ as the output corresponding to the input x.

We often specify a function f by a formula or an equation that tells how to find $f(x)$ from x. Instead of "the function f that assigns to each real number its square," we write "$f(x) = x^2$" or "$y = x^2$". Although the domain of f is not mentioned, it is understood to be the largest set of x for which the formulas make sense. In this example, the domain of f is the set of all real numbers since x^2 is defined for all real x. The domain of the function $f(x) = \sqrt{x}$ is the set of non-negative real numbers.

Remark It is logically incorrect to say "the function $f(x)$", or "the function x^2", or the function "$y = f(x)$". The symbols "$f(x)$", "y", "x^2" represent numbers, the numbers assigned by the function f to the numbers x. A function is not a *number*, but an *assignment* of a number y or $f(x)$ to each number x in a certain domain. Nevertheless, these slight inaccuracies are very common and harmless; we shall not try to avoid them.

Let us illustrate numerical use of functional notation. Keep in mind that $f(x)$ is the *number* assigned to x by the function f. For example, if $f(x) = 2x + 5$, then no matter what number x is given, the function f doubles it and adds 5.

EXAMPLE 1   Let $f(x) = 2x + 5$. Compute

(a) $f(0)$ (b) $f(3)$ (c) $f(-\sqrt{7})$ (d) $f(1/x)$ (e) $f(x - 8)$.

Solution   (a) $f(0) = 2 \cdot 0 + 5 = 5$

(b) $f(3) = 2 \cdot 3 + 5 = 11$

(c) $f(-\sqrt{7}) = 2(-\sqrt{7}) + 5 = -2\sqrt{7} + 5$

(d) $f(1/x) = 2(1/x) + 5 = 2/x + 5$

(e) $f(x - 8) = 2(x - 8) + 5 = 2x - 11$.

Answer   (a) 5 (b) 11 (c) $-2\sqrt{7} + 5$ (d) $2/x + 5$ (e) $2x - 11$

EXAMPLE 2   Let $f(x) = x^2 + 1$. Compute

(a) $f(x_1 + x_2)$ (b) $f(5x)$ (c) $f(x^2)$ (d) $\dfrac{f(3 + h) - f(3)}{h}$

Solution   (a) $f(x_1 + x_2) = (x_1 + x_2)^2 + 1$

(b) $f(5x) = (5x)^2 + 1 = 25x^2 + 1$

(c) $f(x^2) = (x^2)^2 + 1 = x^4 + 1$

(d) $\dfrac{f(3 + h) - f(3)}{h} = \dfrac{[(3 + h)^2 + 1] - [3^2 + 1]}{h}$

$$= \dfrac{6h + h^2}{h} = 6 + h.$$

Answer   (a) $(x_1 + x_2)^2 + 1$ (b) $25x^2 + 1$

   (c) $x^4 + 1$ (d) $6 + h$

Remark   So far we have used only the letter f to denote functions. However, when several functions occur in the same problem, we must denote each one by a different letter.

EXAMPLE 3   Let $f(x) = \dfrac{1}{x^2 + 2}$ and $g(x) = x^3 + 4x$. Show that

(a) $f(-x) = f(x)$ (b) $g(-x) = -g(x)$.

Solution   (a) $f(-x) = \dfrac{1}{(-x)^2 + 2} = \dfrac{1}{x^2 + 2} = f(x)$

(b) $g(-x) = (-x)^3 + 4(-x) = -x^3 - 4x = -(x^3 + 4x) = -g(x)$.

Warning   In practice, most functions are computed by formulas such as $f(x) = x^2$ or $f(x) = 1 + 7x^4$. Yet there are perfectly good functions not given by formulas. Here are a few examples:

(a) $f(x) =$ the largest integer y for which $y \le x$.

(b) $f(x) = \begin{cases} 1 \text{ if } x > 0 \\ 0 \text{ if } x = 0 \\ -1 \text{ if } x < 0. \end{cases}$

(c) $f(x) = 1$ if x is an integer, $f(x) = -1$ if x is not an integer.

(d) $f(x) =$ number of letters in the English spelling of the rational number x in lowest terms. For example, $f(\frac{1}{2}) = 7, f(3) = 5$. (A **rational** number is a real number that is the quotient of two integers.)

Domains

Most functions arising in practice have simple domains. The most common domains are the whole line, an interval (segment) $a \le x \le b$, a "half-line" such as $x \ge 0$ or $x < 2$ or some simple combination of these. Examples:

FUNCTION	DOMAIN
$f(x) = 2x + 1$	all real x (the whole line)
$f(x) = \sqrt{x + 2}$	$x \ge -2$ (half line)
$f(x) = \sqrt{1 - x^2}$	$-1 \le x \le 1$ (interval)
$f(x) = 1/x$	all x except $x = 0$ (union of two half-lines)

Note   In the second example, $\sqrt{x + 2}$ is defined provided $x + 2 \ge 0$, which is equivalent to $x \ge -2$. In the third example, $\sqrt{1 - x^2}$ is defined provided $1 - x^2 \ge 0$, that is, $x^2 \le 1$, which is equivalent to $-1 \le x \le 1$.

EXAMPLE 4   Find the domain of $f(x) = \sqrt{x} + \sqrt{1 - x^2}$.

Solution   The domain is the largest set of real numbers x for which the formula is meaningful. The first term $\sqrt{x}$ is defined for $x \geq 0$. The second term $\sqrt{1 - x^2}$ is defined only for $-1 \leq x \leq 1$. In order for both terms to be defined, we must have $0 \leq x \leq 1$. See Fig. 2.

Answer   The interval $0 \leq x \leq 1$.

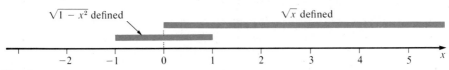

Fig. 2

Exercises

Compute the values of the function

1 $f(x) = 3x - 1$
   (a) $f(1)$ (b) $f(-1)$ (c) $f(0.15)$ (d) $f(-12)$

2 $f(x) = |x - 4|$
   (a) $f(-2)$ (b) $f(2)$ (c) $f(4 - \sqrt{7})$ (d) $f(\sqrt{7} - 4)$

3 $f(x) = 1/x$
   (a) $f(5)$ (b) $f(0.001)$ (c) $f(1/x)$ (d) $f(\tfrac{1}{2}\sqrt{x})$

4 $f(x) = x^2 + x - 1$
   (a) $f(10)$ (b) $f(-0.01)$ (c) $f(-x)$ (d) $f\left(\dfrac{-1 + \sqrt{5}}{2}\right)$.

5 Which of these functions satisfies $f(-x) = f(x)$?
   (a) $f(x) = x^4 + \dfrac{1}{x^4}$ (b) $f(x) = |x|$ (c) $f(x) = 2x + 1$
   (d) $f(x) = (x - 1)^2$.

6 Which of these functions satisfies $f(-x) = -f(x)$?
   (a) $f(x) = 1/x$ (b) $f(x) = x^3$ (c) $f(x) = x^3 + 1$
   (d) $f(x) = x^5 + 7x^3 - 4x$.

7 Which functions satisfy $f(3x) = 3f(x)$?
   (a) $f(x) = x$ (b) $f(x) = 7x$ (c) $f(x) = x + 3$ (d) $f(x) = x^3$.

8 Which functions satisfy $f(x_1 + x_2) = f(x_1) + f(x_2)$?
   (a) $f(x) = ax$ (b) $f(x) = x^2$ (c) $f(x) = 1/x$ (d) $f(x) = \sqrt{x}$.

9 Show that $f(x) = x^n$, where n is any integer, satisfies $f(x_1 x_2) = f(x_1)f(x_2)$. Give an example of a function that does not satisfy this equation.

10 Find an example of a function that satisfies $f(x^2) = [f(x)]^2$ and an example of a function that does not.

Exercises 11–16 concern the function $f(x) = \begin{cases} 1 & \text{if } x > 0 \\ 0 & \text{if } x = 0 \\ -1 & \text{if } x < 0. \end{cases}$

Compute

11 $f(x) + f(-x)$

12 $\frac{1}{2}[f(x) - f(-x)]$

13 $f(2x)$

14 $f[f(x)]$.

15 Show that $xf(x) = |x|$.

16 Show that $f(x)f(y) = f(xy)$.

Find the domain of $f(x)$

17 $f(x) = 3x - 2$

18 $f(x) = -7x + 6$

19 $f(x) = (4x - 5)^3$

20 $f(x) = |7 - x|$

21 $f(x) = 1/(2x - 3)$

22 $f(x) = x/(x + 2)$

23 $f(x) = x/(3x - 5)$

24 $f(x) = x/(x - 1)(x - 3)$

25 $f(x) = \sqrt{x - 6}$

26 $f(x) = \sqrt{5 - 2x}$

27 $f(x) = \sqrt{4 - 9x^2}$

28 $f(x) = \sqrt{15x^2 + 11}$

29 $f(x) = \sqrt{2x - 3}$

30 $f(x) = \dfrac{1}{\sqrt{x + 4}}$

31 $f(x) = \sqrt{\frac{1}{4} - x^2}$

32 $f(x) = \sqrt{x^2 - 9} + \sqrt{9 - x^2}$

33 $f(x) = \sqrt{(x - 1)(x - 4)}$

34 $f(x) = \sqrt{x^3 + 1}$

35 $\sqrt{x^2 - x - 12}$

36 $\sqrt{(x - 1)(x - 2)(x - 3)}$.

Find a function $f(x)$ that describes

37 the number of seconds in x hours

38 the Celsius reading corresponding to $x°$ Fahrenheit

39 the diagonal of a square of side x

40 the time required for a plane to fly 1000 miles at x miles/hr if the last 300 miles are against a 40 mph head-wind

41 the value after one year of a $10,000 investment if x dollars are invested at 6% yearly interest and the rest at 9%

42 the value after 3 years of $1000 invested at x% annual interest, compounded annually

3 GRAPHS OF FUNCTIONS

We associate to each function f its graph, a geometric picture of the function. The **graph** of f is the set of all points in the coordinate plane of the form $(x, f(x))$. To each x in the domain of f corresponds one point (x, y) on its graph, where $y = f(x)$. We sometimes think of $f(x)$ as the "height" of the graph at x, above the x-axis if $f(x) > 0$, below if $f(x) < 0$. See Fig. 1.

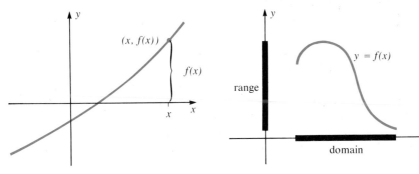

Fig. 1 Graph of f **Fig. 2** Domain and range of f

Geometrically, the domain of f is the set of all points on the x-axis at which f is defined. Recall that another set associated with a function f is its **range.** It is the set of all *values* of f:

Range   The **range** of f is the set of all real numbers $f(x)$, where x is any real number in the domain of f.

We can see both the domain and range of a function f by inspecting its graph (Fig. 2). The domain of f is the part of the x-axis directly above or below the graph. In other words, the domain is the vertical projection of the graph onto the x-axis. The range of f is the set of all "heights" of the graph. In other words, the range is the horizontal projection of the graph onto the y-axis.

If you think of a function as a machine, then the domain is the set of all inputs the machine will accept, and the range is the set of all outputs it produces.

Terminology   It is common to say "the graph of $y = f(x)$" or even "the graph of $f(x)$" instead of "the graph of the function f". For example, if $f(x) = x^2$, we may refer to "the graph of $y = x^2$". Very informal, but common terminology is simply "the graph of x^2".

EXAMPLE 1   Graph the constant function $f(x) = 2$. Show the domain and range of the function.

Solution   This extremely simple function assigns to each number the same number, 2. Its graph consists of all points of the form $(x, 2)$. These points form the horizontal line 2 units above the x-axis (Fig. 3).

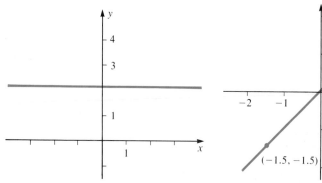

Fig. 3 Graph of $f(x) = 2$
Dom $= \{$all $x\}$ Range $= \{2\}$

Fig. 4 Graph of $f(x) = x$
Dom $= \{$all $x\}$ Range $= \{$all $y\}$

EXAMPLE 2   Graph the function $f(x) = x$. Show the domain and range of the function.

Solution   The graph consists of all points of the form (x, x). Plotting a few of these, we see that they form a straight line through the origin at an angle of 45° with the positive x-axis (Fig. 4).

EXAMPLE 3   Graph $f(x) = |x|$. Show the domain and range of the function.

Solution   If $x \geq 0$, then $|x| = x$. So, on and to the right of the y-axis, the graph is identical to that of $f(x) = x$ in Fig. 4.

If $x < 0$, then $|x| = -x$. So to the left of the y-axis, the graph consists of all points $(x, -x)$, where $x < 0$. Plotting a few of these, we see that they form a straight line through the origin at an angle $135°$ with the positive x-axis. Combining the two parts, we obtain the complete graph (Fig. 5).

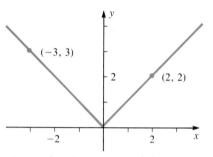

Fig. 5   Graph of $f(x) = |x|$
Dom $= \{$all $x\}$
Range $= \{$all $y \geq 0\}$

Fig. 6   Empirical function

Not only does each function have a graph, but each graph defines a function. By a graph, we mean here a collection of points (x, y) in the plane such that no two of the points have the same first coordinate (only one point can lie above a point on the x-axis). Such a graph automatically defines a function $f(x)$: to each x that occurs as a first coordinate of a point (x, y), it assigns the second coordinate y. Thus, $f(x)$ is the "height" of the graph above x.

The graphical definition of functions is standard procedure in science. For instance, a scientific instrument recording temperature or blood pressure on a graph is defining a function of time (Fig. 6). There is hardly ever an explicit formula for such a function.

Let us give two examples of functions without explicit formulas. The first is the "nearest integer" function defined by

$$I(x) = \begin{cases} \text{nearest integer to } x \text{ if there is } one \text{ such} \\ x \text{ if there are } two \text{ such} \end{cases}$$

For instance, $I(3.49) = 3$, $I(3.51) = 4$, but $I(3.5) = 3.5$. The graph is shown in Fig. 7a, next page.

The second example is the "saw-tooth" function $s(x)$ whose graph is shown in Fig. 7b, next page.

Some Graphing Principles

Suppose we know the graph of $y = f(x)$. Then we can easily find the graph of $y = f(x) + 1$. For each point $(x, f(x) + 1)$ on this graph is one unit higher than the point $(x, f(x))$ on the graph of $y = f(x)$. Therefore, the graph of $y = f(x) + 1$ is one unit higher than that of $y = f(x)$. Similarly, if $c > 0$,

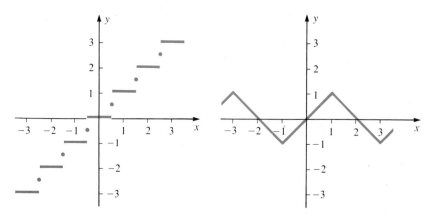

(a) nearest integer function $I(x)$ (b) saw-tooth function $s(x)$

Fig. 7   Functions without explicit formulas

then the graph of $y = f(x) + c$ is c units higher and the graph of $y = f(x) - c$ is c units lower (Fig. 8).

Let $c > 0$. Then the graph of

$\left.\begin{array}{l} y = f(x) + c \\ y = f(x) - c \end{array}\right\}$ is the graph of $y = f(x)$ shifted c units $\left\{\begin{array}{l} \text{upward} \\ \text{downward}. \end{array}\right.$

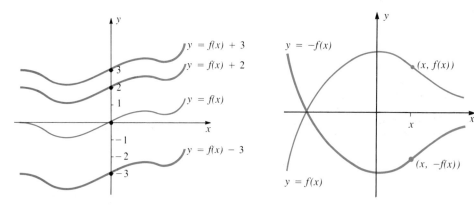

Fig. 8   Graphs of $y = f(x) \pm c$ **Fig. 9**   Graph of $y = -f(x)$

Next we consider the graph of $y = -f(x)$. Each of its points has the form $(x, -y)$, where (x, y) is on the graph of $y = f(x)$. But $(x, -y)$ is the reflection of (x, y) in the x-axis. Therefore the graph of $y = -f(x)$ is the reflection of the graph of $y = f(x)$. See Fig. 9.

The graph of $y = -f(x)$ is the reflection of the graph of $y = f(x)$ in the x-axis.

EXAMPLE 4 Graph $y = 2 - |x|$.

Solution   Figure 5 shows the graph of $y = |x|$. Turn it upside down (reflect in the x-axis) to obtain (Fig. 10a) the graph of $y = -|x|$. Then shift upward 2 units to obtain (Fig. 10b) the graph of $y = -|x| + 2 = 2 - |x|$.

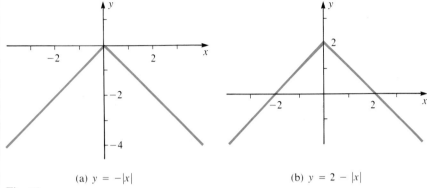

(a) $y = -|x|$ (b) $y = 2 - |x|$

Fig. 10

As our final graphing principle, let us consider the graph of $y = 2f(x)$. Each point of this graph has the form $(x, 2y)$, where (x, y) is on the graph of $y = f(x)$. Therefore, we obtain the graph of $y = 2f(x)$ when we stretch the graph of $y = f(x)$ by a factor of 2 in the y-direction (doubling the ''heights''). Similarly, for $c > 0$ the graph of $y = cf(x)$ is the graph of $y = f(x)$ stretched vertically by a factor c. If $c < 1$, then ''stretched'' must be interpreted as ''shrunk.'' See Fig. 11.

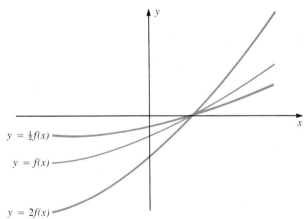

$y = \frac{1}{2}f(x)$

$y = f(x)$

$y = 2f(x)$

Fig. 11 Vertical stretching: $y = cf(x)$

> If $c > 0$, then the graph of $y = cf(x)$ is the graph of $y = f(x)$ stretched vertically by a factor c.

EXAMPLE 5 Graph $y = 2 + \frac{1}{2}s(x)$, where $s(x)$ is the saw-tooth function in Fig. 7b.

Solution   Shrink the graph of $s(x)$ vertically by a factor of $\frac{1}{2}$ to obtain (Fig. 12a) the graph of $y = \frac{1}{2}s(x)$. Then shift upward 2 units to obtain (Fig. 12b) the graph of $y = 2 + \frac{1}{2}s(x)$.

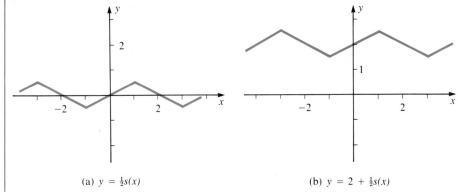

(a) $y = \frac{1}{2}s(x)$ (b) $y = 2 + \frac{1}{2}s(x)$

Fig. 12

Exercises

Graph

1 $f(x) = x + 2$ **2** $f(x) = x - 1$ **3** $f(x) = -x$

4 $f(x) = -x + 1$ **5** $f(x) = -17$ **6** $f(x) = 0.03$

7 $f(x) = 2x$ **8** $f(x) = \frac{1}{2}x - 3$ **9** $f(x) = 2|x|$

10 $f(x) = |x - 1|$ **11** $f(x) = \begin{cases} 0 & x \le 0 \\ 2x & x > 0 \end{cases}$ **12** $f(x) = \begin{cases} x - 1 & x \le 3 \\ 2 & x > 3 \end{cases}$

13 $f(x) = \begin{cases} 1 & x > 0 \\ 0 & x = 0 \\ -1 & x < 0 \end{cases}$ **14** $f(x) = \begin{cases} 1 & \text{if } x \text{ is an integer} \\ -1 & \text{if } x \text{ is not an integer} \end{cases}$

15 $f(x) = \begin{cases} \text{nearest } \textit{even} \text{ integer if there is } \textit{one} \text{ such} \\ x \text{ if there are } \textit{two} \text{ such} \end{cases}$

16 $f(x) = \frac{1}{3}|x| - 1$.

Recall the function $s(x)$ in Fig. 7b. Graph

17 $f(x) = 3s(x)$ **18** $f(x) = -s(x)$

19 $f(x) = |s(x)|$ **20** $f(x) = 1 - 2s(x)$.

21 Graph the function that gives the first class postage on a letter as a function of its weight. (As we go to press, the first ounce or fraction costs 15¢, each additional ounce or fraction, 13¢.)

22 Graph $f(x)$, the distance from the real number x to the nearest integer.

23 A taxi charges 75¢ as soon as you enter, then 10¢ after each $\frac{1}{6}$ mile. Graph $f(x)$, the cost of riding x miles in this cab.

24 A 120-yard high hurdle race has 10 hurdles. The first is 15 yards after the start, and all the rest are spaced at 10-yard intervals. Graph $f(x)$, the number of hurdles remaining after a runner has covered x yards.

Let $[x]$ denote the largest integer n such that $n \le x$. Graph

25 $f(x) = [x]$ **26** $f(x) = [1/x]$ $x > \frac{1}{5}$ **27** $f(x) = [\frac{1}{2}x]$

28 $f(x) = [2x]$ **29** $f(x) = (-1)^{[x]}$
30 $f(x) = $ the remainder when $[x]$ is divided by 3.

4 CONSTRUCTION OF FUNCTIONS

In this section we discuss several standard methods for building new functions out of old ones.

Addition of Functions

Given functions f and g whose domains overlap, their **sum** $f + g$ is the function defined on the common part (intersection) of the domains of f and g by

$$[f + g](x) = f(x) + g(x).$$

Examples

(1) $f(x) = 2x - 3$ $g(x) = x^2 - x - 1$.

f and g are defined for all real x; so is $f + g$, and

$$[f + g](x) = (2x - 3) + (x^2 - x - 1) = x^2 + x - 4.$$

(2) $f(x) = \sqrt{x - 2}$ $g(x) = \sqrt{3 - x}$.

The domain of f is $\{x \geq 2\}$; the domain of g is $\{x \leq 3\}$. The common part of these domains is $\{2 \leq x \leq 3\}$; this set is the domain of $f + g$, and

$$[f + g](x) = \sqrt{x - 2} + \sqrt{3 - x}.$$

A black box interpretation of the sum of functions is shown in Fig. 1.

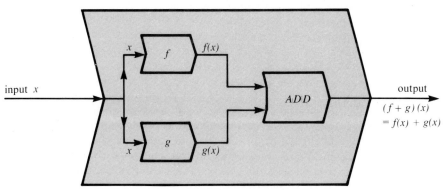

Fig. 1

Multiplication of Functions

Given functions f and g, their **product** fg is defined by

$$[fg](x) = f(x)g(x).$$

For example, if $f(x) = 16x^2$ and $g(x) = (x - 3)^5$, then

$$[fg](x) = 16x^2(x - 3)^5.$$

The domain of fg is the intersection of the domains of f and of g. The reason is precisely the same as for sums: the product is defined wherever both individual factors $f(x)$ and $g(x)$ are defined.

For a black box picture of fg, just replace the adder in Fig. 1 by a multiplier.

Composition of Functions

Given functions f and g, their **composition** $f \circ g$ (also called the **composite function** of f and g) is defined by

$$[f \circ g](x) = f[g(x)].$$

Think of substituting one function into the other, or replacing the variable of f by the function g.

In order for $f[g(x)]$ to be defined, the number $g(x)$ must be in the domain of f. Therefore the domain of $f \circ g$ is the set of all numbers x such that x is in the domain of g *and* $g(x)$ is in the domain of f.

Examples

(1) $f(x) = x^2 + 2x \qquad g(x) = -3x$.

$$[f \circ g](x) = f[g(x)] = [g(x)]^2 + 2[g(x)]$$
$$= (-3x)^2 + 2(-3x)$$
$$= 9x^2 - 6x.$$

Since both g and f are defined for all real x, so is $f \circ g$; the domain of $f \circ g$ is $\{$all $x\}$.

(2) $f(x) = 3x - 4 \qquad g(x) = 2x^2 - x + 1$:

$$[f \circ g](x) = f[g(x)] = 3g(x) - 4$$
$$= 3(2x^2 - x + 1) - 4$$
$$= 6x^2 - 3x - 1.$$

Again, the domain of $f \circ g$ is $\{$all $x\}$.

The black box interpretation is particularly useful for composite functions (Fig. 2). The black boxes for g and for f are connected in series; the output from g is the input for f. The black boxes for f and g, connected in series, make the new (composite) black box for $f \circ g$.

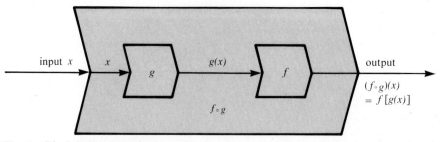

Fig. 2 Black boxes in series

EXAMPLE 1 Let $f(x) = x - 5$ and $g(x) = \sqrt{x}$. Compute and give the domains of

(a) $f \circ g$ (b) $g \circ f$.

Solution   (a) g is defined for all $x \geq 0$ and f is defined for *all* x, so $g(x)$ is in the domain of f for each x in the domain of g. Hence $f \circ g$ is defined for all $x \geq 0$ and

$$[f \circ g](x) = g(x) - 5 = \sqrt{x} - 5 \quad (x \geq 0).$$

(b) f is defined for all x, but g is defined only for all $x \geq 0$. Now $f(x) \geq 0$ only if $x \geq 5$, so the domain of $g \circ f$ is $\{x \geq 5\}$, and

$$[g \circ f](x) = \sqrt{f(x)} = \sqrt{x - 5} \quad (x \geq 5).$$

Answer   (a) $[f \circ g](x) = \sqrt{x} - 5$, domain $\{x \geq 0\}$

(b) $[g \circ f](x) = \sqrt{x - 5}$, domain $\{x \geq 5\}$.

EXAMPLE 2 Let $f(x) = \sqrt{x}$ and $g(x) = \sqrt{25 - x^2} - x$.

(a) Find the domain of $f \circ g$.   (b) Compute $f \circ g$.
(c) Find $[f \circ g](3)$ and $[f \circ g](-4)$.

Solution   (a) The domain of f is $\{x \geq 0\}$ and the domain of g is $\{-5 \leq x \leq 5\}$. For $g(x)$ to be in the domain of f requires

$$\sqrt{25 - x^2} - x \geq 0 \qquad \sqrt{25 - x^2} \geq x.$$

If $-5 \leq x \leq 0$, this is automatically satisfied. If $0 < x \leq 5$, then $\sqrt{25 - x^2} \geq x$ only if

$$25 - x^2 \geq x^2 \qquad 25 \geq 2x^2 \qquad x^2 \leq \tfrac{25}{2} \qquad x \leq \tfrac{5}{2}\sqrt{2}.$$

The domain of $f \circ g$ is the set of all x in the domain of g such that $g(x)$ is in the domain of f. In this case it is the set $\{-5 \leq x \leq \tfrac{5}{2}\sqrt{2}\}$.

(b) $[f \circ g](x) = \sqrt{g(x)} = \sqrt{\sqrt{25 - x^2} - x}$.

(c) $[f \circ g](3) = \sqrt{\sqrt{25 - 9} - 3} = \sqrt{4 - 3} = 1$.

$[f \circ g](-4) = \sqrt{\sqrt{25 - 16} + 4} = \sqrt{3 + 4} = \sqrt{7}$.

Answer   (a) $\{-5 \leq x \leq \tfrac{5}{2}\sqrt{2}\}$ (b) $[f \circ g](x) = \sqrt{\sqrt{25 - x^2} - x}$

(c) 1 $\sqrt{7}$

Exercises

Find $[f + g](x)$ and $[fg](x)$ and their domains

1 $f(x) = 3x + 1$ $g(x) = x - 4$ 2 $f(x) = 2x - 1$ $g(x) = 2x + 3$
3 $f(x) = x^2$ $g(x) = -2x + 1$ 4 $f(x) = x^2 + 1$ $g(x) = -x^2 + x$
5 $f(x) = x$ $g(x) = 1/x$ 6 $f(x) = \sqrt{x}$ $g(x) = \sqrt{1 - x}$
7 $f(x) = \sqrt{1 - x}$ $g(x) = \sqrt{x - 2}$ 8 $f(x) = \dfrac{1}{x^2 + x}$ $g(x) = \dfrac{1}{x^2 - 4}$.

Compute $[f \circ g](x)$ and $[g \circ f](x)$, and give their domains

9 $f(x) = 3x + 1$ $g(x) = x - 2$ **10** $f(x) = 2x - 1$ $g(x) = -x^2 + 3x$

11 $f(x) = 2x^2$ $g(x) = -x - 1$ **12** $f(x) = x + 1$ $g(x) = -x + 1$

13 $f(x) = 2x$ $g(x) = -2x$ **14** $f(x) = x + 3$ $g(x) = 4x + 1$

15 $f(x) = x^2$ $g(x) = 3$ **16** $f(x) = \pi x^2$ $g(x) = 2x + 5$

17 $f(x) = \sqrt{x}$ $g(x) = x^2 - 9$ **18** $f(x) = \sqrt{x}$ $g(x) = \sqrt[3]{x}$

19 $f(x) = 1/x$ $g(x) = x^2 + 2$ **20** $f(x) = \sqrt{x}$ $g(x) = |x|$

21 $f(x) = \dfrac{x - 7}{3}$ $g(x) = 3x + 7$ **22*** $f(x) = \dfrac{2x + 1}{3x + 1}$ $g(x) = \dfrac{x - 1}{-3x + 2}$.

Compute $[f \circ g](x)$

23 $f(x) = x$ and $g(x)$ is any function   **24** $g(x) = x$ and $f(x)$ is any function.

Compute $[f \circ f](x)$

25 $f(x) = 1 - x$ **26** $f(x) = 1/x$ $x \neq 0$.

Does $f \circ g$ make sense?

27 $f(x) = \sqrt{2x - 5}$ $g(x) = 1 - x^2$ **28** $f(x) = g(x) = -\sqrt{x}$.

Suppose the domain of f is the interval $0 \leq x \leq 2$. Find the domain of

29 $f(x - 3)$ **30** $f(-x)$ **31** $f(2x)$ **32** $f(x^2)$.

33 A function f is called **strictly increasing** if whenever x_1 and x_2 are in the domain of f and $x_1 < x_2$, then $f(x_1) < f(x_2)$. Show that the sum of two strictly increasing functions is strictly increasing.

34 (cont.) Show that the composite of two strictly increasing functions is strictly increasing.

35 Let $f(x) = x + 2$. Find a function $g(x)$ such that $[f \circ g](x) = x$.

36 (cont.) Do the same for $f(x) = 2x - 3$.

How do the calculator sequences represent composition of functions? How do the composite functions differ?

37 $\boxed{x^2}$ $\boxed{\sqrt{}}$ and $\boxed{\sqrt{}}$ $\boxed{x^2}$ **38** $\boxed{+}$ 1 $\boxed{=}$ $\boxed{x^2}$ and $\boxed{x^2}$ $\boxed{+}$ 1 $\boxed{=}$.

In the next four exercises you are asked to find the function represented by a black box that is built out of other black boxes. These are all simplified models of **servomechanisms** because part of the output is returned to the input. We use the building blocks in Fig. 3.

(a) subtractor

(b) amplifier
$(0 < \alpha)$

(c) signal splitter
$(0 < c < 1)$

Fig. 3

Express y as a function of x

39

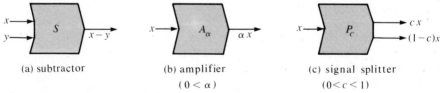

40

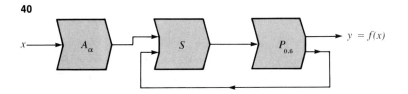

41

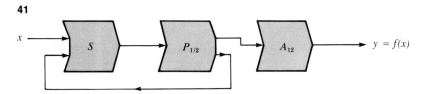

42

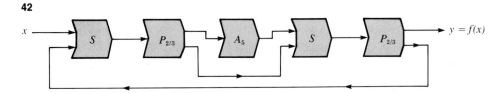

REVIEW EXERCISES

1 Plot the points $(2, 3)$, $(2, -3)$, $(-2, 3)$, and $(-2, -3)$. Show that they are the vertices of a rectangle.

2 Plot all points (x, y) in the plane for which
(a) $y \geq 0$ (b) $-1 \leq x \leq 1$ and y is an integer.

3 Given $f(x) = 3x + 1$
(a) compute $f(0)$, $f(-2)$, $f[f(x)]$
(b) show that $f(a + b) = f(a) + f(b) - 1$.

4 Find a function $f(x)$ that describes the diameter of a circle of area x.

Graph and show the domain and range

5 $f(x) = x - |x|$ **6** $f(x) = \begin{cases} x & \text{if } x \text{ is an integer} \\ -x & \text{otherwise.} \end{cases}$

7 If $f(x) = 2x + 1$ and $g(x) = 3x - 2$, find a formula for $h(x) = 2f(x) - g(x) + f(x)g(x)$.

8 If $f(x) = \sqrt{x}$ and $g(x) = 3 - x$, compute
(a) $[f \circ g](x)$ (b) $[g \circ f](x)$.
In each case state the domain and range of the function.

9 Find a linear function $y = ax + b$ whose graph passes through $(0, 6)$ and is parallel to the graph of $y = -x$.

10 Find a linear function $y = ax + b$ whose graph passes through $(0, 3)$ and $(1, 5)$.

11 Graph the **parabola** $y = x^2$.

12 Graph $y = -2x^2 - 1$.

2 THE SINE AND COSINE FUNCTIONS

1 ANGLES

In this chapter we study the two most basic of the trigonometric functions, the sine function and the cosine function. We begin our work with a review of the geometry of angles and their measurement.

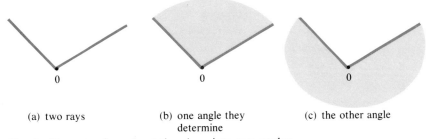

(a) two rays

(b) one angle they determine

(c) the other angle

Fig. 1 Two rays from 0 cut the plane into two angles.

Two rays (half lines) from a point 0 cut the plane into two **angles** (Fig. 1). The point 0 is called the **vertex** of each angle. There is one exception, which we agree to include anyhow: when the rays coincide (Fig. 2a). Then they determine a **whole angle** (Fig. 2b) and a **zero angle** (Fig. 2c).

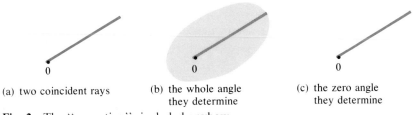

(a) two coincident rays

(b) the whole angle they determine

(c) the zero angle they determine

Fig. 2 The "exception", included anyhow

Each angle has an **initial side** and a **terminal side.** If you stand on the vertex 0 of the angle facing into the angle, its initial side is on your right and its terminal side is on your left (Fig. 3a). Alternatively, imagine a ray from 0 that sweeps the angle *counterclockwise,* from its initial to its terminal side (Fig. 3b).

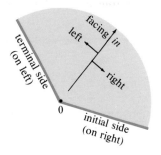

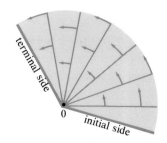

(a) The observer looks into the angle from the vertex.

(b) counterclockwise sweep of the angle

Fig. 3

The common unit of angle *measure* is the **degree** (°). The whole angle is divided into 360 degrees (Fig. 4a). Examples are shown in Figs. 4b and 4c.

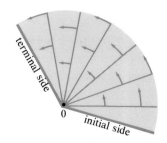

(b) 30° angle

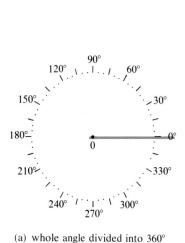

(a) whole angle divided into 360°

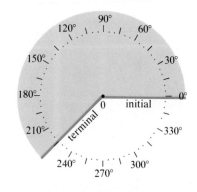

(c) 225° angle

Fig. 4

Angle measure has several important properties.

(1)   Angle measure is *additive:* if an angle of measure $\gamma°$ is divided by a ray from its vertex into two subangles of measure $\alpha°$ and $\beta°$ respectively, then $\gamma° = \alpha° + \beta°$.

(2)   Congruent angles have equal measure. That is, if a transformation of the euclidean plane that preserves all distances (**congruence** transformation) takes an angle of measure $\alpha°$ to an angle of measure $\beta°$, then $\alpha° = \beta°$.

(3)   Similar angles have equal measure. That is, if a transformation of the euclidean plane that stretches all distances by the same positive factor k (a **similarity** transformation) takes an angle of measure $\alpha°$ to an angle of measure $\beta°$, then $\alpha° = \beta°$.

   Recall the theorems: (1) Congruent triangles have equal corresponding angles. (2) Similar triangles have equal corresponding angles.

Terminology   It is customary to refer to two different angles of equal *measure* as "equal angles." This is convenient and harmless. It goes along with the common usage of saying "equal segments" when we mean segments of equal length.

   We shall often say things like "α is an angle and $\alpha = 30°$", when we mean "α is an angle and the *measure* of α equals $30°$". It is just too tedious to drag along the word measure when the context is perfectly clear.

   A **straight angle** is half of a whole angle, that is, an angle of 180°. See Fig. 5a. A **right angle** is half a straight angle, or a quarter of a whole angle, that is, an angle of 90°. See Fig. 5b. Three quarters of a whole angle has no special name; it is an angle of 270°. An example is the angle from the positive x-axis, counterclockwise to the negative y-axis (Fig. 5c).

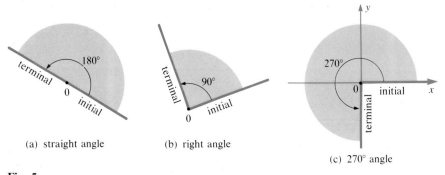

(a) straight angle   (b) right angle

(c) 270° angle

Fig. 5

   An angle is in **standard position** with respect to coordinate axes when its vertex is at the origin and its initial side is along the positive x-axis. The angles in Fig. 4 are in standard position; so is the 270° angle in Fig. 5c.

Exercises

Draw in standard position

1	10°	**2**	80°	**3**	120°	**4**	210°
5	240°	**6**	300°	**7**	330°	**8**	190°
9	160°	**10**	315°	**11**	165°	**12**	255°.

2 SINE AND COSINE

Let θ be an angle in standard position. Let (x, y) be *any* point on its terminal side different from its vertex (Fig. 1a). Let r be the distance from the origin to (x, y). By the distance formula (really the Pythagorean theorem),

$$r = \sqrt{x^2 + y^2}.$$

Of course $r > 0$ because $(x, y) \neq (0, 0)$. Now consider the two ratios

$$\frac{x}{r} \quad \text{and} \quad \frac{y}{r}.$$

These ratios depend *only* on θ, not on the choice of (x, y) on the terminal side. Why? Consider the *point*

$$\left(\frac{x}{r}, \frac{y}{r} \right)$$

whose coordinates are the ratios x/r and y/r in question. On the one hand this point lies on the terminal side because its coordinates are just the coordinates of (x, y) stretched by the positive factor $1/r$. On the other hand, it lies on the **unit circle,** the circle of radius 1 and center $(0, 0)$. This is so because

$$\left(\frac{x}{r} \right)^2 + \left(\frac{y}{r} \right)^2 = \frac{x^2 + y^2}{r^2} = \frac{r^2}{r^2} = 1.$$

Therefore, the point $(x/r, y/r)$ is determined geometrically (Fig. 1b) as the unique intersection of the terminal side with the unit circle.

Thus x/r and y/r depend only on θ, that is, they are *functions* of θ. We call them the **cosine** of θ and the **sine** of θ respectively.

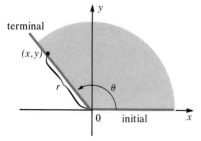

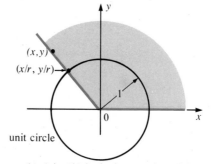

(a) θ is in standard position,
(x,y) on its terminal side,
$r = \sqrt{x^2 + y^2} > 0$.

(b) $(x/r, y/r)$ is the intersection of the terminal side of θ with the unit circle.

Fig. 1

Sine and Cosine   Let θ be an angle in standard position, and let (x, y) be any point on its terminal side, different from the origin. Set $r = \sqrt{x^2 + y^2}$. Then

$$\cos \theta = \frac{x}{r} \qquad \sin \theta = \frac{y}{r}.$$

Thus we have defined two functions of θ with domain $0 \leq \theta \leq 360°$.

Examples

(1)   0°   terminal side = positive *x*-axis,

$(x, y) = (x, 0)$ with $x > 0$ $r = x$

$\dfrac{x}{r} = \dfrac{x}{x} = 1$ $\dfrac{y}{r} = \dfrac{0}{r} = 0$ $\cos 0° = 1$ $\sin 0° = 0$

(2)   90°   terminal side = positive *y*-axis

$(x, y) = (0, y)$ with $y > 0$ $r = y$

$\dfrac{x}{r} = \dfrac{0}{r} = 0$ $\dfrac{y}{r} = \dfrac{y}{y} = 1$ $\cos 90° = 0$ $\sin 90° = 1$

(3)   180°   terminal side = negative *x*-axis

$(x, y) = (x, 0)$ with $x < 0$ $r = |x| = -x$

$\dfrac{x}{r} = \dfrac{x}{-x} = -1$ $\dfrac{y}{r} = \dfrac{0}{r} = 0$ $\cos 180° = -1$ $\sin 180° = 0$

(4)   270°   terminal side = negative *y*-axis

$(x, y) = (0, y)$ with $y < 0$ $r = |y| = -y$

$\dfrac{x}{r} = \dfrac{0}{r} = 0$ $\dfrac{y}{r} = \dfrac{y}{-y} = -1$ $\cos 270° = 0$ $\sin 270° = -1$

(5)   360°   same as 0° $\cos 360° = 1$ $\sin 360° = 0$

The functions sine and cosine have many interesting properties which we shall study soon. For the moment we note only the following.

$$-1 \leq \sin \theta \leq 1 \qquad -1 \leq \cos \theta \leq 1$$
$$\sin^2 \theta + \cos^2 \theta = 1$$

[Here $\cos^2 \theta$ is an abbreviation for $(\cos \theta)^2$, etc.] They are consequences of the fact that $(\cos \theta, \sin \theta) = (x/r, y/r)$ is a point on the *unit* circle, which means $\cos^2 \theta + \sin^2 \theta = 1$. It is clear geometrically that $|\cos \theta| \leq 1$ and $|\sin \theta| \leq 1$, or analytically from $\cos^2 \theta \leq \cos^2 \theta + \sin^2 \theta = 1$, etc.

Further Examples

EXAMPLE 1   Find the cosine and sine of   (a)   45°   (b)   135°.

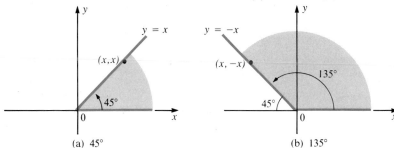

(a) 45°   (b) 135°

Fig. 2

Solution (a) The terminal side of a $45°$ angle in standard position is the portion of the line $y = x$ in the first quadrant (Fig. 2a). Its typical point is $(x, y) = (x, x)$ where $x > 0$. Therefore

$$r = \sqrt{x^2 + y^2} = \sqrt{x^2 + x^2} = x\sqrt{2}$$

$$\cos 45° = \frac{x}{r} = \frac{x}{x\sqrt{2}} = \frac{1}{\sqrt{2}} = \tfrac{1}{2}\sqrt{2}$$

$$\sin 45° = \frac{y}{r} = \frac{x}{x\sqrt{2}} = \frac{1}{\sqrt{2}} = \tfrac{1}{2}\sqrt{2}.$$

(b) Since $135° = 180° - 45°$, the terminal side of a $135°$ angle in standard position is the second quadrant portion of the line $y = -x$. See Fig. 2b. Its typical point is $(x, y) = (x, -x)$ where $x < 0$. Therefore

$$r = \sqrt{x^2 + y^2} = \sqrt{x^2 + x^2} = |x|\sqrt{2} = -x\sqrt{2}$$

$$\cos 135° = \frac{x}{r} = \frac{x}{-x\sqrt{2}} = \frac{-1}{\sqrt{2}} = -\tfrac{1}{2}\sqrt{2}$$

$$\sin 135° = \frac{y}{r} = \frac{-x}{-x\sqrt{2}} = \frac{1}{\sqrt{2}} = \tfrac{1}{2}\sqrt{2}.$$

Answer (a) $\cos 45° = \sin 45° = \tfrac{1}{2}\sqrt{2}$

(b) $\cos 135° = -\tfrac{1}{2}\sqrt{2}$ $\sin 135° = \tfrac{1}{2}\sqrt{2}$

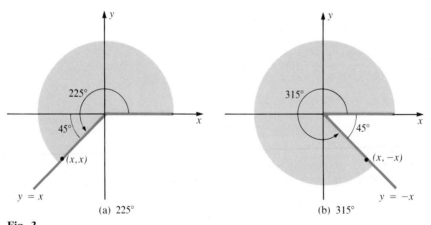

(a) $225°$ (b) $315°$

Fig. 3

EXAMPLE 2 Find the cosine and sine of (a) $225°$ (b) $315°$.

Solution (a) Since $225° = 180° + 45°$, the terminal side of a $225°$ angle in standard position is the third quadrant portion of $y = x$. See Fig. 3a. Its typical point is $(x, y) = (x, x)$ where $x < 0$. Therefore

$$r = \sqrt{x^2 + y^2} = \sqrt{x^2 + x^2} = |x|\sqrt{2} = -x\sqrt{2}$$

$$\cos 225° = \frac{x}{r} = \frac{x}{-x\sqrt{2}} = \frac{-1}{\sqrt{2}} = -\tfrac{1}{2}\sqrt{2}$$

$$\sin 225° = \frac{y}{r} = \frac{x}{-x\sqrt{2}} = \frac{-1}{\sqrt{2}} = = -\tfrac{1}{2}\sqrt{2}.$$

(b) Since $315° = 360° - 45°$, the terminal side of a $315°$ angle in standard position is the fourth quadrant portion of $y = -x$. See Fig. 3b. Its typical point is $(x, y) = (x, -x)$ where $x > 0$. Therefore

$$r = \sqrt{x^2 + y^2} = \sqrt{x^2 + x^2} = x\sqrt{2}$$

$$\cos 315° = \frac{x}{r} = \frac{x}{x\sqrt{2}} = \frac{1}{\sqrt{2}} = \tfrac{1}{2}\sqrt{2}$$

$$\sin 315° = \frac{y}{r} = \frac{-x}{x\sqrt{2}} = \frac{-1}{\sqrt{2}} = -\tfrac{1}{2}\sqrt{2}.$$

Answer   (a) $\cos 225° = \sin 225° = -\tfrac{1}{2}\sqrt{2}$

   (b) $\cos 315° = \tfrac{1}{2}\sqrt{2}$ $\sin 315° = -\tfrac{1}{2}\sqrt{2}$

Graphs

Let us tabulate all the values we have found, rounding to three places ($\tfrac{1}{2}\sqrt{2} \approx 0.707$):

$\theta°$	0	45	90	135	180	225	270	315	360
$\cos \theta°$	1	.707	0	$-.707$	-1	$-.707$	0	.707	1
$\sin \theta°$	0	.707	1	.707	0	$-.707$	-1	$-.707$	0

This gives us enough data for rough graphs of $y = \cos \theta$ and $y = \sin \theta$. See Fig. 4. We shall make a detailed study of these graphs in a later section.

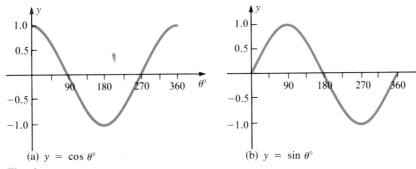

(a) $y = \cos \theta°$ (b) $y = \sin \theta°$

Fig. 4

Calculating Trigonometric Functions

All scientific calculators can estimate $\sin \theta°$ and $\cos \theta°$. They often calculate also in other units of angle measure (*radians*, sometimes *grads*) and you must make sure before calculating that the mode switch (or key) is set to $\boxed{\text{deg}}$. Then you simply enter $\theta°$ and press $\boxed{\text{sin}}$ or $\boxed{\text{cos}}$.

Examples

(1) sin 60°   6 0 | sin | 0.8660254

(2) cos 135°   1 3 5 | cos | -0.7071068

Exercises

Find sin θ and cos θ exactly for an angle θ in standard position whose terminal side passes through

1 $(4,3)$	**2** $(-3,4)$	**3** $(-5,-12)$	**4** $(12,-5)$
5 $(7,24)$	**6** $(-24,-7)$	**7** $(40,-9)$	**8** $(-9,40)$
9 $(-11,-60)$	**10** $(-60,11)$	**11** $(1,\sqrt{2})$	**12** $(-\sqrt{2},-1)$
13 $(\sqrt{2},\sqrt{7})$	**14** $(-\sqrt{7},-\sqrt{2})$	**15** $(\sqrt{3},-\sqrt{5})$	**16** $(\sqrt{5},-\sqrt{3})$.

Estimate by calculator to 4 places

17 sin 17.61°	**18** sin 48.92°	**19** cos 5.21°	**20** cos 78.33°
21 sin 100.62°	**22** sin 164.55°	**23** cos 93.75°	**24** cos 171.82°
25 sin 206.17°	**26** sin 251.39°	**27** cos 199.02°	**28** cos 254.10°
29 sin 279.03°	**30** sin 294.16°	**31** cos 280.40°	**32** cos 355.01°

3 RIGHT TRIANGLES

The angles of a right triangle are its right angle, 90°, and its two complementary acute angles, α and β. Remember that **acute** means $0 < \alpha° < 90°$ and $0 < \beta° < 90°$, and **complementary** means $\alpha° + \beta° = 90°$. The angles α and β are complementary because the sum of the three angles of (any) triangle is 180°, so $\alpha° + \beta° + 90° = 180°$, hence $\alpha° + \beta° = 90°$.

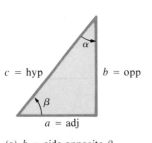

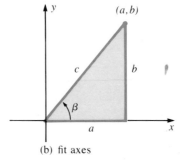

(a) b = side opposite β
    a = side adjacent β
    c = hypotenuse

(b) fit axes

Fig. 1 Right triangle

Take a right triangle, labelled as in Fig. 1a. To find sin β and cos β, we fit axes as in Fig. 1b. (It may be necessary to turn the triangle upside-down first.) Then $(x,y) = (a,b)$ is a point on the terminal side of β, now in standard position, and $r = c$. Hence

$$\sin \beta = \frac{b}{c} \quad \text{and} \quad \cos \beta = \frac{a}{c}.$$

Since b is the length of the leg of the triangle opposite β and a is the length of the leg adjacent to α, we may write:

> For an acute angle β of a right triangle,
>
> $$\sin \beta = \frac{\text{opposite}}{\text{hypotenuse}} \qquad \cos \beta = \frac{\text{adjacent}}{\text{hypotenuse}}.$$

Look at α in Fig. 1a. Its opposite side is a, and its adjacent side is b, so

$$\sin \alpha = \frac{a}{c} = \cos \beta \quad \text{and} \quad \cos \alpha = \frac{b}{c} = \sin \beta.$$

Since $\beta° = 90° - \alpha°$, this proves an important result:

> If α is any acute angle, then
>
> $$\sin \alpha° = \cos(90° - \alpha°) \quad \text{and} \quad \cos \alpha° = \sin(90° - \alpha°).$$

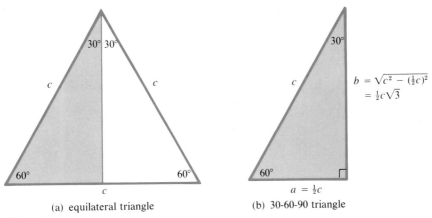

(a) equilateral triangle (b) 30-60-90 triangle

Fig. 2

EXAMPLE 1 Find the cosine and sine of

(a) 60° (b) 30°.

Solution (a) Drop a perpendicular from one vertex of an equilateral triangle of side c to its opposite side (Fig. 2a). It splits the triangle into two 30-60-90 triangles. The side adjacent to the 60° angle (Fig. 2b) has length $a = \frac{1}{2}c$. By the Pythagorean theorem, the opposite side is

$$b = \sqrt{c^2 - a^2} = \sqrt{c^2 - (\tfrac{1}{2}c)^2} = \sqrt{\tfrac{3}{4}c^2} = \tfrac{1}{2}c\sqrt{3}.$$

Therefore

$$\sin 60° = \frac{\text{opp}}{\text{hyp}} = \frac{b}{c} = \frac{\tfrac{1}{2}c\sqrt{3}}{c} = \tfrac{1}{2}\sqrt{3}$$

$$\cos 60° = \frac{\text{adj}}{\text{hyp}} = \frac{a}{c} = \frac{\tfrac{1}{2}c}{c} = \tfrac{1}{2}$$

(b)   Since 30° and 60° are complementary acute angles,

$$\sin 30° = \cos 60° = \tfrac{1}{2} \quad \text{and} \quad \cos 30° = \sin 60° = \tfrac{1}{2}\sqrt{3}.$$

Answer (a) $\sin 60° = \tfrac{1}{2}\sqrt{3}$ $\cos 60° = \tfrac{1}{2}$

(b) $\sin 30° = \tfrac{1}{2}$ $\cos 30° = \tfrac{1}{2}\sqrt{3}$

The 30-60-90 triangle can be used to find sines and cosines of many non-acute angles. Remember that the sides are

short leg: $\tfrac{1}{2}c$ long leg: $\tfrac{1}{2}c\sqrt{3}$ hypotenuse: c.

It helps to take $c = 2$: the sides of a 30-60-90 triangle are in the proportion 1 to $\sqrt{3}$ to 2.

EXAMPLE 2   Find $\sin 150°$ and $\cos 150°$.

Solution   Since $150° = 180° - 30°$, the terminal side of a 150° angle in standard position lies in the second quadrant, at 30° to the negative *x*-axis (Fig. 3, next page). Take the point *P* on this terminal side at distance 2 from $(0,0)$, and drop a perpendicular from *P* to the *x*-axis. A 30-60-90 triangle is formed; its hypotenuse is 2, so its short leg is 1 and its long leg $\sqrt{3}$. Therefore

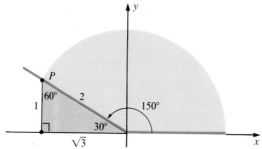

Fig. 3

$P = (-\sqrt{3}, 1)$, being in the *second* quadrant. Consequently

$$\cos 150° = \frac{x}{r} = \frac{-\sqrt{3}}{2} = -\tfrac{1}{2}\sqrt{3} \qquad \sin 150° = \frac{y}{r} = \frac{1}{2} = \tfrac{1}{2}.$$

Answer $\cos 150° = -\tfrac{1}{2}\sqrt{3}$ $\sin 150° = \tfrac{1}{2}$

EXAMPLE 3   Express in terms of $v = \sin 35°$

(a) $\cos 35°$ (b) $\sin 125°$ (c) $\cos 235°$.

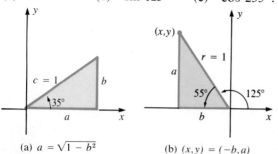

(a) $a = \sqrt{1 - b^2}$ (b) $(x,y) = (-b,a)$ (c) $(x,y) = (-b,-a)$

Fig. 4

Solution (a) Draw $35°$ in standard position and a corresponding 35-55-90 right triangle (Fig. 4a). Take hyp $= 1$. Then

$$v = \sin 35° = \frac{\text{opp}}{\text{hyp}} = \frac{b}{1} = b.$$

By the Pythagorean theorem,

$$a = \text{adj} = \sqrt{1 - b^2} = \sqrt{1 - v^2}$$

so

$$\cos 35° = \frac{\text{adj}}{\text{hyp}} = \frac{\sqrt{1 - v^2}}{1} = \sqrt{1 - v^2}.$$

[Of course, we could have used $\cos^2 35° = 1 - \sin^2 35°$, but we need the triangle for (b) and (c).]

(b) In standard position, the terminal side of $125°$ lies in the second quadrant, $55°$ short of the negative x-axis. Place the triangle from (a) accordingly (Fig. 4b). Clearly $x = -b = -v$ and $y = a = \sqrt{1 - v^2}$, so

$$\sin 125° = \frac{y}{r} = \frac{\sqrt{1 - v^2}}{1} = \sqrt{1 - v^2}.$$

(c) The terminal side of $235°$ lies in the third quadrant, $55°$ forward of the negative x-axis ($270°$). Place the triangle from (a) accordingly (Fig. 4c). Clearly $x = -b = -v$ and $y = -a = -\sqrt{1 - v^2}$, so

$$\cos 235° = \frac{x}{r} = \frac{-v}{1} = -v.$$

Answer (a) $\sqrt{1 - v^2}$ (b) $\sqrt{1 - v^2}$ (c) $-v$

Calculating Angles

Each angle θ of a triangle satisfies $0 < \theta° < 180°$. When you solve a triangle (as we do in Chapter 8) for one of its angles θ, you usually find $\sin \theta$ or $\cos \theta$, from which you must calculate θ. Your calculator does this by the sequences

$$\boxed{\text{INV}} \quad \boxed{\text{sin}} \quad \text{and} \quad \boxed{\text{INV}} \quad \boxed{\text{cos}},$$

but some care must be taken. First let's look at $\boxed{\text{INV}} \boxed{\text{cos}}$. From Fig. 4a, page 29, it is clear that if $-1 < y < 1$, then there is a unique θ in the domain $0 < \theta° < 180°$ such that $\cos \theta° = y$. This is what your calculator estimates.

Examples (set $\boxed{\text{deg}}$ mode.)

(1) $\cos \theta° = 0.42$. 4 2 $\boxed{\text{INV}}$ $\boxed{\text{cos}}$ 65.17

(2) $\cos \theta° = -0.834$. 8 3 4 $\boxed{+/-}$ $\boxed{\text{INV}}$ $\boxed{\text{cos}}$ 146.51

Thus $\cos 65.17° \approx 0.42$ and $\cos 146.51° \approx -0.834$, as you can check directly on your calculator.*

Finding θ from $\sin \theta = y$ is more complicated. As the graph in Fig. 4b, page

---

* The symbol $\approx$ denotes "approximately equal."

29, indicates, if $0 < y < 1$, there are *two* values of θ satisfying $0 < \theta° < 180°$ and $\sin \theta° = y$. Your calculator finds the *smaller* one only, the one satisfying $0 < \theta° < 90°$. The larger one is $180° - \theta°$, as is clear from Fig. 5. (Of course, there is only *one* value of θ satisfying $0 < \theta° < 180°$ and $\sin \theta° = 1$; namely $\theta° = 90°$.)

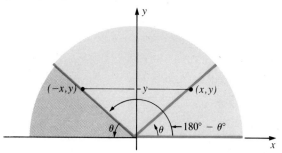

Fig. 5 $\sin(180° - \theta°) = \dfrac{y}{r} = \sin \theta°$

EXAMPLE 4   Estimate to two places all θ satisfying $0 < \theta° < 180°$ and $\sin \theta° = 0.6413$.

Solution   Set the ⃞deg⃞ mode, then key in

.6 4 1 3 ⃞INV⃞   ⃞sin⃞ $\boxed{39.89}$

Hence one answer is 39.89°. The other is

180 − 39.89 = 140.11°

Check   3 9 . 8 9 ⃞sin⃞   .6413

1 4 0 . 1 1 ⃞sin⃞   .6413

Answer   39.89°   140.11°

Remark   Evidently $\cos \theta$ is a strictly decreasing function for $0 < \theta° < 180°$. The sequence ⃞INV⃞ ⃞cos⃞ calculates its *inverse* function. This function is commonly denoted arc cos y. Similarly arc sin y denotes the inverse function of the strictly increasing function $\sin \theta$ on the interval $0 < \theta° < 90°$. We shall study these inverse trigonometric functions in Chapter 4.

Exercises

Find (exact values)

1 $\sin 120°$	**2** $\cos 120°$	**3** $\sin 210°$	**4** $\cos 210°$
5 $\sin 240°$	**6** $\cos 240°$	**7** $\sin 300°$	**8** $\cos 300°$
9 $\sin 330°$	**10** $\cos 330°$.		

Express in terms of $K = \cos 20°$

11 $\sin 20°$	**12** $\sin 70°$	**13** $\cos 160°$	**14** $\sin 200°$
15 $\cos 290°$	**16** $\cos 340°$.		

Estimate to two places the angle θ satisfying $0 < \theta° < 180°$ and $\cos \theta° =$

17 0.3333	**18** 0.6667	**19** 0.8984	**20** 0.75

21 -0.2 **22** -0.31 **23** -0.2179 **24** -0.3964.

Estimate to two places all angles θ satisfying $0 < \theta° < 180°$ and $\sin \theta° =$

25 0.3333 **26** 0.6667 **27** 0.9243 **28** 0.6.

4 THE CIRCULAR FUNCTIONS

We have defined the sine and cosine for any angle θ whose degree measure satisfies $0° \leq \theta° \leq 360°$. Our next task is to enlarge the definitions so that the "circular functions" $\sin \theta$ and $\cos \theta$ are defined for all real numbers θ. It is convenient to measure θ, not in degrees, but in a unit called the **radian.**

Radian Measure

The degree is the unit of angle measure used in practical work, like surveying, navigation, and machine design. Another unit of angle measure called the radian is better for scientific work. Recall what a degree is: it is the unit of angle measure taken so that the whole angle measures $360°$. The **radian** is the unit of angle measure taken so that the whole angle measures 2π radians. Of course π denotes the real number such that a circle of radius r has circumference $2\pi r$. Let's look at a circle of radius r in Fig. 1a. The whole angle is 2π radians and the circumference is $2\pi r$. It follows (Fig. 1b) that a central angle of θ radians subtends (determines) an arc of length $s = r\theta$. In particular (Fig. 1c) a central angle of 1 radian in a circle of radius r subtends an arc of length precisely r.

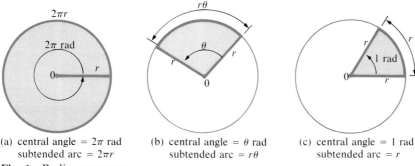

(a) central angle $= 2\pi$ rad
   subtended arc $= 2\pi r$

(b) central angle $= \theta$ rad
   subtended arc $= r\theta$

(c) central angle $= 1$ rad
   subtended arc $= r$

Fig. 1 Radian measure

Since 2π radians equals $360°$, we have π radians equal $180°$ (easier to remember). Thus 1 radian equals $180/\pi$ degrees and $1°$ equals $\pi/180$ radians.

Radian Measure A central angle of 1 radian in a circle of radius r subtends an arc of length r.

$$\pi \text{ radians} = 180° \qquad 2\pi \text{ radians} = 360°$$

$$1 \text{ rad} = \frac{180°}{\pi} \approx 57.29578° \qquad 1° = \frac{\pi}{180} \text{ rad} \approx 1.745329 \times 10^{-2} \text{ rad}$$

Examples

(1) $18° = 18 \cdot \dfrac{\pi}{180} = \tfrac{1}{10}\pi$ radians

(2) $47.69° = 47.69 \cdot \dfrac{\pi}{180} \approx 0.8323$ radians

(3) $\tfrac{2}{9}\pi$ radians $= \tfrac{2}{9}\pi \cdot \dfrac{180°}{\pi} = 40°$

(4) 1.3645 radians $= 1.3645 \cdot \dfrac{180°}{\pi} \approx 78.180°$

It is common practice to omit the unit "radian". Thus we write $\theta = \tfrac{1}{10}\pi$ as an abbreviation for $\theta = \tfrac{1}{10}\pi$ radians. (This convention is justified because the radian is, in certain respects, a natural unit of angle measure, whereas the degree is man-made.)

Radian Measures of Common Angles				
$15° = \tfrac{1}{12}\pi$	$30° = \tfrac{1}{6}\pi$	$45° = \tfrac{1}{4}\pi$	$60° = \tfrac{1}{3}\pi$	$75° = \tfrac{5}{12}\pi$
$90° = \tfrac{1}{2}\pi$	$105° = \tfrac{7}{12}\pi$	$120° = \tfrac{2}{3}\pi$	$135° = \tfrac{3}{4}\pi$	$150° = \tfrac{5}{6}\pi$
$165° = \tfrac{11}{12}\pi$	$180° = \pi$	$210° = \tfrac{7}{6}\pi$	$225° = \tfrac{5}{4}\pi$	$240° = \tfrac{4}{3}\pi$
$270° = \tfrac{3}{2}\pi$	$300° = \tfrac{5}{3}\pi$	$315° = \tfrac{7}{4}\pi$	$330° = \tfrac{11}{6}\pi$	$360° = 2\pi$

The $\boxed{\pi}$ key on your calculator is a shortcut for keying in π. Since $\theta°$ equals $(\pi/180)\theta$ radians, the key sequence

$\boxed{\pi}$ $\boxed{\div}$ $\boxed{1}$ $\boxed{8}$ $\boxed{0}$ $\boxed{\times}$ $\boxed{\theta}$ $\boxed{=}$

converts $\theta°$ to radians.

Note   If your calculator has a constant factor capability (p. 102), you can keep $\pi/180$ as a constant factor and easily do a batch of conversions, like Exs. 21–24 below.

Similarly

$\boxed{1}$ $\boxed{8}$ $\boxed{0}$ $\boxed{\div}$ $\boxed{\pi}$ $\boxed{\times}$ $\boxed{\beta}$ $\boxed{=}$

converts β radians to degrees.

Functions of Radians

From now on (unless otherwise noted) when we write $\sin\theta$ or $\cos\theta$, we mean that θ is in radians.

Examples

$$\sin\tfrac{1}{2}\pi = 1 \qquad \sin\tfrac{1}{6}\pi = \tfrac{1}{2} \qquad \cos\tfrac{3}{4}\pi = -\tfrac{1}{2}\sqrt{2} \qquad \cos\pi = -1.$$

These all come from previous calculations. As quickly as possible, you should learn to think in radians and toss away the crutch of passing through degrees. You should think of a right angle as $\tfrac{1}{2}\pi$, of half of a right angle as $\tfrac{1}{4}\pi$, of one-third of a right angle as $\tfrac{1}{6}\pi$, etc.

The Sine Function

Let us recall the graph of $y = \sin \theta$, only this time with θ in radians, $0 \le \theta \le 2\pi$. The function y increases from 0 to 1 as θ increases from 0 to $\frac{1}{2}\pi$. Then y decreases from 1 to -1 as θ increases from $\frac{1}{2}\pi$ to $\frac{3}{2}\pi$. Finally y increases from -1 to 0 as θ increases from $\frac{3}{2}\pi$ to 2π. See Fig. 2.

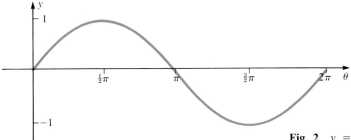

Fig. 2 $y = \sin \theta$ $0 \le \theta \le 2\pi$

Now we are going to do a construction that extends this function $y = \sin \theta$ to be a function on the whole real line, not just on the interval $0 \le \theta \le 2\pi$. We do it simply by shifting the graph both left and right in steps of 2π. See Fig. 3. The pieces match because $\sin 2\pi = \sin 0$, so the next piece always takes off from where the previous one stops.

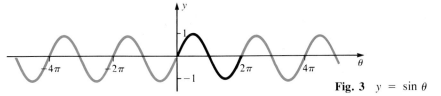

Fig. 3 $y = \sin \theta$ all θ

The result is a function $y = \sin \theta$ defined for all θ, and satisfying the relation

$$\sin(\theta + 2\pi n) = \sin \theta \qquad n \text{ any integer}.$$

This expresses analytically what we did geometrically with the graph. The piece from $2\pi n$ to $2\pi(n + 1)$ is obtained by shifting the original piece (from 0 to 2π) to the right $2\pi n$ units (left if n is negative).

Examples

(1) $\sin \frac{27}{4}\pi = \sin(\frac{3}{4}\pi + 6\pi) = \sin \frac{3}{4}\pi = \frac{1}{2}\sqrt{2}$

(2) $\sin \frac{34}{3}\pi = \sin(\frac{4}{3}\pi + 10\pi) = \sin \frac{4}{3}\pi = -\frac{1}{2}\sqrt{3}$

(3) $\sin(-\frac{19}{6}\pi) = \sin(\frac{5}{6}\pi - 4\pi) = \sin \frac{5}{6}\pi = \frac{1}{2}$.

Your calculator will handle, with reasonable accuracy, $\sin \theta$ for any reasonable value of θ, negative, or greater than 2π, provided of course that it fits on the calculator.

Examples (Set $\boxed{\text{rad}}$ mode.)

(1) $\sin \frac{34}{3}\pi$ $\qquad$ $3 \; 4 \; \boxed{\times} \; \boxed{\pi} \; \boxed{\div} \; 3 \; \boxed{=} \; \boxed{\sin}$ $\qquad$ -0.866025

(2) $\sin(-27.68)$ $2\,7\,.\,6\,8$ $\boxed{+/-}$ $\boxed{\sin}$ -0.559957

Note   Many calculators have a $\boxed{\text{grad}}$ mode. Some countries on the metric system prefer the **grad** as the practical unit of angle measure. There are 100 grads in a right angle, so $\frac{1}{6}\pi = 30° = 33\frac{1}{3}$ grad.

Angles Again

We have defined $\sin\theta$ for each real number θ. Does that mean we have defined angles of all possible sizes, even negative? Yes and no. We can't change geometry; a geometric angle is what it was in Fig. 1, p. 23: one of the regions determined by two rays. But let's think dynamically instead of statically. Let's think of a ray, pivoted at the vertex, that starts on the initial side and sweeps counterclockwise to the terminal side. Figure 4a shows an angle of $\frac{1}{6}\pi$ swept out this way. Now let's suppose that the ray, like a lighthouse beam, sweeps a full revolution and then stops at the same terminal side (Fig. 4b). Then it has swept through an angle of 2π plus $\frac{1}{6}\pi$ more, altogether $\frac{13}{6}\pi$. If instead it rotates two full revolutions, then stops at the terminal side (Fig. 4c), the result is a swept angle of $4\pi + \frac{1}{6}\pi = \frac{25}{6}\pi$. If it sweeps backwards (*clockwise*), that counts as a negative angle (Fig. 4d).

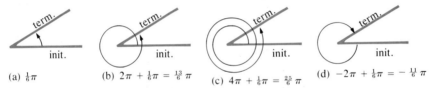

(a) $\frac{1}{6}\pi$ (b) $2\pi + \frac{1}{6}\pi = \frac{13}{6}\pi$ (c) $4\pi + \frac{1}{6}\pi = \frac{25}{6}\pi$ (d) $-2\pi + \frac{1}{6}\pi = -\frac{11}{6}\pi$

Fig. 4   Swept angles—same initial and terminal sides

Our new (dynamic) angle consists of an initial side, a terminal side, and a history of how we got there: counterclockwise or clockwise, and how many full revolutions before stopping. The initial side and one single real number describes all this. For instance, $\theta = -\frac{38}{3}\pi$ tells you the rotation is *clockwise* through 6 full revolutions $(-6 \times 2\pi = -12\pi = -\frac{36}{3}\pi)$ and $-\frac{2}{3}\pi$ more. Its terminal side is in the 3-rd quadrant at angle $\frac{1}{3}\pi$ with the negative x-axis. (Check this.)

Remark   The **revolution** is another unit of angle measurement, commonly used with rotating machinery. It is a whole angle, so

   1 rev $= 360° = 2\pi$ radians .

For instance, a certain electric hand drill idles at 1150 rpm (revolutions per minute). That means in each minute it turns (sweeps) through an angle of 1150 rev $= 2300\pi$.

The Cosine Function

Now let us recall the graph of $y = \cos\theta$, only with θ in radians, $0 \le \theta \le 2\pi$. The function decreases from 1 to -1 as θ increases from 0 to π. Then y increases from -1 back to 1 as θ increases from π to 2π. See Fig. 5.
   We extend this function to a function on the whole real line by shifting the graph

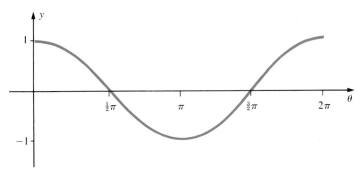

Fig. 5 $y = \cos \theta \qquad 0 \le \theta \le 2\pi$

left and right in steps of 2π. See Fig. 6. The result is a function $y = \cos \theta$ defined for all θ, and satisfying the relation

$$\cos(\theta + 2\pi n) = \cos \theta.$$

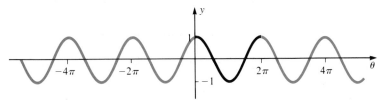

Fig. 6 $y = \cos \theta \qquad$ all θ

Exercises

Convert to exact degrees

1 $\frac{1}{10}\pi$	**2** $\frac{1}{15}\pi$	**3** $\frac{3}{5}\pi$	**4** $\frac{3}{8}\pi$	**5** $\frac{2}{9}\pi$
6 $\frac{1}{90}\pi$	**7** $\frac{13}{12}\pi$	**8** $\frac{17}{12}\pi$	**9** $\frac{19}{12}\pi$	**10** $\frac{23}{12}\pi$.

Convert to radians; express in the form $b\pi$, where b is rational

11 $10°$	**12** $12°$	**13** $36°$	**14** $67.5°$	**15** $54°$
16 $202.5°$	**17** $357°$	**18** $275°$	**19** $228°$	**20** $108°$.

Convert to 4-place radians on a calculator

21 $12.85°$	**22** $10.34°$	**23** $172.91°$	**24** $0.0751°$.

Convert to 2-place degrees on a calculator

25 0.5400	**26** 5.1774	**27** 0.03161	**28** 1.0003.

Find the arc length of an arc of radius 5 subtended by a central angle $\theta =$

29 2.604	**30** 0.1313	**31** $11.65°$	**32** $104.61°$.

Find (exactly) $\sin \theta$ and $\cos \theta$ for $\theta =$

33 3π	**34** $-\pi$	**35** 51π	**36** -7π
37 $\frac{19}{4}\pi$	**38** $\frac{73}{4}\pi$	**39** $-\frac{19}{4}\pi$	**40** $-\frac{73}{4}\pi$
41 $\frac{28}{3}\pi$	**42** $\frac{67}{6}\pi$	**43** $-\frac{28}{3}\pi$	**44** $-\frac{67}{6}\pi$.

Find (exactly) all θ in the interval $-\pi \le \theta \le \pi$ that satisfy

45 $\sin \theta = -\frac{1}{2}$ **46** $\cos \theta = -\frac{1}{2}$ **47** $\cos \theta = \frac{1}{2}\sqrt{2}$ **48** $\sin \theta = \frac{1}{2}\sqrt{2}$

49 $\sin \theta = -1$ **50** $\cos \theta = -1$ **51** $\cos \theta = \frac{1}{2}\sqrt{3}$ **52** $\sin \theta = \frac{1}{2}\sqrt{3}$.

Find all real numbers θ that satisfy

53 $\sin \theta = 0$ **54** $\cos \theta = 0$ **55** $\sin \theta = 1$ **56** $\cos \theta = -1$.

57   (A true story) The instruction manual of a certain calculator gives these directions for calculating $\sin 63°$ and $\cos \frac{1}{4}\pi$:

6 3 $\boxed{\text{sin}}$ $\quad 0.8910065$

$\boxed{\pi}$ $\boxed{\div}$ 4 $\boxed{=}$ $\boxed{\text{cos}}$ 0.9999061

Explain why the writer of the manual should be fired.

58*   Find all real numbers θ such that θ radians and θ degrees have the same terminal side. Equivalently, such that both $\cos \theta = \cos \theta°$ and $\sin \theta = \sin \theta°$.

59   I am not sure what mode my calculator is in, so I key 3 0 $\boxed{\text{sin}}$. The result is a negative number. Conclusion?

60   (cont.) The result is a positive number, not 0.5. Conclusion?

5 PROPERTIES OF SINE AND COSINE

Properties of the Sine Function

Let θ_1 and θ_2 be two real numbers. Suppose the corresponding angles they determine have the same terminal side. Then they differ by an integer multiple of a whole angle, that is, $\theta_2 - \theta_1 = 2\pi n$. Conversely if θ_1 and θ_2 differ by an integer multiple of 2π, then the corresponding angles they sweep out have the same terminal side.

Since $\sin \theta$ was cooked up to satisfy $\sin(\theta + 2\pi n) = \sin \theta$, we conclude that if θ_1 and θ_2 determine the same terminal side, then $\sin \theta_1 = \sin \theta_2$.

Now look again at the graph of $y = \sin \theta$ in Fig. 3, page 37. It looks like an odd function; can we prove it is? We want to prove $\sin(-\theta) = -\sin \theta$. To reach the terminal side of $-\theta$, we do the same motion *clockwise* that we did *counter-*

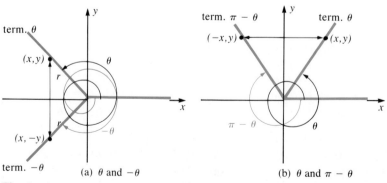

(a) θ and $-\theta$ (b) θ and $\pi - \theta$

Fig. 1

clockwise to reach the terminal side of θ. See Fig. 1a. It is clear from this that the terminal side of $-\theta$ is the reflection in the x-axis of the terminal side of θ. Choose (x, y) on the terminal side of θ, at distance r from $(0, 0)$. Then $(x, -y)$ is on the terminal side of $-\theta$, hence

$$\sin(-\theta) = \frac{-y}{r} = -\frac{y}{r} = -\sin\theta.$$

Look again at the graph of $y = \sin\theta$. It seems to be symmetric in the vertical line $x = \frac{1}{2}\pi$, that is, apparently $\sin(\pi - \theta) = \sin\theta$. We have already verified this relation for the case $0 < \theta < \pi$. But it is true for all θ. Indeed, you rotate as much *clockwise* starting from the negative x-axis to reach the terminal side of $\pi - \theta$ as you rotate *counterclockwise* from the positive x-axis to reach the terminal side of θ. See Fig. 1b. But the negative x-axis is π radians forward of the positive x-axis. Geometrically, this means that the terminal sides of $\pi - \theta$ and θ are symmetric in the y-axis. Now an argument like that of the last paragraph establishes $\sin(\pi - \theta) = \sin\theta$.

$$\sin(-\theta) = -\sin\theta \qquad \sin(\pi - \theta) = \sin\theta$$

From these relations, or by further geometric arguments, other properties of the sine function follow, for example

$$\sin(\theta + \pi) = -\sin\theta \qquad \sin(2\pi - \theta) = -\sin\theta.$$

A Relation Connecting Sine and Cosine

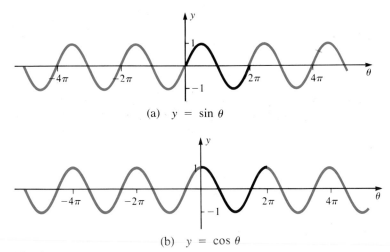

(a) $y = \sin\theta$

(b) $y = \cos\theta$

Fig. 2

The graphs (Fig. 2) of $y = \sin\theta$ and $y = \cos\theta$ look remarkably alike! Apparently if you shift the graph of $y = \sin\theta$ to the left by $\frac{1}{2}\pi$, you get the graph of $y = \cos\theta$. Analytically,

$$\boxed{\sin(\theta + \tfrac{1}{2}\pi) = \cos\theta.}$$

We shall establish this by a geometric argument. Study Fig. 3 and its legend carefully. (You may wish to draw it with the terminal side of θ in other quadrants to be convinced.) If $P = (x, y)$ is on the terminal side of θ at distance r from $(0, 0)$, then $Q = (-y, x)$ is on the terminal side of $\theta + \tfrac{1}{2}\pi$, at the same distance. Therefore

$$\sin(\theta + \tfrac{1}{2}\pi) = \frac{x}{r} = \cos\theta.$$

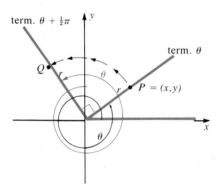

term. $\theta + \tfrac{1}{2}\pi$

term. θ

Q

θ

$P = (x, y)$

θ

Fig. 3 The terminal side of $\theta + \tfrac{1}{2}\pi$ bears the same relation to the positive y-axis that the terminal side of θ bears to the positive x-axis. It also bears the same relation to the *negative* x-axis that term. θ bears to the positive y-axis. Therefore $Q = (-y, x)$.

From the same set-up we also have

$$\cos(\theta + \tfrac{1}{2}\pi) = \frac{-y}{r} = -\frac{y}{r} = -\sin\theta.$$

This also can be derived analytically from the boxed relation above and properties of $\sin\theta$:

$$\cos(\theta + \tfrac{1}{2}\pi) = \sin[(\theta + \tfrac{1}{2}\pi) + \tfrac{1}{2}\pi]$$
$$= \sin(\theta + \pi) = -\sin\theta.$$

Because of the relation $\cos\theta = \sin(\theta + \tfrac{1}{2}\pi)$ expressing $\cos\theta$ in terms of sine, every property of the sine function implies a property of the cosine function. For instance, $\sin(\pi - \theta) = \sin\theta$ can be used as follows:

$$\cos(-\theta) = \sin(-\theta + \tfrac{1}{2}\pi) = \sin[\pi - (\tfrac{1}{2}\pi + \theta)]$$
$$= \sin(\tfrac{1}{2}\pi + \theta) = \cos\theta.$$

Therefore $\cos\theta$ is an even function. Similarly, we can derive other relations involving the cosine. Let us summarize the main such properties of both sine and cosine.

Properties of Sine and Cosine

(1) $-1 \le \sin \theta \le 1 \qquad -1 \le \cos \theta \le 1$

(2) $\sin^2 \theta + \cos^2 \theta = 1$

(3) $\sin(\theta + 2\pi n) = \sin \theta \qquad \cos(\theta + 2\pi n) = \cos \theta$

(4) sin is odd, cosine is even:

$\sin(-\theta) = -\sin \theta \qquad \cos(-\theta) = \cos \theta$

(5) $\sin(\pi - \theta) = \sin \theta \qquad \cos(\pi - \theta) = -\cos \theta$

(6) $\sin(\theta + \frac{1}{2}\pi) = \cos \theta \qquad \cos(\theta + \frac{1}{2}\pi) = -\sin \theta$

(7) $\sin(\theta + \pi) = -\sin \theta \qquad \cos(\theta + \pi) = -\cos \theta$

(8) $\sin(\frac{1}{2}\pi - \theta) = \cos \theta \qquad \cos(\frac{1}{2}\pi - \theta) = \sin \theta$

Remark You should memorize (1)–(4). The first two are easy; the next two are easy if you remember the graphs, which you should. Relations (5)–(8) are a little harder to remember, but you only need (8); the others all follow easily, for example

$$\sin(\theta + \tfrac{1}{2}\pi) = \sin[\tfrac{1}{2}\pi - (-\theta)] = \cos(-\theta) = \cos \theta, \text{ etc.}$$

Actually there is only one relation (8), since both say that if $\alpha + \beta = \frac{1}{2}\pi$, then $\sin \alpha = \cos \beta$. In the case of acute angles, this is an easy-to-remember statement about complementary angles (in a right triangle). Actually, (5)–(8) are all statements about symmetries in the graphs of sine and cosine; they can be recalled as needed by looking at the graphs, or by a quick sketch of angles and their terminal sides. The big box isn't as big a deal as it first appears.

EXAMPLE 1 Express in terms of $y = \sin \frac{1}{8}\pi$

(a) $\sin \frac{9}{8}\pi$ (b) $\cos(\frac{51}{8}\pi)$ (c) $\cos(-\frac{23}{8}\pi)$

Solution (a) Use (7):

$$\sin \tfrac{9}{8}\pi = \sin(\tfrac{1}{8}\pi + \pi) = -\sin \tfrac{1}{8}\pi = -y.$$

(b) Use (3) and (8):

$$\cos(\tfrac{51}{8}\pi) = \cos(\tfrac{3}{8}\pi + 6\pi) = \cos \tfrac{3}{8}\pi$$
$$= \cos(\tfrac{1}{2}\pi - \tfrac{1}{8}\pi) = \sin \tfrac{1}{8}\pi = y.$$

(c) Use (3), (4), (5), and (2):

$$\cos(-\tfrac{23}{8}\pi) = \cos(-\tfrac{7}{8}\pi - 2\pi) = \cos(-\tfrac{7}{8}\pi) = \cos \tfrac{7}{8}\pi$$
$$= \cos(\pi - \tfrac{1}{8}\pi) = -\cos \tfrac{1}{8}\pi$$
$$= -\sqrt{1 - \sin^2 \tfrac{1}{8}\pi} = -\sqrt{1 - y^2}.$$

Answer (a) $-y$ (b) y (c) $-\sqrt{1 - y^2}$

A Graph Construction

Finally, we describe a geometric construction for the graph of $y = \sin\theta$. Of course, the $\cos\theta$ graph is free once we have the sine. This construction is quite accurate for practical purposes; needed are graph paper, compass, ruler, and protractor (not really needed).

First we draw the θ- and y-axes and mark scales on each. Given θ, we wish to construct the corresponding point $(\theta, \sin\theta)$ on the graph of $y = \sin\theta$.

To the left, out of the way, we draw an auxiliary y-axis, an auxiliary x-axis collinear with the θ-axis, and a unit circle (Fig. 4). Given θ, we draw a corresponding angle in standard position. Its terminal side intersects the unit circle in $(\cos\theta, \sin\theta)$. We project this point horizontally; where it meets the vertical through θ (on the right-hand graph) is the desired point $(\theta, \sin\theta)$.

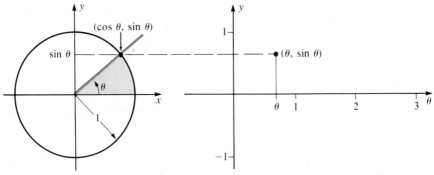

Fig. 4

Let's try it, using $\frac{1}{12}\pi = 15°$ increments for θ. We mark the points that divide the unit circle into 24 equal parts of $\frac{1}{24}(2\pi) = \frac{1}{12}\pi$ each, and project each point horizontally. This gives us the accurate graph of Fig. 5.

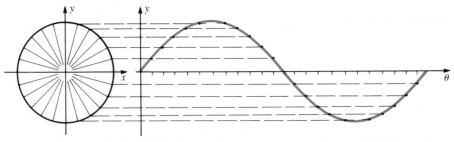

Fig. 5   The construction (15° increments)

Exercises

Express in terms of $\sin\theta$ or $\cos\theta$

1 $\sin(\theta - 3\pi)$ **2** $\cos(\theta + 5\pi)$ **3** $\cos(\theta + \frac{5}{2}\pi)$
4 $\sin(\theta - \frac{7}{2}\pi)$ **5** $\sin(\frac{9}{2}\pi - \theta)$ **6** $\cos(\theta - \frac{17}{2}\pi)$.

Express in terms of $y = \sin \frac{1}{12}\pi$

7 $\sin \frac{13}{12}\pi$ **8** $\sin \frac{5}{12}\pi$ **9** $\cos\left(-\frac{11}{12}\pi\right)$ **10** $\cos\left(-\frac{37}{12}\pi\right)$.

Express in terms of $x = \cos \frac{1}{10}\pi$

11 $\cos \frac{11}{10}\pi$ **12** $\cos \frac{19}{10}\pi$ **13** $\sin \frac{7}{5}\pi$ **14** $\sin\left(-\frac{12}{5}\pi\right)$.

Find all θ in the interval $0 \le \theta \le 2\pi$ that satisfy

15 $\cos \theta \ge 0$ **16** $\sin \theta \ge \frac{1}{2}$ **17** $\sin \theta \ge \cos \theta$

18 $\cos \theta \ge \sin \theta$ **19*** $\sin \theta + \cos \theta \ge 1$ **20*** $\sin \theta - \cos \theta \ge 1$.

21 Prove $\cos\left(\theta + \frac{1}{2}\pi\right) + \cos\left(\theta - \frac{1}{2}\pi\right) = 0$.

22* Prove $\sin \frac{1}{15}\pi + \sin \frac{3}{15}\pi + \sin \frac{5}{15}\pi + \cdots + \sin \frac{29}{15}\pi = 0$.

23 The graph of $y = \sin \theta$ appears symmetric with respect to the line $\theta = \frac{9}{2}\pi$. Prove it so.

24 The graph of $y = \sin \theta$ appears symmetric with respect to the point $(3\pi, 0)$. Prove it so.

REVIEW EXERCISES

Find exact values of $\sin \theta$ and $\cos \theta$ for θ in standard position with its terminal side through

1 $(\sqrt{5}, \sqrt{7})$ **2** $(\sqrt{7}, -\sqrt{5})$.

Estimate by calculator to 5 places

3 $\sin 32.756°$ **4** $\cos 4.1012$.

Express in terms of $S = \sin 27°$

5 $\cos 117°$ **6** $\sin 243°$.

7 Convert $\frac{3}{7}\pi$ to degrees.

8 Convert $113°$ to $b\pi$ radians, b rational.

Find all θ with $0 \le \theta < 2\pi$ such that

9 $\sin \theta = \cos \theta$ **10** $\sin \theta = -\cos \theta$.

3 FURTHER TRIGONOMETRIC FUNCTIONS

1 THE TANGENT

To begin our work on the tangent function, we temporarily return to angles in standard position measured in degrees. If θ is such an angle, then $0° \leq \theta° \leq 360°$. The tangent is undefined if $\theta° = 90°$ or $\theta° = 270°$.

Tangent Let θ be an angle in standard position, but not 90° or 270°. Let (x, y) be any point on its terminal side, different from $(0, 0)$. Then

$$\tan \theta = \frac{y}{x}.$$

Alternatively,

$$\tan \theta = \frac{\sin \theta}{\cos \theta}.$$

In words, $\tan \theta$ is the *slope* of the terminal side.

Note that

$$\frac{\sin \theta}{\cos \theta} = \frac{y/r}{x/r} = \frac{y}{x},$$

so the ratio y/x depends only on θ, not on the choice of the point (x, y) on the terminal side. The exceptions, 90° and 270°, are angles whose terminal sides coincide with one side or the other of the y-axis, hence the denominator $x = 0$, and y/x is undefined. (Alternatively, $\cos \theta = 0$.) Finally, recall that the slope of the terminal side, a non-vertical line through $(0, 0)$, is

$$\text{slope} = \frac{y - 0}{x - 0} = \frac{y}{x} = \tan \theta.$$

If we think of the terminal side as the graph of a function $y = f(x)$, then this function is the linear function $y = (\tan \theta)x$.

Examples

(1) $\tan 45° = \dfrac{\sin 45°}{\cos 45°} = \dfrac{\frac{1}{2}\sqrt{2}}{\frac{1}{2}\sqrt{2}} = 1$

(2) $\tan 150° = \dfrac{\sin 150°}{\cos 150°} = \dfrac{\frac{1}{2}}{-\frac{1}{2}\sqrt{3}} = \dfrac{-1}{\sqrt{3}} = -\frac{1}{3}\sqrt{3}$

If β is an acute angle of a right triangle (Fig. 1, page 30), then

$$\tan \beta = \frac{\text{opposite}}{\text{adjacent}}.$$

EXAMPLE 1 Express in terms of $T = \tan 35°$

(a) $\cos 35°$ (b) $\sin 145°$ (c) $\tan 305°$.

Solution (a) Draw $35°$ in standard position and a corresponding 35-55-90 right triangle (Fig. 1a). Take $\text{adj} = 1$. Then

$$T = \tan 35° = \frac{\text{opp}}{\text{adj}} = \frac{b}{1} = b.$$

By the Pythagorean theorem,

$$c = \text{hyp} = \sqrt{1 + b^2} = \sqrt{1 + T^2}$$

so

$$\cos 35° = \frac{\text{adj}}{\text{hyp}} = \frac{1}{\sqrt{1 + T^2}}.$$

(b) The terminal side of $145°$ lies in the second quadrant, $35°$ short of the negative x-axis. Place the triangle from (a) accordingly (Fig. 1b). Clearly $x = -1$ and $y = b = T$, so

$$\sin 145° = \frac{y}{r} = \frac{T}{\sqrt{1 + T^2}}.$$

(c) The terminal side of $305°$ lies in the fourth quadrant, $35°$ forward of the negative y-axis. Place the triangle from (a) accordingly (Fig. 1c). Clearly $x = b = T$ and $y = -1$, so

$$\tan 305° = \frac{y}{x} = \frac{-1}{T}.$$

Answer (a) $\dfrac{1}{\sqrt{1 + T^2}}$ (b) $\dfrac{T}{\sqrt{1 + T^2}}$ (c) $\dfrac{-1}{T}$

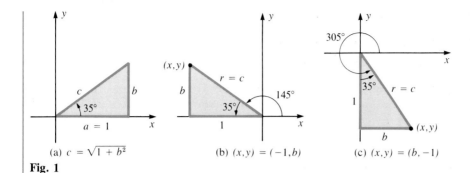

(a) $c = \sqrt{1 + b^2}$ (b) $(x,y) = (-1, b)$ (c) $(x,y) = (b, -1)$

Fig. 1

To estimate $\tan \theta°$ on a calculator, use the $\boxed{\tan}$ key. (First set the mode to $\boxed{\text{deg}}$.)

Examples

(1) $\tan 27.42°$ 2 7 . 4 2 $\boxed{\tan}$ *0.5 188*

(2) $\tan 310°$ 3 1 0 $\boxed{\tan}$ *- 1.19 17536*

The Tangent Function

So far $\tan \theta°$ is defined for $0 \leq \theta° \leq 360°$, not $90°$ or $270°$. Translating to radians, $\tan \theta$ is defined for $0 \leq \theta \leq 2\pi$, $\theta \neq \frac{1}{2}\pi$, $\theta \neq \frac{3}{2}\pi$. We have three ways to look at $\tan \theta$ when θ is in standard position. First, $\tan \theta = y/x$ where (x, y) is any point on the terminal side of θ different from $(0, 0)$. Second, $\tan \theta$ is the slope of its terminal side. Third, $\tan \theta = (\sin \theta)/(\cos \theta)$.

Any of the three can be used to extend $\tan \theta$ to the whole real axis, except for those θ whose terminal sides are vertical; the odd multiples of $\frac{1}{2}\pi$:

$$\pm \tfrac{1}{2}\pi \quad \pm \tfrac{3}{2}\pi \quad \pm \tfrac{5}{2}\pi \quad \cdots$$

> **Tangent** The function $\tan \theta$ is defined for all real θ except $\pm \frac{1}{2}\pi$, $\pm \frac{3}{2}\pi$, $\pm \frac{5}{2}\pi$, $\cdots$. If θ is in standard position, then $\tan \theta$ is the slope of the terminal side of θ, that is, $\tan \theta = y/x$ where (x, y) is on the terminal side and $x \neq 0$. Alternatively,
>
> $$\tan \theta = \frac{\sin \theta}{\cos \theta}.$$

Examples

(1) $\tan \frac{1}{4}\pi = \tan 45° = 1$

(2) $\tan \frac{1}{6}\pi = \tan 30° = \frac{1}{3}\sqrt{3}$

(3) $\tan \frac{17}{3}\pi = \dfrac{\sin \frac{17}{3}\pi}{\cos \frac{17}{3}\pi} = \dfrac{\sin(-\frac{1}{3}\pi + 6\pi)}{\cos(-\frac{1}{3}\pi + 6\pi)} = \dfrac{\sin(-\frac{1}{3}\pi)}{\cos(-\frac{1}{3}\pi)}$

$$= \frac{-\frac{1}{2}\sqrt{3}}{\frac{1}{2}} = -\sqrt{3}.$$

(4) $\tan\left(-\frac{3}{4}\pi\right) = \dfrac{\sin\left(-\frac{3}{4}\pi\right)}{\cos\left(-\frac{3}{4}\pi\right)} = \dfrac{-\sin\frac{1}{4}\pi}{-\cos\frac{1}{4}\pi} = 1$.

Properties of the Tangent Function

The tangent function has two immediate properties, both obvious from the slope definition, but we'll prove them analytically anyhow. First, $\tan\theta$ is an odd function:

$$\tan(-\theta) = \frac{\sin(-\theta)}{\cos(-\theta)} = \frac{-\sin\theta}{\cos\theta} = -\tan\theta.$$

Second, $\tan\theta$ is unchanged when θ is replaced by $\theta \pm \pi$:

$$\tan(\theta \pm \pi) = \frac{\sin(\theta \pm \pi)}{\cos(\theta \pm \pi)} = \frac{-\sin\theta}{-\cos\theta} = \tan\theta.$$

More generally, if n is a whole number, then

$$\tan(\theta \pm n\pi) = \tan\theta.$$

For instance,

$$\tan(\theta + 2\pi) = \tan(\theta + \pi + \pi) = \tan(\theta + \pi) = \tan\theta$$

$$\tan(\theta - 3\pi) = \tan(\theta - \pi - \pi - \pi) = \tan(\theta - \pi - \pi)$$

$$= \tan(\theta - \pi) = \tan\theta.$$

$$\boxed{\tan(-\theta) = -\tan\theta \qquad \tan(\theta \pm n\pi) = \tan\theta}$$

Remark $\tan(\theta + \pi) = \tan\theta$ is a surprise; our first guess would be only $\tan(\theta + 2\pi) = \tan\theta$, the way $\sin\theta$ and $\cos\theta$ behave.

Examples

(1) $\tan\frac{17}{3}\pi = \tan\left(-\frac{1}{3}\pi + 6\pi\right) = \tan\left(-\frac{1}{3}\pi\right) = -\tan\frac{1}{3}\pi = -\sqrt{3}$.

(2) $\tan\left(-\frac{3}{4}\pi\right) = \tan\left(-\frac{3}{4}\pi + \pi\right) = \tan\frac{1}{4}\pi = 1$.

Complementary Angles

The formulas (8) on p. 43 for sin and cos of $\frac{1}{2}\pi - \theta$ imply something about tan:

$$\boxed{\tan\left(\tfrac{1}{2}\pi - \theta\right) = \frac{1}{\tan\theta}.}$$

This is true because

$$\tan\left(\tfrac{1}{2}\pi - \theta\right) = \frac{\sin\left(\frac{1}{2}\pi - \theta\right)}{\cos\left(\frac{1}{2}\pi - \theta\right)} = \frac{\cos\theta}{\sin\theta} = \frac{1}{\tan\theta}.$$

For acute angles, the formula has a direct right triangle interpretation (Fig. 2). In the triangle, α and β are complementary, and

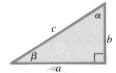

Fig. 2

$$\tan\beta = \frac{\text{opp}}{\text{adj}} = \frac{b}{a} = \frac{1}{a/b} = \frac{1}{\tan\alpha}.$$

Exercises

Find (exact values)

1	$\tan 0°$	**2**	$\tan 180°$	**3**	$\tan 30°$	**4**	$\tan 60°$
5	$\tan 120°$	**6**	$\tan 135°$	**7**	$\tan 210°$	**8**	$\tan 225°$
9	$\tan 240°$	**10**	$\tan 300°$	**11**	$\tan 315°$	**12**	$\tan 330°$.

Express in terms of $K = \cos 20°$

13 $\tan 160°$ **14** $\tan 290°$.

Express in terms of $T = \tan 75°$

15	$\tan 15°$	**16**	$\cos 15°$	**17**	$\sin 165°$	**18**	$\tan 165°$
19	$\tan 195°$	**20**	$\sin 195°$	**21**	$\sin 255°$	**22**	$\cos 255°$.

Estimate to 4 places by calculator

23 $\tan 17.61°$ **24** $\tan 72.41°$ **25** $\tan 100.61°$ **26** $\tan 275.73°$.

Find $\tan \theta$ (exact) for $\theta =$

27	$-\pi$	**28**	7π	**29**	$-\frac{1}{4}\pi$	**30**	$-\frac{1}{6}\pi$
31	$-\frac{1}{3}\pi$	**32**	$-\frac{4}{3}\pi$	**33**	$\frac{13}{3}\pi$	**34**	$\frac{11}{4}\pi$
35	$\frac{100}{3}\pi$	**36**	$-\frac{101}{3}\pi$.				

Find all real θ satisfying

37 $\tan \theta = 0$ **38** $\tan \theta = -1$ **39** $\tan \theta = -\dfrac{1}{\sqrt{3}}$

40 $\tan \theta = -\sqrt{3}$ **41** $\tan \theta > 0$ **42** $\tan \theta > 1$.

Prove

43 $\tan(\pi - \theta) = -\tan \theta$ **44** $\tan(\theta + \frac{1}{2}\pi) = \dfrac{-1}{\tan \theta}$

45 $(1 + \tan^2 \theta)\cos^2 \theta = 1$ **46** $\dfrac{1 - \tan^2 \theta}{1 + \tan^2 \theta} = \cos^2 \theta - \sin^2 \theta$.

Find all possible values for $\tan \theta$ if

47 $\sin \theta = \frac{3}{5}$ **48** $\cos \theta = -\frac{5}{13}$

49 $\cos \theta = \frac{1}{2}$ **50** $\sin \theta = -\frac{1}{2}$.

Estimate by calculator to 6 significant figures

51	$\tan 4.000$	**52**	$\tan(-3.521)$	**53**	$\tan 10.03$
54	$\tan(-15.19)$	**55**	$\tan \frac{11}{7}$	**56**	$\tan \frac{128}{81}$.

2 GRAPHS AND APPROXIMATIONS

Graph of $\tan \theta$

Let us start on the interval $-\frac{1}{2}\pi < \theta < \frac{1}{2}\pi$. There $y = \tan \theta$ is a strictly increasing odd function. By interpreting $\tan \theta$ as the slope of the terminal side of θ, we see that as θ increases from $-\frac{1}{2}\pi$ to 0 to $\frac{1}{2}\pi$, the function $y = \tan \theta$ increases from $-\infty$ to 0 to $+\infty$. This gives us a fair picture (Fig. 1a). Next, be-cause $\tan(\theta + n\pi) = \tan \theta$, we can shift this portion left and right in steps of π to obtain the complete graph (Fig. 1b).

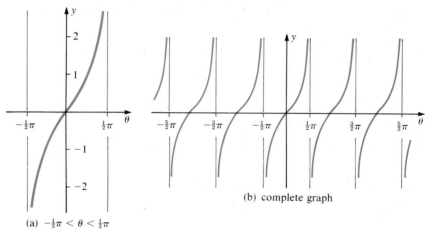

(a) $-\tfrac{1}{2}\pi < \theta < \tfrac{1}{2}\pi$

(b) complete graph

Fig. 1 $y = \tan \theta$

EXAMPLE 1   Find all solutions of $\tan \theta = 1$.

Solution   We know one solution: $\theta = \tfrac{1}{4}\pi$. From the graph, every other solution differs from this one by an integer multiple of π.

Answer $\tfrac{1}{4}\pi + n\pi$ n any integer

A Graph Construction

We modify the construction on page 44 for $y = \sin \theta$ to obtain an accurate construction of the graph of $y = \tan \theta$. We need the vertical line $x = 1$ and the point $(1, \tan \theta)$ where the terminal side of θ meets it (Fig. 2).

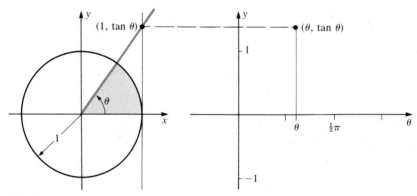

Fig. 2

We apply the construction, with $\tfrac{1}{12}\pi$ increments in θ. See Fig. 3 for one branch of $y = \tan \theta$.

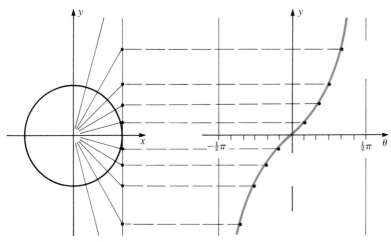

Fig. 3 The construction (15° increments)

Approximations Near $\theta = 0$

We now examine $y = \sin\theta$, $y = \cos\theta$, and $y = \tan\theta$ with a microscope near $\theta = 0$. We begin with a drawing (Fig. 4a) of a small θ. It is geometrically clear that

$$\sin\theta \longrightarrow 0 \quad \text{and} \quad \cos\theta \longrightarrow 1 \quad \text{as} \quad \theta \longrightarrow 0.$$

Thus for our first approximation we can write

$$\sin\theta \approx 0 \quad \text{and} \quad \cos\theta \approx 1 \quad \text{for} \quad \theta \approx 0.$$

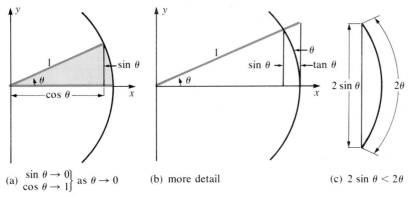

(a) $\left.\begin{matrix} \sin\theta \to 0 \\ \cos\theta \to 1 \end{matrix}\right\}$ as $\theta \to 0$ (b) more detail   (c) $2\sin\theta < 2\theta$

Fig. 4 Small θ

We fine-tune these estimates, starting with a more detailed drawing (Fig. 4b). The arc has length θ; the outside vertical has length $\tan\theta$. Look at the arc (Fig. 4c, left) along the unit circle from the terminal sides of $-\theta$ and θ. Its length 2θ certainly exceeds the corresponding chord length $2\sin\theta$, hence

$$2\sin\theta < 2\theta \qquad \sin\theta < \theta.$$

Now look at the outer right triangle; its legs are 1 and $\tan \theta$, so its area is $\frac{1}{2} \tan \theta$. It includes the sector (of radius 1 and central angle θ) of area $\frac{1}{2} r^2 \theta = \frac{1}{2} \theta$. Hence

$$\tfrac{1}{2}\theta < \tfrac{1}{2} \tan \theta \qquad \theta < \tan \theta = \frac{\sin \theta}{\cos \theta} \qquad \theta \cos \theta < \sin \theta$$

All together,

$$\theta \cos \theta < \sin \theta < \theta.$$

Since $\cos \theta \longrightarrow 1$ as $\theta \longrightarrow 0$, and $\sin \theta$ is trapped between $\theta \cos \theta$ and θ, we conclude that

$$\sin \theta \approx \theta \quad \text{as} \quad \theta \approx 0.$$

We have pulled ourselves up by our bootstraps from $\sin \theta \approx 0$ to $\sin \theta \approx \theta$. (Our figures have $\theta > 0$, but the conclusion holds for $\theta < 0$ also; this is clear from the geometry, or because both $\sin \theta$ and θ are odd functions.)

Let us test $\sin \theta \approx \theta$ for some small θ:

θ	.00	.05	.10	.15	.20	.25	.30
$\sin \theta$	.0000	.0500	.0998	.1494	.1987	.2474	.2955

These values support our estimate. Now let us test smaller θ, this time to 8 places.

θ	.01	.02	.03	.04	.05
$\sin \theta$	.00999983	.01999867	.02999550	.03998933	.04997917

The numerical evidence for $\sin \theta \approx \theta$ is very strong!

Now we can improve our approximation $\cos \theta \approx 1$ as $\theta \approx 0$. First

$$\cos \theta = \sqrt{1 - \sin^2 \theta} \approx \sqrt{1 - \theta^2} \qquad \text{for} \quad \theta \approx 0.$$

For a simpler estimate, we use the approximation*

$$\sqrt{1 - \theta^2} \approx 1 - \tfrac{1}{2}\theta^2 \qquad \text{for} \quad \theta \approx 0.$$

From this,

$$\cos \theta \approx \sqrt{1 - \theta^2} \approx 1 - \tfrac{1}{2}\theta^2 \qquad \text{for} \quad \theta \approx 0.$$

Here is some strong numerical evidence:

θ	.01	.02	.03	.04	.05
$\cos \theta$	.99995000	.99980001	.99955003	.99920011	.99875026
$1 - \tfrac{1}{2}\theta^2$	.99995	.99980	.99955	.99920	.99875

---

* It follows from

$$(1 - \tfrac{1}{2}\theta^2)^2 = 1 - \theta^2 + \tfrac{1}{4}\theta^4 \approx 1 - \theta^2$$

because $\tfrac{1}{4}\theta^4$ is negligible compared to θ^2. Hence

$$\sqrt{1 - \theta^2} \approx \sqrt{(1 - \tfrac{1}{2}\theta^2)^2} = 1 - \tfrac{1}{2}\theta^2.$$

Now let's estimate $\tan \theta$ for $\theta \approx 0$. It's easy to start:

$$\tan \theta = \frac{\sin \theta}{\cos \theta} \approx \frac{0}{1} = 0 \qquad \text{for} \quad \theta \approx 0.$$

Better is

$$\tan \theta = \frac{\sin \theta}{\cos \theta} \approx \frac{\theta}{1} = \theta \qquad \text{for} \quad \theta \approx 0.$$

(It doesn't improve the estimate perceptibly to use $\cos \theta \approx 1 - \frac{1}{2}\theta^2$ in the denominator.) Let's check some calculator values:

θ	.01	.05	.10
$\tan \theta$	.01000033	.05004171	.10033467

The agreement is excellent. Now we summarize our findings:

Approximations Near $\theta = 0$

$\sin \theta \approx \theta \qquad \cos \theta \approx 1 - \frac{1}{2}\theta^2 \qquad \tan \theta \approx \theta$

Remark   It is easy to deduce from the discussion that $(\sin \theta)/\theta \longrightarrow 1$ as $\theta \longrightarrow 0$ but $\theta \neq 0$. This result is important in calculus.

Approximation Near $\theta = \frac{1}{2}\pi$

We know that $\tan \theta \longrightarrow \infty$ as $\theta \longrightarrow \frac{1}{2}\pi -$. How fast? No problem. As $\theta \longrightarrow \frac{1}{2}\pi -$, then $\frac{1}{2}\pi - \theta \longrightarrow 0+$. But (complementary angles)

$$\tan \theta = \frac{1}{\tan(\frac{1}{2}\pi - \theta)} \approx \frac{1}{\frac{1}{2}\pi - \theta} \qquad \text{as} \quad \theta \longrightarrow \frac{1}{2}\pi -.$$

It may be clearer to write this estimate in the form (substituting $\frac{1}{2}\pi - \theta$ for θ):

$$\tan(\tfrac{1}{2}\pi - \theta) \approx \frac{1}{\theta} \qquad \text{as} \quad \theta \longrightarrow 0+$$

We test some calculator values:

θ	.01	.005	.001
$\tan(\frac{1}{2}\pi - \theta)$	99.996667	199.99833	999.99967
$1/\theta$	100	200	1000

Not bad!

Exercises

Tabulate for $\theta = .002, .004, .006, .008, .010$

1 θ and $\sin \theta$

2 θ, $\cos \theta$, and $1 - \frac{1}{2}\theta^2$.

3   Up to approximately what positive value of θ is the estimate $\sin \theta \approx \theta$ accurate to 3 places?

4 (cont.) Answer the same question for the estimate $\cos \theta \approx 1 - \frac{1}{2}\theta^2$.

Justify the estimate

5 $\sin 1° \approx \dfrac{\pi}{180}$

6 $\cos \theta \approx 1 - \frac{1}{2}\sin^2 \theta$, θ small.

7 Show, for $\theta°$ small and positive, that
$$\tan(90° - \theta°) \approx \frac{180}{\pi\theta}.$$

8 (cont.) Find approximately a first quadrant angle $\theta°$ such that $\tan \theta° = 100$.

9 Show that $\tan(\theta - \frac{1}{2}\pi) \approx -1/\theta$ for $\theta \longrightarrow 0+$.

10 Show that $\tan(\theta + \frac{1}{2}\pi) \approx -1/\theta$ for $\theta \longrightarrow 0+$.

11 Prove the estimate
$\frac{1}{2}(\sin \theta + \tan \theta) \approx \theta$ for $\theta \approx 0$.
Verify that it is more accurate than both $\sin \theta \approx \theta$ and $\tan \theta \approx \theta$ for $\theta = 0.1, 0.05, 0.01, 0.005$. Use 8 places if possible.

12 (cont.) Now compare also the estimate
$\frac{2}{3}\sin \theta + \frac{1}{3}\tan \theta \approx \theta$
for the same values. Which is better?

3   COTANGENT, SECANT, AND COSECANT

We now define three more trigonometric functions. They, in a sense, complete the story on what are called direct trigonometric functions.

Cotangent, Secant, Cosecant   Let θ be an angle in standard position and (x, y) a point on its terminal side, different from $(0, 0)$. Set $r = \sqrt{x^2 + y^2}$. Then

$$\cot \theta = \frac{x}{y} \ (y \neq 0) \qquad \sec \theta = \frac{r}{x} \ (x \neq 0) \qquad \csc \theta = \frac{r}{y} \ (y \neq 0).$$

Equivalently,

$$\cot \theta = \frac{\cos \theta}{\sin \theta} = \frac{1}{\tan \theta} \qquad \theta \neq n\pi, \quad n \text{ an integer}$$

$$\sec \theta = \frac{1}{\cos \theta} \qquad \theta \neq (n + \tfrac{1}{2})\pi, \quad n \text{ an integer}$$

$$\csc \theta = \frac{1}{\sin \theta} \qquad \theta \neq n\pi, \quad n \text{ an integer}$$

Examples

(1) $\cot \frac{1}{6}\pi = \dfrac{1}{\tan \frac{1}{6}\pi} = \dfrac{1}{\frac{1}{3}\sqrt{3}} = \sqrt{3}$

(2) $\sec \frac{3}{4}\pi = \dfrac{1}{\cos \frac{3}{4}\pi} = \dfrac{1}{-\frac{1}{2}\sqrt{2}} = -\sqrt{2}$

(3) $\csc\left(-\frac{5}{6}\pi\right) = \dfrac{1}{\sin\left(-\frac{5}{6}\pi\right)} = \dfrac{1}{-\frac{1}{2}} = -2$

Properties

Since cot, sec, csc are the reciprocals of tan, cos, sin respectively, properties of the latter functions translate readily into properties of the former. We first note a batch of relations concerning shifting and parity (even or odd).

$$\begin{cases} \cot(\theta + n\pi) = \cot\theta \\ \cot(-\theta) = -\cot\theta \end{cases} \qquad \begin{cases} \sec(\theta \pm 2\pi n) = \sec\theta \\ \sec(-\theta) = \sec\theta \end{cases}$$

$$\begin{cases} \csc(\theta \pm 2\pi n) = \csc\theta \\ \csc(-\theta) = -\csc\theta \end{cases}$$

Next, we know that $\tan(\frac{1}{2}\pi - \theta) = 1/\tan\theta$ (complementary angles), hence $\cot\theta = \tan(\frac{1}{2}\pi - \theta)$. Also $\sin(\frac{1}{2}\pi - \theta) = \cos\theta$, so taking reciprocals, $\csc(\frac{1}{2}\pi - \theta) = \sec\theta$. Replacing θ by $\frac{1}{2}\pi - \theta$ in these relations leads to two similar ones; all together:

$$\begin{cases} \tan(\frac{1}{2}\pi - \theta) = \cot\theta \\ \cot(\frac{1}{2}\pi - \theta) = \tan\theta \end{cases} \qquad \begin{cases} \sec(\frac{1}{2}\pi - \theta) = \csc\theta \\ \csc(\frac{1}{2}\pi - \theta) = \sec\theta \end{cases}$$

Examples $\frac{1}{3}\pi + \frac{1}{6}\pi = \frac{1}{2}\pi$, so

(1) $\tan\frac{1}{3}\pi = \cot\frac{1}{6}\pi = \sqrt{3}$ (2) $\sec\frac{1}{3}\pi = \csc\frac{1}{6}\pi = 2$

(3) $\tan\frac{1}{6}\pi = \cot\frac{1}{3}\pi = \frac{1}{3}\sqrt{3}$ (4) $\sec\frac{1}{6}\pi = \csc\frac{1}{3}\pi = \dfrac{2}{\sqrt{3}} = \frac{2}{3}\sqrt{3}$.

Graphs

We construct the graph of $y = \cot\theta$ by using the relation

$$\cot\theta = \tan(\tfrac{1}{2}\pi - \theta) = \tan[-(\theta - \tfrac{1}{2}\pi)].$$

We first draw $y = \tan\theta$, then reflect it in the y-axis; the result is $y = \tan(-\theta)$. Then we shift $\frac{1}{2}\pi$ to the right. The result is $y = \tan(\frac{1}{2}\pi - \theta) = \cot\theta$. See Fig. 1, next page.

The complete graph is drawn in Fig. 2, next page. Note the vertical asymptote* at $\theta = n\pi$ for each integer n.

To graph $y = \sec\theta$ and $y = \csc\theta$ we use the relations

$$\sec\theta = \frac{1}{\cos\theta} \qquad \text{and} \qquad \csc\theta = \frac{1}{\sin\theta}$$

---

* A vertical line $\theta = a$ is called a **vertical asymptote** of the graph $y = f(\theta)$ if $y \longrightarrow \infty$ or $y \longrightarrow -\infty$ as $x \longrightarrow a$ from the right or from the left. (We write $x \longrightarrow a+$ to mean x approaches a from the right; that is, $x > a$. Similarly, $x \longrightarrow a-$ means $x \longrightarrow a$ and $x < a$.)

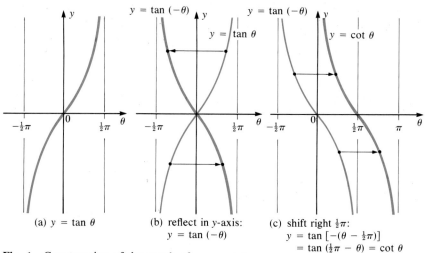

(a) $y = \tan \theta$

(b) reflect in y-axis:
$y = \tan (-\theta)$

(c) shift right $\frac{1}{2}\pi$:
$y = \tan [-(\theta - \frac{1}{2}\pi)]$
$= \tan (\frac{1}{2}\pi - \theta) = \cot \theta$

Fig. 1   Construction of the graph of $y = \cot \theta$

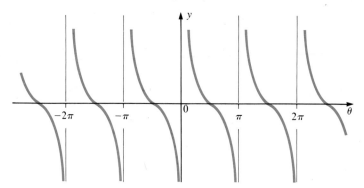

Fig. 2 $y = \cot \theta$

and plot points (Fig. 3 and Fig. 4). Note the vertical asymptotes where the functions are undefined. Note also that $\csc (\theta + \frac{1}{2}\pi) = \sec \theta$ because $\sin (\theta + \frac{1}{2}\pi) = \cos \theta$, so the curve $y = \sec \theta$ is obtained by shifting the curve $y = \csc \theta$ to the left by $\frac{1}{2}\pi$.

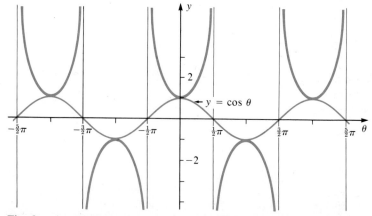

Fig. 3 $y = \sec \theta$

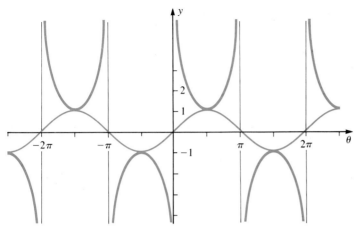

Fig. 4 $y = \csc \theta$

EXAMPLE 1   Find all values of θ for which $\csc \theta = 2$.

Solution   If $\csc \theta = 2$, then

$$\frac{1}{\sin \theta} = 2 \qquad \sin \theta = \tfrac{1}{2}.$$

For $0 \le \theta \le 2\pi$, there are two such values: $\theta = \tfrac{1}{6}\pi$ and $\theta = \tfrac{5}{6}\pi$. By the graph in Fig. 4, the only other possibilities are these plus multiples of 2π.

Answer $\tfrac{1}{6}\pi + 2\pi n \qquad \tfrac{5}{6}\pi + 2\pi n \qquad$ (n any integer)

Values of cot θ, sec θ, csc θ

Calculators do not have keys for these functions. But cot θ, sec θ, csc θ are the reciprocals of tan θ, cos θ, sin θ, so their values can be estimated using $\boxed{\tan}$, $\boxed{\cos}$, $\boxed{\sin}$, and the reciprocal key $\boxed{1/x}$.

Examples   (Set $\boxed{\text{rad}}$ mode).

(1)   cot 0.483   . 4 8 3 $\boxed{\tan}$ $\boxed{1/x}$ 1.9068324

(2)   csc (-0.107) . 1 0 7 $\boxed{+/-}$ $\boxed{\sin}$ $\boxed{1/x}$ -9.3636516

(3)   csc 3π 3 $\boxed{\times}$ $\boxed{\pi}$ $\boxed{=}$ $\boxed{\sin}$ $\boxed{1/x}$ ERROR

(There is a grain of salt in the last example. On a TI-59, we indeed get an error condition. But on a Sharp EL-5806, we get $\sin 3\pi \approx -4 \times 10^{-10}$, so $\csc 3\pi \approx -2.5 \times 10^9$. Small machine errors are likely with calculators; don't forget it.)

Exercises

Find cot θ, sec θ, and csc θ (exact) for $\theta =$

1 $\tfrac{1}{4}\pi$	**2** $\tfrac{2}{3}\pi$	**3** $\tfrac{5}{3}\pi$	**4** $\tfrac{7}{6}\pi$
5 $-\tfrac{1}{3}\pi$	**6** $-\tfrac{2}{3}\pi$	**7** $-\tfrac{3}{4}\pi$	**8** $-\tfrac{13}{6}\pi$
9 $\tfrac{29}{3}\pi$	**10** $\tfrac{53}{6}\pi$.		

Find all θ for which

11 $\sec \theta = 2$ **12** $\csc \theta = -2$ **13** $\cot \theta = \sqrt{3}$

14 $\cot \theta = \frac{1}{3}\sqrt{3}$ **15** $\sec \theta = -1$ **16** $\csc \theta = 1$.

Estimate on a calculator to 6 significant figures

17 $\cot 0.3113$ **18** $\cot(-1.0144)$ **19** $\sec(2.500)$

20 $\sec(-0.0439)$ **21** $\csc(-3.4124)$ **22** $\csc(0.1158)$.

23 Justify the estimate $\cot \theta \approx 1/\theta$ for θ small.

24 (cont.) Tabulate $1/\theta - \cot \theta$ for $\theta = 0.10, 0.08, 0.06, 0.04, 0.02; 0.01; 0.005$.

25 (cont.) Examine the table for 0.1 and 0.01. Do you see a pattern? If so, test it on the other entries in the table.

26 Justify the estimate $\sec \theta \approx 1 + \frac{1}{2}\theta^2$ for θ small. Tabulate $\sec \theta$ and $1 + \frac{1}{2}\theta^2$ for $\theta = 0.1, 0.05, 0.01, 0.005, 0.001$.

27 Justify the estimate $\csc \theta \approx 1/\theta$ for θ small. Test for $\theta = 0.1, 0.05, 0.01, 0.005, 0.001$.

28 In view of Exs. 23 and 27, you would guess that $\csc \theta - \cot \theta$ is not too large if θ is small. Show numerically that $\csc \theta - \cot \theta \approx \frac{1}{2}\theta$ appears to be true.

Find all θ satisfying

29 $\tan \theta \geq \cot \theta$ **30*** $\tan \theta + \cot \theta \geq 2$.

The quarter circle has radius 1. Express each length in terms of θ

31 $\overline{AE}$ **32** $\overline{BC}$

33 $\overline{OE}$ **34** $\overline{OC}$

35 $\overline{OF}$ **36** $\overline{FP}$

37 $\overparen{AP}$ **38** $\overparen{PB}$.

4 PERIODIC FUNCTIONS

A function $f(\theta)$ with domain {all θ} is called **periodic of period p** provided p is a positive constant and

$$f(\theta + p) = f(\theta)$$

for all θ.

Examples

(1) $f(\theta) = \sin \theta$ is periodic of period 2π

(2) $f(\theta) = \cos \theta$ is periodic of period 10π (also of period 2π)

(3) $f(\theta) = 5\cos^2 3\theta - \sin^3 7\theta$ is periodic of period 2π.

If $f(\theta)$ is periodic of period p, then

$$f(\theta + 2p) = f(\theta + p + p) = f(\theta + p) = f(\theta)$$
$$f(\theta + 3p) = f(\theta + 2p + p) = f(\theta + 2p) = f(\theta)$$

and in general

$$f(\theta + np) = f(\theta) \qquad n \text{ any whole number.}$$

On the other side,

$$f(\theta - p) = f(\theta - p + p) = f(\theta),$$

and as with positive multiples,

$$f(\theta - np) = f(\theta).$$

> If $f(\theta)$ is periodic of period p, then
>
> $$f(\theta + np) = f(\theta) \qquad n \text{ any integer}.$$

Constructing Periodic Functions

Let $f(\theta)$ be periodic of period p. The graph of $f(\theta)$ on an interval of length p such as $a \le \theta < a + p$ or $a < \theta \le a + p$ is called a **cycle** of the function. (Note that exactly one end-point is included.) For instance, the graph of $y = \sin \theta$ on $0 \le \theta < 2\pi$ is a cycle of $\sin \theta$; so is its graph on $-\pi < \theta \le \pi$. What is important about a cycle is that it completely determines the periodic function. This is so because given any θ, there is an integer n such that $\theta + np$ is in the interval; then $f(\theta) = f(\theta + np)$.

This reasoning can be turned around; what we have in mind is the process by which we defined $\sin \theta$ once we had it for $0 \le \theta < 2\pi$. Given any interval $a \le \theta < a + p$ (or $a < \theta \le a + p$) of length p and a function $f(\theta)$ on this interval, there is a unique extension of this function to a periodic function of period p on the whole θ-axis. Its graph is simply the graph on $a \le \theta < a + p$ shifted left and right by steps of p.

Examples (1) $f(\theta) = \theta$ for $0 \le \theta < 1, p = 1$.

The corresponding periodic function of period 1 is graphed in Fig. 1.

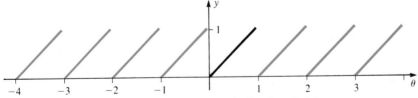

Fig. 1 $y = \theta$ for $0 \le \theta < 1$ and $y(\theta)$ periodic of period 1

(2) $f(\theta) = \theta^2$ for $-1 \le \theta < 1, p = 2$.

The corresponding periodic function of period 2 is graphed in Fig. 2.

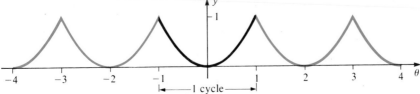

Fig. 2 $y = \theta^2$ for $-1 \le \theta < 1$ and $y(\theta)$ periodic of period 2

(3) $f(\theta) = \begin{cases} \sin\theta & \text{for} \quad 0 \le \theta \le \pi \\ 0 & \text{for} \quad \pi < \theta < 2\pi \end{cases} \quad p = 2\pi.$

The corresponding periodic function is graphed in Fig. 3. It is what a rectified alternating electric current (half wave rectifier) looks like, current vs. time.

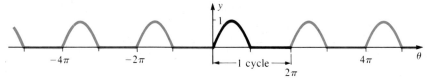

Fig. 3 $y = \begin{cases} \sin\theta & \text{for } 0 \le \theta < \pi \\ 0 & \text{for } \pi \le \theta < 2\pi \end{cases}$ and $y(\theta)$ periodic of period 2π

Fundamental Period

Let $f(\theta)$ be a periodic function of period p. Then $f(\theta)$ is periodic of period $2p$ since

$$f(\theta + 2p) = f(\theta).$$

It is also periodic of period $3p$, or period $4p$, etc. Thus each of the numbers

$$p, 2p, 3p, 4p, \cdots$$

is a period of $f(\theta)$. There may be other periods, for instance, possibly $\frac{1}{3}p$ is also a period of $f(\theta)$. For instance $p = 6\pi$ is a period of $f(\theta) = \cos\theta$. So are $12\pi, 18\pi, \cdots$. But so also is $\frac{1}{3}p = 2\pi$. This number 2π is the *least* (positive) period of $f(\theta) = \cos\theta$, as is obvious from its graph.

If it exists, the *least* (positive) period p of a periodic function $f(\theta)$ is called the **fundamental period** of $f(\theta)$.

A function can be periodic without having a fundamental period: any constant function is so, and there are others (Ex. 14). While a certain function may have a fundamental period, it may not be easy to find it and prove it so. The advantage of having the fundamental period rather than some multiple is that you have to study the function in detail on a smaller interval.

For instance, consider $f(\theta) = \sin^3\theta \cos^5\theta$. Obviously 2π is a period of $f(\theta)$. It is less obvious, but true, that π is a period of $f(\theta)$. That π is *the fundamental period* of $f(\theta)$ is far from obvious; we probably need a graph to be convinced. [Actually, we can do it this way. First we note that $f(\theta) = 0$ only for $\theta = \frac{1}{2}\pi n$. Therefore the only candidate for a period smaller than π is $\frac{1}{2}\pi$. But $f(\frac{1}{4}\pi) > 0$ and $f(\frac{1}{4}\pi + \frac{1}{2}\pi) < 0$, so $\frac{1}{2}\pi$ is not a period; π is the fundamental period.]

Generally, if you are asked to find a period of a function, you should find one as small as you possibly can. But finding the fundamental period, if there is one, might be very difficult. Try it for

$$f(\theta) = 1.4\sin 3\theta + 1.2\sin 5\theta - 2.3\cos 6\theta$$

for instance. Here 2π is surely a period; is there a smaller one? That's hard!

Singularities

We would like to say that $\tan \theta$ is periodic of period π because

$$\tan(\theta \pm n\pi) = \tan \theta.$$

We are not supposed to do so because the domain of $\tan \theta$ falls short of the whole θ-axis. The points $\theta = (n + \frac{1}{2})\pi$ are singularities of $\tan \theta$, points where it cannot be defined.

Usually one says so anyhow, taking a looser attitude towards periodic functions. Some points can be omitted, but whenever θ_0 is omitted, then all $\theta_0 + np$, where p is the period, must also be omitted.

So don't be surprised to hear that $f(\theta) = \sec \theta$ is periodic of period 2π, etc.

Exercises

Graph

1 $y = \begin{cases} 1 & 0 \le \theta < 1 \\ -1 & 1 \le \theta < 2 \end{cases}$ period 2

2 $y = \sqrt{1 - \theta^2} \ -1 \le \theta < 1$ period 2

3 $y = \begin{cases} \theta & 0 \le \theta < 2 \\ 4 - \theta & 2 \le \theta < 4 \end{cases}$ period 4

4 $y = |\theta| \ -1.5 \le \theta < 1.5$ period 3.

Show that $f(\theta)$ is periodic by finding a period

5 $\sin \frac{1}{2}\theta - \cos \frac{1}{5}\theta$ **6** $\sin \frac{1}{3}\theta + \cos \frac{1}{4}\theta$

7 $\cos 2\pi\theta$ **8** $\sin \pi\theta$

9 $|\cos \theta|$ **10** $\sin^2 \theta$

11 $\sin 2\theta \cos^3 2\theta$ **12** $\sin(\theta - \frac{1}{6}\pi) + \cos(\theta - \frac{1}{5}\pi)$.

13 Suppose that, to graph $y = \sin 20\pi\theta$, you plot points for $\theta = 0, 0.1, 0.2, 0.3, \cdots, 2.0$. What happens and why?

14* Suppose $f(\theta) = 0$ if θ is irrational and $f(\theta) = 1$ if θ is rational. Show that $f(\theta)$ is periodic of period p for each positive rational number p, but $f(\theta)$ has no fundamental period.

5   FURTHER GRAPHS

Graphs of $y = \sin b\theta$

Let's start with $y = \sin 2\theta$. This is just like $y = \sin \theta$, except that we have changed the scale (stretched) on the θ-axis. As θ goes from 0 to π, then 2θ goes from 0 to 2π, so $y = \sin 2\theta$ has a cycle on $0 \le \theta < \pi$. It is periodic of period π because

$$\sin 2(\theta + \pi) = \sin(2\theta + 2\pi) = \sin 2\theta.$$

The graph is sketched in Fig. 1, next page.

Similarly, the graph of $y = \sin 3\theta$ makes one cycle on the interval $0 \le \theta < \frac{2}{3}\pi$ and three cycles on the interval $0 \le \theta < 2\pi$. The graph of $y = \sin\frac{1}{2}\theta$

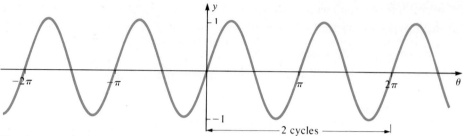

Fig. 1 $y = \sin 2\theta$

makes one cycle on the interval $0 \le \theta < 4\pi$, a half-cycle on the interval $0 \le \theta < 2\pi$. See Fig. 2.

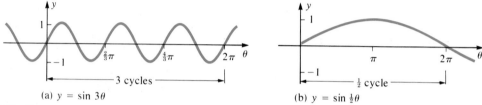

(a) $y = \sin 3\theta$ (b) $y = \sin \frac{1}{2}\theta$

Fig. 2

In general, for each $b > 0$, the graph of $y = \sin b\theta$ is the graph of $y = \sin \theta$ compressed (or stretched) horizontally. The period of $y = \sin b\theta$ is $2\pi/b$:

$$\sin b(\theta + 2\pi/b) = \sin(b\theta + 2\pi) = \sin b\theta.$$

Similar statements apply to $y = \cos b\theta$.

Graph of $y = a \sin(b\theta - c) + d$

Once we note that

$$\sin(b\theta - c) = \sin b\left(\theta - \frac{c}{b}\right),$$

it is clear that the graph of

$$y = a \sin(b\theta - c) + d$$

comes from the graph of $y = \sin b\theta$ by our usual routine of stretching, reflecting, and shifting horizontally and vertically. The same holds if sin is replaced by cos.

EXAMPLE 1 Graph

(a) $y = -2 \cos 3\theta$ (b) $y = 5 \sin(\frac{1}{3}\theta - \pi) + 10$.

Solution (a) Start with $y = \cos \theta$. First compress the θ-axis by a factor of 3, then stretch the y-axis by a factor of 2, and finally reflect in the θ-axis for the sign change (Fig. 3a).

(b) Rewrite the equation in the form

$$y = 5 \sin \tfrac{1}{3}(\theta - 3\pi) + 10.$$

The graph is obtained from that of $y = 5 \sin \frac{1}{3}\theta$ by shifting 3π units to the right and 10 units up. Now $\sin \frac{1}{3}\theta$ oscillates $\frac{1}{3}$ as fast as $\sin \theta$; a cycle has length $3 \cdot 2\pi = 6\pi$. Also $y = 5 \sin \frac{1}{3}\theta$ oscillates between the levels $y = 5$ and $y = -5$. The shifted graph oscillates between the levels $y = 5 + 10 = 15$ and $y = -5 + 10 = 5$. See Fig. 3b.

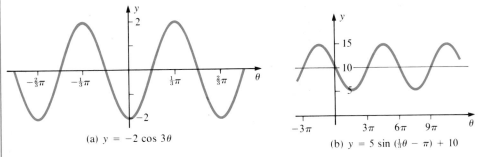

(a) $y = -2 \cos 3\theta$

(b) $y = 5 \sin (\frac{1}{3}\theta - \pi) + 10$

Fig. 3

EXAMPLE 2 A point P moves uniformly (with constant angular speed) counterclockwise around the circle of radius 4 and center $(0,0)$. It starts at $(0,4)$ at time $t = 0$, and its angular speed is 10 revolutions per second. Express its x-coordinate in terms of t (in seconds) and graph the resulting function in the t,x-plane.

Solution From Fig. 4a, we see that $x = 4 \cos \theta$. But we are asked for x in terms of t. So we should try to express θ in terms of t.

Given: the motion is uniform. That means θ is a linear function of t. Because $\theta = \frac{1}{2}\pi$ when $t = 0$, we may write $\theta = at + \frac{1}{2}\pi$. The constant a is the constant rate of θ (angular speed) in rad/sec. Thus

$$a = 10 \text{ rev/sec} = 10(2\pi) \text{ rad/sec} = 20\pi \text{ rad/sec}.$$

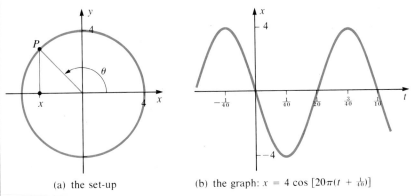

(a) the set-up

(b) the graph: $x = 4 \cos [20\pi(t + \frac{1}{40})]$

Fig. 4

Therefore $\theta = 20\pi t + \frac{1}{2}\pi$,

$$x = 4 \cos \theta = 4 \cos(20\pi t + \frac{1}{2}\pi) = 4 \cos [20\pi(t + \frac{1}{40})].$$

This is the desired function; its graph (Fig. 4b) is a cosine curve, shifted left $\frac{1}{40}$ and with the scales adjusted.

Remark   Suppose a point moves uniformly around a circle, like P in Example 2. The oscillatory motion of its projection on any line is called **simple harmonic motion.** Such motion is very common in nature. For example, imagine a weight hanging on the end of a spring attached to the ceiling. If pulled down a little from its rest position and released, the spring will oscillate up and down. The motion is simple harmonic motion along a vertical line.

Suppose you must graph a periodic function $y = f(\theta)$ of period p. Then graph it carefully on any segment of the θ-axis of length p. After that, or before that, it simply repeats itself.

E_{XAMPLE} 3   Graph $y = \cos 2\theta + 2 \cos \theta$.

Solution   The function is periodic of period 2π. We tabulate enough values for $-\pi \leq \theta < \pi$ to make an accurate graph on that interval. Actually, we can cut our calculations in half by noting that $y(\theta)$ is an even function. Therefore, we need only tabulate values for $0 \leq \theta \leq \pi$:

θ	0	$\frac{1}{10}\pi$	$\frac{2}{10}\pi$	$\frac{3}{10}\pi$	$\frac{4}{10}\pi$	$\frac{5}{10}\pi$	$\frac{6}{10}\pi$	$\frac{7}{10}\pi$	$\frac{8}{10}\pi$	$\frac{9}{10}\pi$	π
y	3.0	2.7	1.9	0.9	-0.2	-1.0	-1.4	-1.5	-1.3	-1.1	-1.0

We plot these points, reflect in the y-axis, then join by a smooth curve. By periodicity we can draw as much of the graph as we want (Fig. 5).

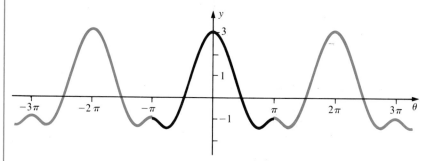

Fig. 5 $y = \cos 2\theta + 2 \cos \theta$

Other Trigonometric Graphs

Functions such as $f(\theta) = \theta \sin \theta$ and $f(\theta) = \theta + \cos \theta$ involve trigonometric functions but are not periodic. Let us graph one such function.

E_{XAMPLE} 4   Graph $y = \theta \sin \theta$.

Solution   Since $\theta \sin \theta$ is even, we can concentrate on $\theta \geq 0$. Ordinarily $\sin \theta$ oscillates between the horizontal lines $y = 1$ and $y = -1$. But here the values of $\sin \theta$ are magnified by the factor θ. Therefore the graph oscillates between the lines $y = \theta$ and $y = -\theta$. We draw these lines as guides.

Now we plot some crucial points. Since $y = 0$ where $\sin \theta = 0$, the graph meets the θ-axis at $\theta = 0, \pi, 2\pi, \cdots$. Halfway between these values, at $\theta = \frac{1}{2}\pi, \frac{3}{2}\pi, \frac{5}{2}\pi, \cdots$ we have $\sin \theta = \pm 1$, so $y = \pm \theta$. For these values of θ, the curve alternately reaches its upper and lower guidelines. One final piece of information: $\sin \theta \approx \theta$ near $\theta = 0$, so $y \approx \theta^2$ for $\theta \approx 0$.

This is enough information to sketch the graph for $\theta \geq 0$. We reflect in the y-axis to extend the graph to $\theta < 0$. See Fig. 6.

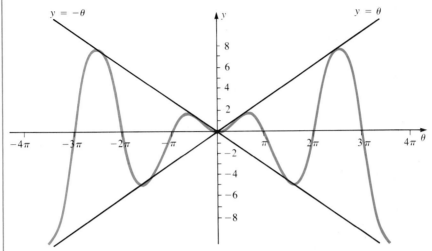

Fig. 6 $y = \theta \sin \theta$

Exercises

Graph $y =$

1 $\sin 4\theta$	**2** $\sin 3\theta$	**3** $\cos 3\theta$
4 $\cos 4\theta$	**5** $\sin 2\pi\theta$	**6** $\cos \pi\theta$
7 $3 \sin 2\theta$	**8** $-3 \cos 2\theta$	**9** $1 + \cos \theta$
10 $2 - \sin \theta$	**11** $11 + 3 \cos \theta$	**12** $-6 + 2 \sin \theta$
13 $|\sin \theta|$	**14** $|\cos \theta|$	**15** $\sin^2 \theta$
16 $\cos^2 \theta$	**17** $\sin(\theta + \frac{1}{6}\pi)$	**18** $\cos(\theta + \frac{1}{4}\pi)$
19 $\frac{1}{2} \sin(\theta + \frac{1}{3}\pi)$	**20** $2 \cos(\theta - \frac{1}{6}\pi)$	**21** $\sin(3\theta - \frac{1}{3}\pi)$
22 $2 \cos 4(\theta - \frac{1}{6}\pi)$	**23** $-4 \sin 3\theta$	**24** $-2 \cos(\theta + \frac{1}{4}\pi)$.

Solve Example 2 for the following data

	Center	Radius	Initial position	Angular speed
25	$(0,0)$	3	$(-3,0)$	1 rev/sec
26	$(0,0)$	3	$(0,-3)$	30 rev/min
27	$(1,0)$	1	$(2,0)$	10 rev/sec
28	$(-3,0)$	2	$(-3,2)$	5 rev/sec

Graph $y =$

29 $\sin 2\theta - \sin \theta$	**30** $\sin \theta + \cos 2\theta$
31 $\sin \theta + \cos \theta$	**32** $\sin \theta - 2 \cos \theta$
33 $3 \cot \theta$	**34** $\csc 2\theta + 1$
35 $\tan \theta + \cot \theta$	**36** $\tan \theta - \cot \theta$
37 $\sec \theta - \csc \theta$	**38** $\sec \theta + \csc \theta$
39 $\theta \cos \theta$	**40** $\theta^2 \sin \theta$
41 $\theta + \cos \theta$	**42** $\theta + \sin \theta$
43 $|\theta| \sin \theta$	**44** $2^{-\theta/2\pi} \sin \theta$.

45 Show graphically that $\cos \theta = \theta$ has only one solution, and locate it approximately.

46 Show graphically that $\pi \cos 2\theta = \theta$ has exactly five solutions, and locate them approximately.

47 Locate graphically all solutions of $\sin 2\theta = \cos \theta$ in the interval $0 \le \theta \le 2\pi$.

48 Locate graphically all solutions of $\theta = 3\pi \sin \theta$.

REVIEW EXERCISES

Find

1 $\tan\left(-\frac{23}{4}\pi\right)$ exactly

2 $\tan \frac{2}{7}\pi$ to 6 places.

Calculate to 6 places

3 $\sec^2 1.385 - \tan^2 1.385$

4 $\csc^2 4.998 - \cot^2 4.998$.

Find all θ for which

5 $\sec \theta = \frac{2}{3}\sqrt{3}$

6 $\cot \theta = -1$.

Graph

7 $y = 2 \sin \theta - \cos 2\theta$

8 $y = \sin \theta + \tan \theta$.

9 Find the smallest positive angle θ on your calculator such that $\cos \theta \ne 1$.

10 Find the largest angle $\theta° < 90°$ on your calculator such that $\sin \theta° \ne 1$.

4 IDENTITIES

1 BASIC IDENTITIES

Because the six trigonometric functions are closely related to each other, there are a multitude of identities interconnecting them; we begin with three very basic ones:

$$\sin^2 \theta + \cos^2 \theta = 1$$
$$1 + \tan^2 \theta = \sec^2 \theta$$
$$1 + \cot^2 \theta = \csc^2 \theta.$$

The first identity, already discussed on p. 27, is an immediate consequence of the definition of $\sin \theta$ and $\cos \theta$:

$$\cos^2 \theta + \sin^2 \theta = \left(\frac{x}{r}\right)^2 + \left(\frac{y}{r}\right)^2 = \frac{x^2 + y^2}{r^2} = \frac{r^2}{r^2} = 1.$$

To derive the second identity, divide both sides of the first by $\cos^2 \theta$:

$$\frac{\sin^2 \theta}{\cos^2 \theta} + \frac{\cos^2 \theta}{\cos^2 \theta} = \frac{1}{\cos^2 \theta} \qquad \tan^2 \theta + 1 = \sec^2 \theta.$$

To derive the third, divide both sides of the first by $\sin^2 \theta$:

$$\frac{\sin^2 \theta}{\sin^2 \theta} + \frac{\cos^2 \theta}{\sin^2 \theta} = \frac{1}{\sin^2 \theta} \qquad 1 + \cot^2 \theta = \csc^2 \theta.$$

Remark The identity $1 + \tan^2 \theta = \sec^2 \theta$ is valid wherever both sides are defined, that is, for all θ except $\pm\frac{1}{2}\pi$, $\pm\frac{3}{2}\pi$, $\pm\frac{5}{2}\pi$, $\cdots$. Remember that in general an identity is an equation that is valid for all values of the variable(s) for which all terms in the equation are defined.

In addition to the three relations in the box, there are the relations that define tan, cot, sec, and csc:

$$\tan \theta = \frac{\sin \theta}{\cos \theta} \qquad \cot \theta = \frac{1}{\tan \theta} = \frac{\cos \theta}{\sin \theta}$$

$$\sec \theta = \frac{1}{\cos \theta} \qquad \csc \theta = \frac{1}{\sin \theta}.$$

Now we have the tools for deriving and proving many other identities.

EXAMPLE 1   Show that

(a) $\cos \theta \tan \theta = \sin \theta$ (b) $\cot \theta \sec \theta = \csc \theta$.

Solution   (a) $\cos \theta \tan \theta = \cos \theta \dfrac{\sin \theta}{\cos \theta} = \sin \theta$.

(b) $\cot \theta \sec \theta = \dfrac{\cos \theta}{\sin \theta} \dfrac{1}{\cos \theta} = \dfrac{1}{\sin \theta} = \csc \theta$.

EXAMPLE 2   Show that

(a) $\tan^2 \theta + \tan^4 \theta = \tan^2 \theta \sec^2 \theta$ (b) $\dfrac{\cos^2 \theta}{1 - \cos^2 \theta} = \cot^2 \theta$.

Solution   (a)   Factor the left-hand side and use $1 + \tan^2 \theta = \sec^2 \theta$:

$$\tan^2 \theta + \tan^4 \theta = \tan^2 \theta (1 + \tan^2 \theta) = \tan^2 \theta \sec^2 \theta.$$

(b)   From $\sin^2 \theta + \cos^2 \theta = 1$ follows $1 - \cos^2 \theta = \sin^2 \theta$. Hence

$$\frac{\cos^2 \theta}{1 - \cos^2 \theta} = \frac{\cos^2 \theta}{\sin^2 \theta} = \left(\frac{\cos \theta}{\sin \theta} \right)^2 = \cot^2 \theta.$$

Functions in Terms of Other Functions

Suppose we are given an acute θ with $\sin \theta = \frac{3}{4}$. Can we find $\cos \theta$? Yes; from $\cos^2 \theta = 1 - \sin^2 \theta$,

$$\cos \theta = \sqrt{1 - \sin^2 \theta} = \sqrt{(1 - (\tfrac{3}{4})^2} = \sqrt{\tfrac{7}{16}} = \tfrac{1}{4}\sqrt{7}.$$

Watch out, though. If θ is in the *second* quadrant, then $\cos \theta$ is *negative*, so

$$\cos \theta = -\sqrt{1 - \sin^2 \theta} = -\tfrac{1}{4}\sqrt{7}.$$

In general,

$$\sin \theta = \pm\sqrt{1 - \cos^2 \theta} \quad \text{and} \quad \cos \theta = \pm\sqrt{1 - \sin^2 \theta},$$

where the choice of $+$ or $-$ depends on the quadrant of θ. Similarly,

$$\tan \theta = \pm\sqrt{\sec^2 \theta - 1} \qquad \sec \theta = \pm\sqrt{1 + \tan^2 \theta}$$

$$\cot \theta = \pm\sqrt{\csc^2 \theta - 1} \qquad \csc \theta = \pm\sqrt{1 + \cot^2 \theta}.$$

Again, the choice of sign depends on the quadrant of θ.

EXAMPLE 3 Suppose $\tan \theta = -2$ and θ is an angle in the second quadrant. Find

(a) $\sec \theta$ (b) $\cos \theta$ (c) $\sin \theta$.

Solution (a) In the second quadrant, $\sec \theta < 0$. Therefore,

$$\sec \theta = -\sqrt{1 + \tan^2 \theta} = -\sqrt{1 + (-2)^2} = -\sqrt{5}.$$

(b) $\cos \theta = \dfrac{1}{\sec \theta} = -\dfrac{1}{\sqrt{5}}.$

(c) In the second quadrant, $\sin \theta > 0$. Therefore

$$\sin \theta = \sqrt{1 - \cos^2 \theta} = \sqrt{1 - \tfrac{1}{5}} = \sqrt{\tfrac{4}{5}} = \frac{2}{\sqrt{5}}.$$

Another way:

$$\sin \theta = \cos \theta \frac{\sin \theta}{\cos \theta} = \cos \theta \tan \theta = \left(-\frac{1}{\sqrt{5}}\right)(-2) = \frac{2}{\sqrt{5}}.$$

Answer (a) $-\sqrt{5}$ (b) $-\dfrac{1}{\sqrt{5}}$ (c) $\dfrac{2}{\sqrt{5}}$

EXAMPLE 4 For θ in the second quadrant, express in terms of $\tan \theta$:

(a) $\sec \theta$ (b) $\cos \theta$ (c) $\sin \theta$.

Solution This is the same problem as Example 3, except that $\tan \theta$ is not given a specific value. We solve it in the same way.

(a) $\sec \theta = -\sqrt{1 + \tan^2 \theta}$

(b) $\cos \theta = \dfrac{1}{\sec \theta} = \dfrac{-1}{\sqrt{1 + \tan^2 \theta}}$

(c) $\sin \theta = \cos \theta \tan \theta = \dfrac{-\tan \theta}{\sqrt{1 + \tan^2 \theta}}$

Answer (a) $-\sqrt{1 + \tan^2 \theta}$ (b) $\dfrac{-1}{\sqrt{1 + \tan^2 \theta}}$ (c) $\dfrac{-\tan \theta}{\sqrt{1 + \tan^2 \theta}}$

EXAMPLE 5 Suppose θ is in the fourth quadrant. Express the other five trig functions in terms of $\csc \theta$.

Solution Note that $\csc \theta < 0$ in the fourth quadrant (actually $\csc \theta < -1$). We have first

$$\sin \theta = \frac{1}{\csc \theta}.$$

Since $\cos \theta > 0$ in the fourth quadrant,

$$\cos \theta = \sqrt{1 - \sin^2 \theta} = \sqrt{1 - \frac{1}{\csc^2 \theta}} = \sqrt{\frac{\csc^2 \theta - 1}{\csc^2 \theta}} = \frac{\sqrt{\csc^2 \theta - 1}}{-\csc \theta}.$$

The minus sign is chosen so $\cos \theta > 0$. (Don't forget that $\csc \theta < 0$ in this example.)

Having $\sin \theta$ and $\cos \theta$, it is now easy to express the remaining three functions in terms of $\csc \theta$:

$$\tan \theta = \frac{\sin \theta}{\cos \theta} = \left(\frac{1}{\csc \theta}\right) \bigg/ \left(\frac{\sqrt{\csc^2 \theta - 1}}{-\csc \theta}\right) = \frac{-1}{\sqrt{\csc^2 \theta - 1}}$$

$$\cot \theta = \frac{1}{\tan \theta} = -\sqrt{\csc^2 \theta - 1} \qquad \sec \theta = \frac{1}{\cos \theta} = \frac{-\csc \theta}{\sqrt{\csc^2 \theta - 1}}$$

Answer

$\sin \theta$	$\cos \theta$	$\tan \theta$	$\cot \theta$	$\sec \theta$
$\dfrac{1}{\csc \theta}$	$-\dfrac{\sqrt{\csc^2 \theta - 1}}{\csc \theta}$	$\dfrac{-1}{\sqrt{\csc^2 \theta - 1}}$	$-\sqrt{\csc^2 \theta - 1}$	$\dfrac{-\csc \theta}{\sqrt{\csc^2 \theta - 1}}$

A Geometric Method

Let $f(\theta)$ be any one of the six functions $\sin \theta, \cdots, \csc \theta$. Then $f(\theta)$ equals a quotient of two of the three numbers x, y, r, where (x, y) lies on the terminal side of θ. By setting the denominator equal to ± 1, according to the quadrant of θ, we easily express x, y, r all in terms of $f(\theta)$. Then any other of the six trig functions, another quotient, is readily expressed in terms of $f(\theta)$.

EXAMPLE 6   Express $\sec \theta$ in terms of $\csc \theta$ for θ in the first quadrant.

Solution   For the first quadrant, all we need is a right triangle (Fig. 1). Since $\csc \theta = \text{hyp}/\text{opp}$, we choose opp $= 1$. Then hyp $= \csc \theta$ and

$$\text{adj} = \sqrt{\text{hyp}^2 - \text{opp}^2} = \sqrt{\csc^2 \theta - 1}.$$

Therefore $\sec \theta = \dfrac{\text{hyp}}{\text{adj}} = \dfrac{\csc \theta}{\sqrt{\csc^2 \theta - 1}}.$

Answer $\sec \theta = (\csc \theta)/\sqrt{\csc^2 \theta - 1}$.

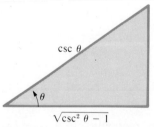

Fig. 1

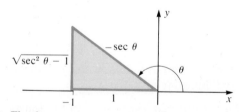

Fig. 2

EXAMPLE 7   Express $\csc \theta$ in terms of $\sec \theta$ for θ in the second quadrant.

Solution   In the second quadrant, $x < 0$ and $y > 0$. Since $\sec \theta = r/x$, we choose $x = -1$. Then $r = -\sec \theta$ and

$$y = \sqrt{r^2 - x^2} = \sqrt{\sec^2 \theta - 1}.$$

See Fig. 2. Therefore $\quad \csc \theta = \dfrac{r}{y} = \dfrac{-\sec \theta}{\sqrt{\sec^2 \theta - 1}}.$

Answer $\quad \csc \theta = -(\sec \theta)/\sqrt{\sec^2 \theta - 1}$

Exercises

Show that

1 $\sin \theta \sec \theta = \tan \theta$

2 $\sec \theta \cot \theta = \csc \theta$

3 $\tan \theta \csc \theta = \sec \theta$

4 $\cos \theta \csc \theta = \cot \theta$

5 $\dfrac{\sin^2 \theta}{1 - \sin^2 \theta} = \tan^2 \theta$

6 $\dfrac{1 - \cos^2 \theta}{\tan^2 \theta} = \cos^2 \theta$

7 $\dfrac{\sec^2 \theta}{\sec^2 \theta - 1} = \dfrac{1}{\sin^2 \theta}$

8 $\dfrac{\cot^2 \theta + 1}{\cot^2 \theta} = \dfrac{1}{\cos^2 \theta}$

9 $\dfrac{\sin \theta}{1 - \sin^2 \theta} = \tan \theta \sec \theta$

10 $\dfrac{1 - \cos^2 \theta}{\cos \theta} = \sin \theta \tan \theta$

11 $\sec^4 \theta - \sec^2 \theta = \sec^2 \theta \tan^2 \theta$

12 $\cos^3 \theta - \cos \theta = -\sin^2 \theta \cos \theta$

13 $1 - (\cos \theta - \sin \theta)^2 = 2 \sin \theta \cos \theta$

14 $1 - \sin \theta \cos \theta \tan \theta = \cos^2 \theta$

15 $\dfrac{\cos^2 \theta}{1 + \sin \theta} = 1 - \sin \theta$

16 $\dfrac{\sec \theta - 1}{\tan^2 \theta} = \dfrac{1}{\sec \theta + 1}.$

Find the exact values of the remaining trigonometric functions

17 $\cos \theta = -\frac{4}{5}$ quadrant 3

18 $\sin \theta = -\frac{3}{5}$ quadrant 4

19 $\tan \theta = 3$ quadrant 1

20 $\cot \theta = \frac{1}{2}$ quadrant 1

21 $\sec \theta = -2$ quadrant 2

22 $\csc \theta = 3$ quadrant 2

23 $\cot \theta = -\frac{3}{2}$ quadrant 4

24 $\tan \theta = 4$ quadrant 3.

Express the first function in terms of the second; adjust the sign so the formula is correct in the given quadrant

25 $\sin \theta$ in terms of $\cot \theta$ quadrant 2

26 $\sin \theta$ in terms of $\cot \theta$ quadrant 4

27 $\csc \theta$ in terms of $\cot \theta$ quadrant 1

28 $\csc \theta$ in terms of $\cot \theta$ quadrant 3

29 $\sec \theta$ in terms of $\cot \theta$ quadrant 3

30 $\cos \theta$ in terms of $\cot \theta$ quadrant 2

31 $\tan \theta$ in terms of $\sin \theta$ quadrant 4

32 $\sec \theta$ in terms of $\sin \theta$ quadrant 1

33 $\csc \theta$ in terms of $\cos \theta$ quadrant 2

34 $\tan \theta$ in terms of $\cos \theta$ quadrant 3.

2 PROVING IDENTITIES

Because there are so many relations among the trigonometric functions, two trigonometric expressions may look quite different, yet be identical. For example, $\tan \theta$ is identical to

$$\frac{\sin \theta}{\cos \theta} \qquad \frac{1}{\cot \theta} \qquad \sin \theta \sec \theta \qquad \cot \theta (\sec^2 \theta - 1)$$

and to many other expressions. If, for instance, the expression $(\sec^2 \theta - 1) \cot \theta$ were involved in a numerical computation, it would save work to replace it by $\tan \theta$.

Suppose we are given two algebraic expressions A and B in the basic trigonometric functions $\sin \theta$, $\cos \theta$, etc., and we want to prove the identity $A = B$ (for all values of θ for which the expressions are defined). If the identity is simple enough, we can sometimes convert A into B by a sequence of steps.

EXAMPLE 1   Prove $\cos \theta + \sin \theta \tan \theta = \sec \theta$.

Solution

$$\cos \theta + \sin \theta \tan \theta = \cos \theta + \sin \theta \frac{\sin \theta}{\cos \theta}$$

$$= \cos \theta + \frac{\sin^2 \theta}{\cos \theta} = \frac{\cos^2 \theta + \sin^2 \theta}{\cos \theta}$$

$$= \frac{1}{\cos \theta} = \sec \theta.$$

In a more complicated identity, it may be hard to transform one side into the other. A systematic procedure for proving such an identity is to express everything in A and B in terms of sines and cosines only, then to simplify. If the results are the same, then $A = B$ is proved.

EXAMPLE 2   Prove $\dfrac{\csc \theta + 1}{\cot \theta} = \dfrac{1}{\sec \theta - \tan \theta}$.

Solution   Express the left-hand side in terms of sines and cosines:

$$A = \frac{\csc \theta + 1}{\cot \theta} = \frac{\dfrac{1}{\sin \theta} + 1}{\dfrac{\cos \theta}{\sin \theta}} = \frac{1 + \sin \theta}{\cos \theta}$$

The right-hand side is

$$B = \frac{1}{\sec \theta - \tan \theta} = \frac{1}{\dfrac{1}{\cos \theta} - \dfrac{\sin \theta}{\cos \theta}} = \frac{\cos \theta}{1 - \sin \theta}.$$

To prove $A = B$, we must show that

$$\frac{1 + \sin \theta}{\cos \theta} = \frac{\cos \theta}{1 - \sin \theta}, \quad \text{that is,} \quad (1 + \sin \theta)(1 - \sin \theta) = \cos^2 \theta.$$

But $(1 + \sin \theta)(1 - \sin \theta) = 1 - \sin^2 \theta = \cos^2 \theta$, so indeed $A = B$.

One way to prove an identity $A = B$ is to prove a string of equalities

$$A = B_1 = B_2 = \cdots = B_n = B.$$

Actually, this is a string of equivalent identities

$$A = B_1 \qquad A = B_2 \cdots A = B_n \qquad A = B$$

that transform A into B one step at a time. This is how we did Example 1.

This method is fine if you can do it. However, for more complicated identities, as in Example 2, it can be difficult to apply. In general, you can just as well prove any equivalent relation such as $A + C = B + C$ or $(A + C)D = (B + C)D$, provided $D \neq 0$. In fact, if you can find a chain of equivalent relations

$$A = B, \quad A_1 = B_1, \quad A_2 = B_2, \quad \cdots, \quad A_n = B_n,$$

and the last one is an identity, then so is the first, and the proof is complete.

Warning Be sure that each step is *reversible*. Whatever reasoning leads from $A_j = B_j$ to $A_{j+1} = B_{j+1}$ must work in reverse gear; that is, the truth of $A_{j+1} = B_{j+1}$ must imply the truth of $A_j = B_j$, not just the other way. In particular, squaring both sides is invalid; for instance the chain

$$-\sin\theta = \sqrt{1 - \cos^2\theta}, \quad \sin^2\theta = 1 - \cos^2\theta, \quad \sin\theta = \sqrt{1 - \cos^2\theta}$$

certainly requires some qualification.

A convenient shorthand is to abbreviate $\cos\theta$ by c and $\sin\theta$ by s. This not only saves a lot of writing, but also decreases the chance of error.

EXAMPLE 3 Prove $\dfrac{1 - \cos\theta}{1 + \cos\theta} = (\csc\theta - \cot\theta)^2$.

Solution Express everything in terms of $c = \cos\theta$ and $s = \sin\theta$; then form a chain of equivalent relations:

$$\frac{1 - c}{1 + c} = \left(\frac{1}{s} - \frac{c}{s}\right)^2 \qquad \frac{1 - c}{1 + c} = \frac{(1 - c)^2}{s^2}$$

$$s^2(1 - c) = (1 + c)(1 - c)^2 = (1 + c)(1 - c)(1 - c)$$

$$= (1 - c^2)(1 - c) = s^2(1 - c).$$

Converting to sines and cosines is not the only approach, nor necessarily the quickest in every case.

EXAMPLE 4 Prove $\dfrac{\tan\alpha - \tan\beta}{\cot\alpha - \cot\beta} = -\tan\alpha\tan\beta$.

Solution Here it is easiest to express the cotangents in terms of tangents. The chain of relations is

$$\tan\alpha - \tan\beta = -\tan\alpha\tan\beta(\cot\alpha - \cot\beta)$$

$$= -\tan\alpha\tan\beta\left(\frac{1}{\tan\alpha} - \frac{1}{\tan\beta}\right)$$

$$= -\tan\beta + \tan\alpha.$$

Each student develops his or her own style for proving identities. Some like to start with the left-hand side and transform it into the right-hand side. Others start with the more complicated looking side and modify it. We recommend a middle course; fiddle around a little with both sides and try to find a middle meeting ground. Sometimes if you want to prove $A/B = C/D$, it pays to clear of fractions first, then attempt to prove the *equivalent* identity $AD = BC$.

Exercises

Simplify

1 $\sin^4 \theta - \cos^4 \theta$

2 $(\sin \theta + \cos \theta)^2 - 2 \sin \theta \cos \theta$

3 $\dfrac{1 - \cos^2 \theta}{\sin \theta}$

4 $\dfrac{1 + \tan^2 \theta}{\tan^2 \theta}$.

5 $(\sec^2 \theta - \tan^2 \theta)^3$

6 $\tan^4 \theta \cot^5 \theta$

7 $\cos^2 \theta + \cos^2(\tfrac{1}{2}\pi - \theta)$

8 $\cot \theta \sec \theta$.

Prove the identity

9 $\tan \theta + \cot \theta = \sec \theta \csc \theta$

10 $2 \cos^2 \theta - 1 = 1 - 2 \sin^2 \theta$

11 $(1 - \sin \theta)(1 + \csc \theta) = \cos \theta \cot \theta$

12 $(\cos \theta + 1)(\sec \theta - 1) = \sin \theta \tan \theta$

13 $\sec \theta - \cos \theta = \sin \theta \tan \theta$

14 $\csc \theta = \sin \theta + \cos \theta \cot \theta$

15 $\sec^2 \theta - \csc^2 \theta = \tan^2 \theta - \cot^2 \theta$

16 $\sin^2 \theta \cot^2 \theta + \cos^2 \theta \tan^2 \theta = 1$

17 $\sec^2 \theta + \csc^2 \theta = \sec^2 \theta \csc^2 \theta$

18 $\tan^2 \theta - \sin^2 \theta = \tan^2 \theta \sin^2 \theta$

19 $\sec^4 \theta - \tan^4 \theta = 1 + 2 \tan^2 \theta$

20 $\sin^4 \theta + \cos^4 \theta + 2 \sin^2 \theta \cos^2 \theta = 1$

21 $\cot^4 \theta + \cot^2 \theta = \csc^4 \theta - \csc^2 \theta$

22 $\cos^6 \theta + 3 \cos^2 \theta \sin^2 \theta + \sin^6 \theta = 1$

23 $(\cos \theta - \sin \theta)^2 + (\cos \theta + \sin \theta)^2 = 2$

24 $(x \cos \theta + y \sin \theta)^2 + (-x \sin \theta + y \cos \theta)^2 = x^2 + y^2$

25 $\dfrac{1 - \tan \theta}{1 + \tan \theta} = \dfrac{\cot \theta - 1}{\cot \theta + 1}$

26 $\dfrac{1}{1 - \sin \theta} - \dfrac{1}{1 + \sin \theta} = 2 \tan \theta \sec \theta$

27 $\dfrac{\tan \theta}{\sec \theta - \cos \theta} = \dfrac{\sec \theta}{\tan \theta}$

28 $\dfrac{\sin \theta}{\csc \theta + \cot \theta} = 1 - \cos \theta$

29 $\sec \theta + \csc \theta = (\tan \theta + \cot \theta)(\cos \theta + \sin \theta)$

30 $\dfrac{1 + \cos \theta}{\sin \theta} + \dfrac{\sin \theta}{1 + \cos \theta} = 2 \csc \theta$

31 $\dfrac{\tan \theta - \sin \theta \cos \theta}{\sec \theta} = \sin^3 \theta$

32 $\dfrac{1 + 2 \sin \theta \cos \theta}{\cos^2 \theta} = (1 + \tan \theta)^2$

33 $\cos^4 \theta - \cos^2 \theta = \sin^4 \theta - \sin^2 \theta$

34 $\dfrac{\cos^3 \theta - \sin^3 \theta}{\cos \theta - \sin \theta} = 1 + \sin \theta \cos \theta$

35 $(1 - \tan \theta)(1 - \cot \theta) = 2 - \dfrac{1}{\sin \theta \cos \theta}$

36 $\tan \theta - \cot \theta = \dfrac{1 - 2 \cos^2 \theta}{\sin \theta \cos \theta}$

37 $\dfrac{\cos \theta}{1 - \tan \theta} + \dfrac{\sin \theta}{1 - \cot \theta} = \sin \theta + \cos \theta$

38 $\dfrac{\tan^2 \theta + 1}{\tan^2 \theta - 1} = \dfrac{\sec^2 \theta + \csc^2 \theta}{\sec^2 \theta - \csc^2 \theta}$

39 $\dfrac{\cos \alpha - \sin \beta}{\cos \beta - \sin \alpha} = \dfrac{\cos \beta + \sin \alpha}{\cos \alpha + \sin \beta}$

40 $\dfrac{\tan \alpha + \tan \beta}{\cot \alpha + \cot \beta} = \dfrac{\tan \alpha \tan \beta - 1}{1 - \cot \alpha \cot \beta}$.

3   THE ADDITION LAWS

The **addition laws** for the sine and cosine are formulas that express $\sin(\alpha + \beta)$ and $\cos(\alpha + \beta)$ in terms of $\sin \alpha$, $\sin \beta$, $\cos \alpha$, and $\cos \beta$:

$$\sin(\alpha + \beta) = \sin \alpha \cos \beta + \cos \alpha \sin \beta$$
$$\cos(\alpha + \beta) = \cos \alpha \cos \beta - \sin \alpha \sin \beta.$$

These addition formulas are very basic in mathematics and should be memorized. Before we discuss a proof, let us note the equivalent forms, obtained by substituting $-\beta$ for β and using $\cos(-\beta) = \cos \beta$ and $\sin(-\beta) = -\sin \beta$:

$$\sin(\alpha - \beta) = \sin \alpha \cos \beta - \cos \alpha \sin \beta$$
$$\cos(\alpha - \beta) = \cos \alpha \cos \beta + \sin \alpha \sin \beta.$$

(The reasoning is reversible: the same substitution $\beta \longrightarrow -\beta$ takes the formulas of the second box into those of the first.)

Another point to note is that the addition law for the cosine implies that for the sine (and vice versa). For suppose $\cos(\alpha - \beta) = \cos \alpha \cos \beta + \sin \alpha \sin \beta$ is true for all α and β. Replace α by $\frac{1}{2}\pi - \alpha$ and use the relations

$$\cos(\tfrac{1}{2}\pi - \theta) = \sin \theta \qquad \sin(\tfrac{1}{2}\pi - \theta) = \cos \theta.$$

We have

$$\cos(\tfrac{1}{2}\pi - \alpha - \beta) = \cos(\tfrac{1}{2}\pi - \alpha)\cos \beta + \sin(\tfrac{1}{2}\pi - \alpha)\sin \beta$$

hence

$$\sin(\alpha + \beta) = \sin \alpha \cos \beta + \cos \alpha \sin \beta.$$

Thus we shall have proved both addition laws when we have proved the cosine one.

Proof of the Cosine Addition Law

Take α and β in standard position (Fig. 1a, next page). Their terminal sides meet the unit circle in

$$A = (\cos \alpha, \sin \alpha) \quad \text{and} \quad B = (\cos \beta, \sin \beta).$$

By the distance formula, the chord length $\overline{AB}$ satisfies

$$\overline{AB}^2 = (\cos \alpha - \cos \beta)^2 + (\sin \alpha - \sin \beta)^2$$
$$= (\cos^2 \alpha + \sin^2 \alpha) + (\cos^2 \beta + \sin^2 \beta)$$
$$- 2(\cos \alpha \cos \beta + \sin \alpha \sin \beta),$$

hence

$$\overline{AB}^2 = 2 - 2(\cos \alpha \cos \beta + \sin \alpha \sin \beta).$$

Now let us *rotate* the triangle $0AB$ about 0 through angle $-\beta$. See Fig. 1b, next page. The resulting triangle $0A'B'$, congruent to $0AB$, has

$$A' = (\cos(\alpha - \beta), \sin(\alpha - \beta)) \quad \text{and} \quad B' = (1, 0).$$

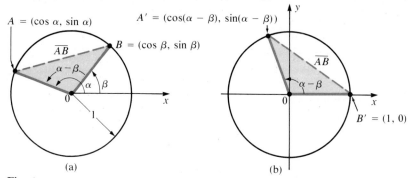

Fig. 1

In other words, $0A'$ is the terminal side of $\alpha' = \alpha - \beta$ and $0B'$ is the terminal side of $\beta' = 0$. By the formula for chord length, with α and β replaced by α' and β',

$$\overline{A'B'}^2 = 2 - 2\cos(\alpha - \beta).$$

But $\overline{A'B'} = \overline{AB}$ because rotation preserves length. Hence

$$2 - 2\cos(\alpha - \beta) = 2 - 2(\cos\alpha\cos\beta - \sin\alpha\sin\beta),$$

so

$$\cos(\alpha - \beta) = \cos\alpha\cos\beta - \sin\alpha\sin\beta.$$

Addition Law for the Tangent

Addition formulas for the other functions can be derived from those for sine and cosine. The most important is the formula for tangent:

$$\boxed{\tan(\alpha + \beta) = \frac{\tan\alpha + \tan\beta}{1 - \tan\alpha\tan\beta} \qquad (\tan\alpha\tan\beta \neq 1).}$$

Proof

$$\tan(\alpha + \beta) = \frac{\sin(\alpha + \beta)}{\cos(\alpha + \beta)} = \frac{\sin\alpha\cos\beta + \cos\alpha\sin\beta}{\cos\alpha\cos\beta - \sin\alpha\sin\beta}.$$

Divide numerator and denominator by $\cos\alpha\cos\beta$:

$$\tan(\alpha + \beta) = \frac{\dfrac{\sin\alpha}{\cos\alpha} + \dfrac{\sin\beta}{\cos\beta}}{1 - \dfrac{\sin\alpha\sin\beta}{\cos\alpha\cos\beta}} = \frac{\tan\alpha + \tan\beta}{1 - \tan\alpha\tan\beta}.$$

EXAMPLE 1 Find exact values for

(a) $\cos 75°$ (b) $\tan 75°$.

Solution (a) Since $75 = 45 + 30$, we can use the addition laws to express $\cos 75°$ in terms of $\sin 45°$, $\cos 45°$, $\sin 30°$, and $\cos 30°$:

$$\cos 75° = \cos(45° + 30°) = \cos 45° \cos 30° - \sin 45° \sin 30°$$
$$= (\tfrac{1}{2}\sqrt{2})(\tfrac{1}{2}\sqrt{3}) - (\tfrac{1}{2}\sqrt{2})(\tfrac{1}{2}) = \tfrac{1}{4}\sqrt{6} - \tfrac{1}{4}\sqrt{2}.$$

(b)   By the addition law for tangent,

$$\tan 75° = \frac{\tan 45° + \tan 30°}{1 - \tan 45° \tan 30°} = \frac{1 + \tfrac{1}{3}\sqrt{3}}{1 - 1\cdot\tfrac{1}{3}\sqrt{3}} = \frac{3 + \sqrt{3}}{3 - \sqrt{3}}.$$

For a neater answer, we rationalize the denominator:

$$\tan 75° = \frac{3 + \sqrt{3}}{3 - \sqrt{3}} = \frac{3 + \sqrt{3}}{3 - \sqrt{3}}\frac{3 + \sqrt{3}}{3 + \sqrt{3}} = \frac{9 + 6\sqrt{3} + 3}{9 - 3}$$
$$= \frac{12 + 6\sqrt{3}}{6} = 2 + \sqrt{3}.$$

Answer   (a) $\tfrac{1}{4}(\sqrt{6} - \sqrt{2})$ (b) $2 + \sqrt{3}$

EXAMPLE 2   Suppose α and β are angles in the first quadrant, $\tan \alpha = \tfrac{1}{2}$ and $\tan \beta = \tfrac{1}{3}$. Find $\alpha + \beta$.

Solution   By the addition law for tangent,

$$\tan(\alpha + \beta) = \frac{\tan \alpha + \tan \beta}{1 - \tan \alpha \tan \beta} = \frac{\tfrac{1}{2} + \tfrac{1}{3}}{1 - \tfrac{1}{2}\cdot\tfrac{1}{3}} = \frac{\tfrac{5}{6}}{\tfrac{5}{6}} = 1.$$

Thus $\tan(\alpha + \beta) = 1$. Therefore $\alpha + \beta$ is one of the angles $\tfrac{1}{4}\pi + 2\pi n$ or $\tfrac{5}{4}\pi + 2\pi n$. Which one? Well, α and β are first quadrant angles, so $0 < \alpha + \beta < \pi$. The only possibility in this range is $\alpha + \beta = \tfrac{1}{4}\pi$.

Answer $\tfrac{1}{4}\pi$

Double-Angle Formulas

There are important special cases of the addition formulas that express $\sin 2\theta$, $\cos 2\theta$, and $\tan 2\theta$ in terms of $\sin \theta$, $\cos \theta$, and $\tan \theta$. In the formulas for $\sin(\alpha + \beta)$, $\cos(\alpha + \beta)$, and $\tan(\alpha + \beta)$, set $\alpha = \beta = \theta$. The results are the **double-angle formulas:**

$$\sin 2\theta = 2 \sin \theta \cos \theta \qquad \cos 2\theta = \cos^2 \theta - \sin^2 \theta$$
$$\tan 2\theta = \frac{2 \tan \theta}{1 - \tan^2 \theta}.$$

The second formula has two useful alternative forms. First we replace $\cos^2 \theta$ by $1 - \sin^2 \theta$:

$$\cos 2\theta = \cos^2 \theta - \sin^2 \theta = (1 - \sin^2 \theta) - \sin^2 \theta = 1 - 2 \sin^2 \theta.$$

Next we replace $\sin^2 \theta$ by $1 - \cos^2 \theta$:

$$\cos 2\theta = \cos^2 \theta - \sin^2 \theta = \cos^2 \theta - (1 - \cos^2 \theta) = 2 \cos^2 \theta - 1.$$

$$\cos 2\theta = 1 - 2 \sin^2 \theta = 2 \cos^2 \theta - 1.$$

EXAMPLE 3   Suppose θ is an angle in the second quadrant and $\sin \theta = \frac{3}{5}$. Find

(a) $\cos 2\theta$ (b) $\sin 2\theta$ (c) the quadrant of 2θ.

Solution   (a)   Use the double angle formula for cosine, preferably the form that involves only $\sin \theta$:

$$\cos 2\theta = 1 - 2 \sin^2 \theta = 1 - 2(\tfrac{3}{5})^2 = 1 - \tfrac{18}{25} = \tfrac{7}{25}.$$

(b)   The double angle formula for sine requires both $\sin \theta$ and $\cos \theta$. Since θ is in the second quadrant, $\cos \theta < 0$. Therefore

$$\cos \theta = -\sqrt{1 - \sin^2 \theta} = -\sqrt{1 - (\tfrac{3}{5})^2} = -\sqrt{1 - \tfrac{9}{25}}$$

$$= -\sqrt{\tfrac{16}{25}} = -\tfrac{4}{5}.$$

Now apply the double angle formula:

$$\sin 2\theta = 2 \sin \theta \cos \theta = 2(\tfrac{3}{5})(-\tfrac{4}{5}) = -\tfrac{24}{25}.$$

(c)   Since $\cos 2\theta > 0$ and $\sin 2\theta < 0$, the terminal side of 2θ lies in the fourth quadrant.

Answer   (a) $\frac{7}{25}$ (b) $-\frac{24}{25}$ (c) 4-th quadrant

EXAMPLE 4   Express $\sin 3\theta$ in terms of $\sin \theta$ and $\cos \theta$.

Solution   Write $3\theta = \theta + 2\theta$, and use the addition laws to express $\sin 3\theta$ in terms of $\sin \theta$, $\cos \theta$, $\sin 2\theta$, and $\cos 2\theta$,

$$\sin 3\theta = \sin(\theta + 2\theta) = \sin \theta \cos 2\theta + \cos \theta \sin 2\theta.$$

Now apply the double-angle formulas for $\sin 2\theta$ and $\cos 2\theta$:

$$\sin 3\theta = \sin \theta [\cos^2 \theta - \sin^2 \theta] + \cos \theta [2 \sin \theta \cos \theta]$$

$$= 3 \sin \theta \cos^2 \theta - \sin^3 \theta.$$

If you prefer the answer in terms of $\sin \theta$ alone, you can replace $\cos^2 \theta$ by $1 - \sin^2 \theta$.

Answer $\sin 3\theta = 3 \sin \theta \cos^2 \theta - \sin^3 \theta = 3 \sin \theta - 4 \sin^3 \theta.$

Remarkable!   The addition laws for sine and cosine are perhaps the most striking formulas in elementary mathematics. The only formulas you may have seen before vaguely resembling them are the addition laws for exponential and logarithm functions:

$$a^{x+y} = a^x a^y \qquad \log_a x + \log_a y = \log_a (xy).$$

(See Chapter 12, Sections 2 and 3.) But the sine and cosine addition laws express $\sin(\alpha + \beta)$ and $\cos(\alpha + \beta)$ in terms of *both* sines and cosines of α and β.

Note that all the formulas of this section are consequences of the addition laws.

Exercises

Use the addition laws to compute

1   $\cos(\theta + 2\pi)$ and $\sin(\theta + 2\pi)$ 2 $\cos(\theta + \pi)$ and $\sin(\theta + \pi)$

3   $\cos(\theta + \tfrac{1}{2}\pi)$ and $\sin(\theta + \tfrac{1}{2}\pi)$ 4 $\cos(\tfrac{1}{2}\pi - \theta)$ and $\sin(\tfrac{1}{2}\pi - \theta)$.

Express

5 $\cos(\theta + \frac{1}{4}\pi)$ and $\sin(\theta + \frac{1}{4}\pi)$ in terms of $\cos \theta$ and $\sin \theta$

6 $\cos(\theta + \frac{1}{6}\pi)$ and $\sin(\theta + \frac{1}{6}\pi)$ in terms of $\cos \theta$ and $\sin \theta$

7 $\cot(\alpha + \beta)$ in terms of $\cot \alpha$ and $\cot \beta$

8 $\tan(\alpha - \beta)$ in terms of $\tan \alpha$ and $\tan \beta$.

Find exactly

9 $\sin 75°$ **10** $\tan 105°$ **11** $\cos 15°$ **12** $\sin 15°$

13 $\sin 9° \cos 21° + \cos 9° \sin 21°$ **14** $\sin 81° \cos 36° - \cos 81° \sin 36°$

15 $\cos \frac{3}{10}\pi \cos \frac{1}{20}\pi + \sin \frac{3}{10}\pi \sin \frac{1}{20}\pi$

16 $\cos 0.3\pi \cos 0.7\pi - \sin 0.3\pi \sin 0.7\pi$.

17 Suppose $0 < \alpha < \frac{1}{2}\pi$, $\pi < \beta < \frac{3}{2}\pi$, $\tan \alpha = \frac{1}{3}$ and $\tan \beta = \frac{1}{2}$. Find $\alpha + \beta$.

18 Suppose α, β, and γ are in the first quadrant, $\tan \alpha = \frac{1}{3}$, $\tan \beta = \frac{1}{7}$, and $\tan \gamma = \frac{1}{2}$, Show that $\alpha + \beta = \gamma$.

19 Suppose $0 < \alpha < \frac{1}{2}\pi$, $\cos \alpha = \frac{1}{4}$, $\frac{1}{2}\pi < \beta < \pi$, and $\sin \beta = \frac{1}{5}$. Find $\sin(\beta - \alpha)$.

20 Suppose $\frac{1}{2}\pi < \alpha < \pi$, $\sin \alpha = \frac{2}{3}$, $\pi < \beta < \frac{3}{2}\pi$, and $\tan \beta = 2$. Find $\cos(\alpha + \beta)$.

21 Suppose θ is in the second quadrant and $\cos \theta = -\frac{1}{3}$. Find $\sin 2\theta$ and $\cos 2\theta$.

22 Suppose θ is in the fourth quadrant and $\sin \theta = -\frac{2}{3}$. Find $\sin 2\theta$ and $\cos 2\theta$.

23 Suppose $\tan \theta = 4$. Find $\cos 2\theta$ and $\tan 2\theta$.

24 Suppose $\tan \theta = 2$. Find $\sin 2\theta$ and $\tan 2\theta$.

Express

25 $\cos 3\theta$ in terms of $\cos \theta$ **26** $\tan 3\theta$ in terms of $\tan \theta$

27 $\csc 3\theta$ in terms of $\csc \theta$ **28** $\sec 3\theta$ in terms of $\sec \theta$

[Use Example 3.]   [Use Example 25.]

Use double-angle formulas to prove

29 $\sin 4\theta = 4 \sin \theta \cos \theta - 8 \sin^3 \theta \cos \theta$

30 $\cos 4\theta = 8 \cos^4 \theta - 8 \cos^2 \theta + 1$

31 $\cot 2\theta + \csc 2\theta = \cot \theta$ **32** $\csc 2\theta - \cot 2\theta = \tan \theta$

33 $\tan \theta + \cot \theta = 2 \csc 2\theta$ **34** $\sin \theta \cos \theta \cos 2\theta \cos 4\theta = \frac{1}{8} \sin 8\theta$.

35 If $0 < \theta < \frac{1}{4}\pi$, which is larger, $\tan 2\theta$ or $2 \tan \theta$?

36* If $\frac{1}{4}\pi < \theta < \frac{1}{3}\pi$, show that $\sin \theta < \sin 2\theta < \sqrt{2} \sin \theta$.

Prove

37 $\cot 2\theta = \dfrac{\cot^2 \theta - 1}{2 \cot \theta}$ **38** $\sec 2\theta = \dfrac{\sec^2 \theta}{2 - \sec^2 \theta}$

39 $\cot(\alpha + \beta) = \dfrac{\cot \alpha \cot \beta - 1}{\cot \alpha + \cot \beta}$ **40** $\tan 4\theta = \dfrac{4 \tan \theta - 4 \tan^3 \theta}{1 - 6 \tan^2 \theta + \tan^4 \theta}$

41 $\dfrac{1 + \cos 2\theta}{1 - \cos 2\theta} = \cot^2 \theta$ **42** $\dfrac{\sin 2\theta + \sin \theta}{\cos 2\theta + \cos \theta + 1} = \tan \theta$

43 $\dfrac{\sec \theta - \csc \theta}{\sec \theta + \csc \theta} = \tan 2\theta - \sec 2\theta$ **44** $\dfrac{\cot \theta - \tan \theta}{\cot \theta + \tan \theta} = \cos 2\theta$.

4   FURTHER IDENTITIES

Half-Angle Formulas

From the double-angle formulas, we can obtain half-angle formulas that express $\sin \frac{1}{2}\theta$, $\cos \frac{1}{2}\theta$, and $\tan \frac{1}{2}\theta$ in terms of $\sin \theta$ and $\cos \theta$. We start with the double-angle formulas

$$\cos 2\theta = 2 \cos^2 \theta - 1 = 1 - 2 \sin^2 \theta$$

and solve for $\sin \theta$ and $\cos \theta$:

$$\sin \theta = \pm\sqrt{\frac{1 - \cos 2\theta}{2}} \qquad \cos \theta = \pm\sqrt{\frac{1 + \cos 2\theta}{2}},$$

where the sign depends on the quadrant. Replacing θ by $\frac{1}{2}\theta$, we obtain the half-angle formulas:

$$\sin \tfrac{1}{2}\theta = \pm\sqrt{\frac{1 - \cos \theta}{2}} \qquad \cos \tfrac{1}{2}\theta = \pm\sqrt{\frac{1 + \cos \theta}{2}},$$

where the sign of each is chosen according to the quadrant of $\frac{1}{2}\theta$.

We could obtain a formula for $\tan \frac{1}{2}\theta$ by dividing the expressions for $\sin \frac{1}{2}\theta$ and $\cos \frac{1}{2}\theta$, but we avoid square roots by using a little trick:

$$\tan \tfrac{1}{2}\theta = \frac{\sin \frac{1}{2}\theta}{\cos \frac{1}{2}\theta} = \frac{\sin \frac{1}{2}\theta}{\cos \frac{1}{2}\theta} \cdot \frac{2 \cos \frac{1}{2}\theta}{2 \cos \frac{1}{2}\theta}.$$

By the double-angle formulas, the numerator is $\sin 2(\frac{1}{2}\theta) = \sin \theta$ and the denominator is

$$2 \cos^2 (\tfrac{1}{2}\theta) = 1 + [2 \cos^2 (\tfrac{1}{2}\theta) - 1] = 1 + \cos 2(\tfrac{1}{2}\theta) = 1 + \cos \theta.$$

Therefore

$$\tan \tfrac{1}{2}\theta = \frac{\sin \theta}{1 + \cos \theta}.$$

This half-angle formula for the tangent has a geometric interpretation. In Fig. 1a, the inscribed angle is $\frac{1}{2}\theta$, half the corresponding central angle. From the right triangle (Fig. 1b), $\tan \frac{1}{2}\theta = \sin \theta/(1 + \cos \theta)$.

Identities expressing $\sin \theta$ and $\cos \theta$ in terms of $\tan \frac{1}{2}\theta$ are useful in calculus:

$$\sin \theta = \frac{2 \tan \frac{1}{2}\theta}{1 + \tan^2 \frac{1}{2}\theta} \qquad \cos \theta = \frac{1 - \tan^2 \frac{1}{2}\theta}{1 + \tan^2 \frac{1}{2}\theta}.$$

The proofs are easy:

$$\sin 2\theta = 2 \sin \theta \cos \theta = \frac{2 \sin \theta \cos \theta}{\sin^2 \theta + \cos^2 \theta} = \frac{2\dfrac{\sin \theta}{\cos \theta}}{\dfrac{\sin^2 \theta}{\cos^2 \theta} + 1}$$

$$= \frac{2 \tan \theta}{1 + \tan^2 \theta}.$$

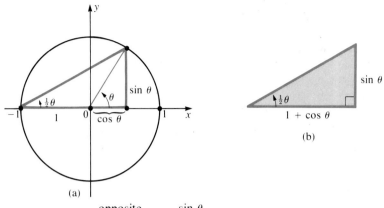

Fig. 1 $\tan \frac{1}{2}\theta = \dfrac{\text{opposite}}{\text{adjacent}} = \dfrac{\sin \theta}{1 + \cos \theta}$

$$\cos 2\theta = \cos^2 \theta - \sin^2 \theta = \frac{\cos^2 \theta - \sin^2 \theta}{\cos^2 \theta + \sin^2 \theta} = \frac{1 - \dfrac{\sin^2 \theta}{\cos^2 \theta}}{1 + \dfrac{\sin^2 \theta}{\cos^2 \theta}}$$

$$= \frac{1 - \tan^2 \theta}{1 + \tan^2 \theta}.$$

Now replace θ by $\frac{1}{2}\theta$.

EXAMPLE 1   Find exact values for   (a) $\cos 105°$ (b) $\tan \frac{1}{12}\pi$.

Solution   (a) Since $105° = \frac{1}{2}(210°)$ and $210° = 180° + 30°$, use the half-angle formula for cosine, with the negative sign because $105°$ is in the second quadrant:

$$\cos 105° = -\sqrt{\frac{1 + \cos 210°}{2}} = -\sqrt{\frac{1 - \frac{1}{2}\sqrt{3}}{2}} = -\tfrac{1}{2}\sqrt{2 - \sqrt{3}}.$$

(b)   Since $\frac{1}{12}\pi = \frac{1}{2}(\frac{1}{6}\pi)$, use the formula for $\tan \frac{1}{2}\theta$:

$$\tan \tfrac{1}{12}\pi = \frac{\sin \frac{1}{6}\pi}{1 + \cos \frac{1}{6}\pi} = \frac{\frac{1}{2}}{1 + \frac{1}{2}\sqrt{3}} = \frac{1}{2 + \sqrt{3}}.$$

Rationalizing the denominator gives us $\tan \frac{1}{12}\pi = 2 - \sqrt{3}$.

Answer   (a) $-\frac{1}{2}\sqrt{2 - \sqrt{3}}$ (b) $2 - \sqrt{3}$

Sums of Sines and Cosines

We give two useful applications of the addition formulas. The first is to express any function

$$f(\theta) = a \cos \theta + b \sin \theta \qquad [(a, b) \neq (0, 0)]$$

as a constant times a single cosine or sine.

To do this, we choose an angle θ_0 in standard position whose terminal side passes through (a, b). Then

$$\cos \theta_0 = \frac{a}{\sqrt{a^2 + b^2}} \qquad \sin \theta_0 = \frac{b}{\sqrt{a^2 + b^2}}.$$

(The angle θ_0 is only determined up to a multiple of 2π. We can insist that $0 \leq \theta_0 < 2\pi$; then θ_0 is unique.)

Now we have

$$a \cos \theta + b \sin \theta$$

$$= \sqrt{a^2 + b^2} \left(\frac{a}{\sqrt{a^2 + b^2}} \cos \theta + \frac{b}{\sqrt{a^2 + b^2}} \sin \theta \right)$$

$$= \sqrt{a^2 + b^2} (\cos \theta_0 \cos \theta + \sin \theta_0 \sin \theta) = \sqrt{a^2 + b^2} \cos(\theta - \theta_0).$$

If $a^2 + b^2 > 0$, then

$$a \cos \theta + b \sin \theta = \sqrt{a^2 + b^2} \cos(\theta - \theta_0),$$

where θ_0 is the angle satisfying $0 \leq \theta_0 < 2\pi$ and

$$\cos \theta_0 = \frac{a}{\sqrt{a^2 + b^2}} \qquad \sin \theta_0 = \frac{b}{\sqrt{a^2 + b^2}}.$$

EXAMPLE 2   Express $-\cos \theta + \sqrt{3} \sin \theta$ as a constant times a cosine.

Solution   Here $a = -1$, $b = \sqrt{3}$, and $a^2 + b^2 = 4$, so we write

$$-\cos \theta + \sqrt{3} \sin \theta = 2(-\tfrac{1}{2} \cos \theta + \tfrac{1}{2}\sqrt{3} \sin \theta)$$

We choose θ_0 to satisfy $0 \leq \theta_0 < 2\pi$ and

$$\cos \theta_0 = -\tfrac{1}{2} \qquad \sin \theta_0 = \tfrac{1}{2}\sqrt{3}$$

Clearly $\theta_0 = \tfrac{2}{3}\pi$.

Answer $2 \cos(\theta - \tfrac{2}{3}\pi)$.

Remark   Suppose instead we choose $\theta_0 = \tfrac{1}{6}\pi$. Then $\sin \theta_0 = \tfrac{1}{2}$ and $\cos \theta_0 = \tfrac{1}{2}\sqrt{3}$, so the given function is

$$2(-\sin \theta_0 \cos \theta + \cos \theta_0 \sin \theta) = 2 \sin(\theta - \theta_0).$$

In general, each function $a \cos \theta + b \sin \theta$ can be expressed as a constant multiple of a cosine or a sine.

EXAMPLE 3   Prove that $|-\cos \theta + \sqrt{3} \sin \theta| \leq 2$ for all values of θ.

Solution   By the result of Example 2,

$$|-\cos \theta + \sqrt{3} \sin \theta| = |2 \cos(\theta - \tfrac{2}{3}\pi)| = 2|\cos(\theta - \tfrac{2}{3}\pi)| \leq 2 \cdot 1 = 2.$$

Products of Sines and Cosines

The second application of the addition formulas is to express products of sines and cosines as sums. First we add the formulas

$$\sin(\alpha + \beta) = \sin \alpha \cos \beta + \cos \alpha \sin \beta$$

$$\sin(\alpha - \beta) = \sin \alpha \cos \beta - \cos \alpha \sin \beta,$$

to obtain

$$\sin(\alpha + \beta) + \sin(\alpha - \beta) = 2 \sin \alpha \cos \beta.$$

Next we add and subtract the formulas

$$\cos(\alpha + \beta) = \cos \alpha \cos \beta - \sin \alpha \sin \beta$$

$$\cos(\alpha - \beta) = \cos \alpha \cos \beta + \sin \alpha \sin \beta,$$

to obtain

$$\cos(\alpha + \beta) + \cos(\alpha - \beta) = 2 \cos \alpha \cos \beta$$

$$\cos(\alpha - \beta) - \cos(\alpha + \beta) = 2 \sin \alpha \sin \beta.$$

We interpret the results as formulas for products of sines and cosines:

$$\sin \alpha \sin \beta = \tfrac{1}{2}[\cos(\alpha - \beta) - \cos(\alpha + \beta)]$$

$$\cos \alpha \cos \beta = \tfrac{1}{2}[\cos(\alpha + \beta) + \cos(\alpha - \beta)]$$

$$\sin \alpha \cos \beta = \tfrac{1}{2}[\sin(\alpha + \beta) + \sin(\alpha - \beta)].$$

Examples

(1) $\sin 75° \sin 15° = \tfrac{1}{2}(\cos 60° - \cos 90°) = \tfrac{1}{2}(\tfrac{1}{2} - 0) = \tfrac{1}{4}$

(2) $\sin 105° \cos 75° = \tfrac{1}{2}(\sin 180° + \sin 30°) = \tfrac{1}{2}(0 + \tfrac{1}{2}) = \tfrac{1}{4}$

(3) $\cos 105° \cos 75° = \tfrac{1}{2}(\cos 180° + \cos 30°) = \tfrac{1}{2}(-1 + \tfrac{1}{2}\sqrt{3})$

Sometimes we go the other way and express the sum or difference of two sines or two cosines as a product. We write

$$\alpha = \tfrac{1}{2}(\alpha + \beta) + \tfrac{1}{2}(\alpha - \beta) = \gamma + \delta$$

$$\beta = \tfrac{1}{2}(\alpha + \beta) - \tfrac{1}{2}(\alpha - \beta) = \gamma - \delta.$$

where $\gamma = \tfrac{1}{2}(\alpha + \beta)$ and $\delta = \tfrac{1}{2}(\alpha - \beta)$. Then

$$\sin \alpha + \sin \beta = \sin(\gamma + \delta) + \sin(\gamma - \delta)$$

$$= 2 \sin \gamma \cos \delta$$

$$= 2 \sin \tfrac{1}{2}(\alpha + \beta) \cos \tfrac{1}{2}(\alpha - \beta).$$

In this way we derive four formulas:

$$\sin \alpha + \sin \beta = 2 \sin \tfrac{1}{2}(\alpha + \beta) \cos \tfrac{1}{2}(\alpha - \beta)$$

$$\sin \alpha - \sin \beta = 2 \cos \tfrac{1}{2}(\alpha + \beta) \sin \tfrac{1}{2}(\alpha - \beta)$$

$$\cos \alpha + \cos \beta = 2 \cos \tfrac{1}{2}(\alpha + \beta) \cos \tfrac{1}{2}(\alpha - \beta)$$

$$\cos \alpha - \cos \beta = -2 \sin \tfrac{1}{2}(\alpha + \beta) \sin \tfrac{1}{2}(\alpha - \beta).$$

The last two batches of formulas have important applications in situations where one vibration is imposed on another, such as in the modulation of radio signals, and in the phenomenon of beats in acoustics. See the next section.

Exercises

Show that

1 $\sin 22.5° = \frac{1}{2}\sqrt{2 - \sqrt{2}}$ **2** $\tan 22.5° = \sqrt{2} - 1$ **3** $\tan 67.5° = \sqrt{2} + 1$.

4 Compute $\cos 15°$ two ways: by half-angle formulas and by addition laws. Conclude that
$$\sqrt{2 + \sqrt{3}} = \frac{1}{2}(\sqrt{6} + \sqrt{2}).$$

Prove

5 $\tan \frac{1}{2}\theta = \dfrac{1 - \cos \theta}{\sin \theta}$ **6** $\cot \frac{1}{2}\theta = \dfrac{1 + \cos \theta}{\sin \theta}$.

Express each combination as a positive constant times a cosine

7 $4 \cos \theta + 3 \sin \theta$ **8** $3 \cos \theta - 4 \sin \theta$
9 $\cos \theta - 2 \sin \theta$ **10** $\cos \theta + 2 \sin \theta$
11 $2 \cos \theta + 3 \sin \theta$ **12** $-2 \cos \theta - 3 \sin \theta$
13 $3 \cos \theta - \sin \theta$ **14** $5 \sin \theta$.

15 Prove that $|\sin \theta + \cos \theta| \leq \sqrt{2}$.
16 (cont.) Find all θ for which $|\sin \theta + \cos \theta| = \sqrt{2}$.
17 Find the maximum value of $\cos \theta - 3 \sin \theta$.
18 Find the minimum value of $4 \sin \theta + 3 \cos \theta$.

Express each function as half a sum (difference) of sines or cosines

19 $\sin \theta \sin 2\theta$ **20** $\cos 2\theta \cos \theta$
21 $\sin 3\theta \cos 4\theta$ **22** $\sin 101\theta \sin 100\theta$.

Express each function as a product of constants, sines, and cosines

23 $\sin \theta + \sin 2\theta$ **24** $\cos 2\theta - \cos \theta$
25 $\sin 5\theta - \sin 4\theta$ **26** $\sin 5\theta + \sin 4\theta$
27 $\cos 6\theta - \cos 5\theta$ **28** $\cos \theta + \cos 3\theta$.

Prove

29 $\sin 75° + \sin 15° = \frac{1}{2}\sqrt{6}$ **30** $\sin 75° - \sin 15° = \frac{1}{2}\sqrt{2}$
31 $\cos 105° + \cos 15° = \frac{1}{2}\sqrt{2}$ **32** $\cos 105° - \cos 15° = -\frac{1}{2}\sqrt{6}$.

33 From Exs. 29 and 30, find $\sin 75°$ and $\sin 15°$.
34 From Exs. 31 and 32, find $\cos 105°$ and $\cos 15°$.

Prove

35 $\tan \alpha + \tan \beta = \sin(\alpha + \beta)/\cos \alpha \cos \beta$
36 $\cos^4 \theta = \frac{1}{8}(3 + 4 \cos 2\theta + \cos 4\theta)$ [*Hint* Use half-angle formulas twice.]
37 $\sin^4 \theta = \frac{1}{8}(3 - 4 \cos 2\theta + \cos 4\theta)$
38 $\sin^2 \alpha - \sin^2 \beta = \sin(\alpha + \beta)\sin(\alpha - \beta)$
39 $\cos^2 \alpha - \sin^2 \beta = \cos(\alpha + \beta)\cos(\alpha - \beta)$

40 $1 + 2 \cos \theta + 2 \cos 2\theta + 2 \cos 3\theta = \dfrac{\sin \frac{7}{2}\theta}{\sin \frac{1}{2}\theta}$.

[*Hint* Start by multiplying the left side by $\sin \frac{1}{2}\theta$.]

41 $\sin \theta \sin 2\theta \cos 2\theta = \frac{1}{4}(\cos 3\theta - \cos 5\theta)$
42 $\sin \theta \sin 2\theta \sin 3\theta = \frac{1}{4}(\sin 2\theta + \sin 4\theta - \sin 6\theta)$.

43 Set $t = \tan \frac{1}{2}\theta$. Express $\sin 2\theta$ in terms of t. [*Hint* Use half-angle formulas.]
44 Set $t = \tan \frac{1}{2}\theta$. Express $\cos 2\theta$ in terms of t. [*Hint* Use half-angle formulas.]

Prove

45 $\cot \alpha - \cot \beta = \dfrac{\sin (\beta - \alpha)}{\sin \alpha \sin \beta}$

46 $\dfrac{\tan \frac{1}{2}(\alpha + \beta)}{\tan \frac{1}{2}(\alpha - \beta)} = \dfrac{\sin \alpha + \sin \beta}{\sin \alpha - \sin \beta}$

47 $\dfrac{\sin \alpha - \sin \beta}{\cos \alpha + \cos \beta} = \tan \frac{1}{2}(\alpha - \beta)$

48 $\dfrac{\sin \alpha + \sin \beta}{\cos \alpha - \cos \beta} = \cot \frac{1}{2}(\beta - \alpha)$

49 $\dfrac{\sin \alpha + \sin \beta}{\cos \alpha + \cos \beta} = \tan \frac{1}{2}(\alpha + \beta)$

50 $\dfrac{\sin \alpha - \sin \beta}{\cos \alpha - \cos \beta} = -\cot \frac{1}{2}(\alpha + \beta).$

51 Verify numerically that $\sin 40° + \sin 20° = \cos 10°$.

52 (cont.) Now prove the relation is true, not just true to 8 or 10 decimal places.

53 From our work on pp. 53–54, we know that $\sin \theta < \theta < \tan \theta$ for $0 < \theta < \frac{1}{2}\pi$. Apply this with $\theta = \frac{1}{6}\pi$ to prove $3 < \pi < 3.47$.

54* (cont.) Use $\theta = \frac{1}{12}\pi$ and half angle formulas to prove

$$12\sqrt{\tfrac{1}{2}(1 - \tfrac{1}{2}\sqrt{3})} < \pi < 12/(2 + \sqrt{3}),$$

hence $3.105 < \pi < 3.216$.

55* (cont.) Continue estimating π with $\theta = \frac{1}{24}\pi$.

56* (cont.) Use the same method to estimate π, choosing $\theta = \frac{1}{16}\pi$. (You should conclude that on a 4-banger calculator with $\boxed{\sqrt{}}$, you can calculate π with high accuracy. Also, $\sqrt{x}$ can be calculated by hand to any degree of accuracy. Hence so can π!)

5 VIBRATIONS [Optional]

In this section we shall describe some real life applications of trigonometric functions and identities.

Sines and cosines describe many of the important types of vibrations that occur in the physical world, such as vibrations of strings, fluctuations in pressures, currents, voltages, etc. Mathematically, such vibrations are given by functions of the type

$$P = A \sin(2\pi\nu t - \phi).$$

The constants A, ν, and ϕ have names:

    A **amplitude** ν **frequency** ϕ **phase angle**

A fourth constant of the vibration is $T = \dfrac{1}{\nu}$ **period**

The function P is periodic with period T. Its frequency ν is measured in cycles per second (called **hertz,** and abbreviated Hz). Note that P can also be expressed as

$$P = A \cos(2\pi\nu t - \phi - \tfrac{1}{2}\pi),$$

so a vibration can be described by either a sine or a cosine depending on how the phase angle ϕ is adjusted. We won't worry much about ϕ, although in some applications it is important. Usually, we can start the clock so that $\phi = 0$.

To see a vibration, clamp a couple of inches of the tip of a bread knife onto a table with the heel of one hand, most of the blade and the handle projecting from the edge of the table, as shown in Fig. 1. With your other hand, pull the handle down a little and release it sharply, with a "ping." Then the knife will vibrate, and

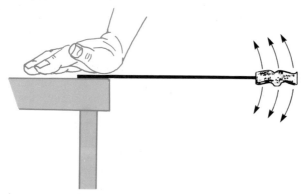

Fig. 1   Vibrating bread knife

the position of the end of the handle will vary from its rest position according to a formula of the type

$$y = A \sin(2\pi \nu t - \phi).$$

For a typical knife, ν will be between 8 and 12 Hz, depending on how much of the knife projects. Hence an up-down-up cycle will take between $\frac{1}{12}$ and $\frac{1}{8}$ sec.

Beats

When a tuning fork is set in motion, it causes a rapid vibration in air pressure given by the formula

$$P = A \sin 2\pi \nu t.$$

The ear detects the disturbance, which it hears as a musical note. For a middle C fork, $\nu \approx 261.63$ Hz, a much faster vibration than that of our bread knife.

Actually, a note produced by a musical instrument is not "pure" but a combination (superposition) of several pure notes. The corresponding formula is of the type

$$P = A_1 \sin 2\pi \nu_1 t + A_2 \sin 2\pi \nu_2 t + A_3 \sin 2\pi \nu_3 t + \cdots$$

where $\nu_1 < \nu_2 < \nu_3 < \cdots$, and the numbers $A_2, A_3, \cdots$ are small compared to A_1. Thus, to the basic pure tone $A_1 \sin 2\pi \nu_1 t$ are added small amounts of higher tones (overtones or harmonics). These overtones give the note its particular quality; they help the ear distinguish between a note on a violin and the same note played on a clarinet.

Suppose two pure musical notes of equal loudness are played together. What do we hear? If their frequencies μ and ν are sufficiently different, we hear two distinct notes (a chord). But if μ and ν are close together, say $|\mu - \nu| \leq 10$, our ears cannot distinguish between the notes, and we hear one note. But we hear that note in an unexpected way; its volume is not steady but pulses up and down several times a second. This is called the phenomonon of **beats.**

To hear beats, try this experiment on a guitar. Tune the low E-string up to F♯

and tune the A-string down to F♯. Then slightly flatten either of these. Pluck the two open strings together and the beats will be unmistakable. The closer in tune with each other the strings are, the slower the beats.

How can we explain beats? Believe it or not, by a trigonometric identity. When the two notes are sounded together, they cause a vibration in air pressure described by

$$P = A \sin 2\pi\mu t + A \sin 2\pi\nu t.$$

(We assume equal amplitudes for simplicity.) Now, there is an identity that expresses a sum of sines as a product (p. 85):

$$P = A(\sin 2\pi\mu t + \sin 2\pi\nu t) = 2A \cos 2\pi\sigma t \sin 2\pi\rho t,$$

where

$$\rho = \tfrac{1}{2}(\mu + \nu) \qquad \sigma = \tfrac{1}{2}(\mu - \nu).$$

Let us write

$$P = V(t)\sin 2\pi\rho t, \quad \text{where} \quad V(t) = 2A \cos 2\pi\sigma t.$$

The ear hears a note of frequency ρ and loudness $V(t)$. Since μ and ν are close, σ is small so $V(t)$ is a slowly fluctuating function. In fact, $V(t)$ varies from $2A$ (loud) to 0 (silence) to $-2A$ (loud) to 0 (silence) and back to $2A$ (loud) as t varies over a full period $0 \le t \le 1/\sigma$. See Fig. 2. The ear does not distinguish between positive and negative loudness; hence in one period of $V(t)$ there are two complete cycles of the beat:

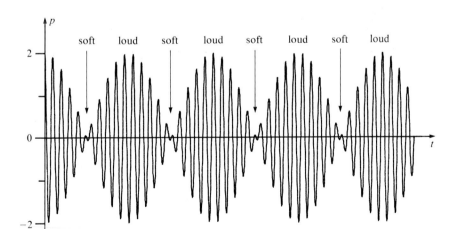

Fig. 2 $P = A(\cos 2\pi\mu t + \cos 2\pi\nu t)$ where $\mu \approx \nu$.
(In this computer-graphics example, $A = 1$, $\mu = \tfrac{21}{16}$, and $\nu = \tfrac{19}{16}$.)

Thus the **beat frequency** is $|2\sigma| = |\mu - \nu|$.

To illustrate these ideas, suppose you tune the A-string of a guitar to 92.5 Hz and the E-string to 90.5 Hz. Then

$$\rho = \tfrac{1}{2}(92.5 + 90.5) = 91.5 \qquad \sigma = \tfrac{1}{2}(92.5 - 90.5) = 1$$

and

$$P = A\,[\sin 2\pi(92.5)t + \sin 2\pi(90.5)t]$$

$$= (2A\cos 2\pi t)\sin 2\pi(91.5)t.$$

The ear will hear the note with pitch given by 91.5 Hz and loudness varying from $2A$ to 0 to $2A$, with two beats per second.

Remark Here is an experiment that makes the beat phenomenon visual. You will need two C-clamps (or a vise), two hacksaw blades, and a small rubber band (Fig. 3). Clamp one blade so the part projecting can be pinged with an audible note. (About 3–4 inches will project.) Clamp the second blade parallel to the first, about 2 inches away, and adjust it to exactly the same note. Then snap the rubber band over the two blades, about halfway. Finally, ping one blade and watch the "beat" move back and forth from one blade to the other.

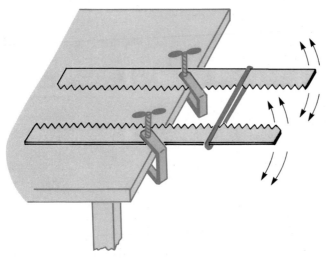

Fig. 3 "Beating" hacksaw blades

Amplitude Modulation

Let us describe an application of trigonometric identities to AM radio. The current generated by an AM transmitter is a pure vibration

$$i = A\sin 2\pi\mu t.$$

The **carrier frequency** (for commercial broadcast stations) lies in the interval

$$550\ \text{khz} \le \mu \le 1600\ \text{khz},$$

where khz means kilohertz, one thousand Hz. When the station is broadcasting its radio signal, but no sound, the amplitude A is constant, $A = A_0$. The constant A_0 is proportional to the power of the station. Sound information is added to the signal by varying (modulating) A, hence the name **amplitude modulation** or AM. To broadcast a sound signal $s(t)$, the amplitude A_0 is modulated by a fraction of its value proportional to $s(t)$. Thus the amplitude of the modulated signal is

$$A = A(t) = A_0[1 + ms(t)] \qquad (0 \le m \le 1)$$

For example, suppose middle C is being broadcast by a station at 980 on your radio dial. Then the modulated current is

$$i = i(t) = A_0(1 + m \sin 2\pi\nu t)\sin 2\pi\mu t,$$

where

$$\mu = 980,000 \text{ Hz} \quad \text{and} \quad \nu \approx 261.63 \text{ Hz}.$$

(Note that ν is very small compared to μ.) The modulated signal is actually the superposition (sum) of three pure vibrations—this is where trigonometric identities come in. We write

$$i = A_0[\sin 2\pi\mu t + m \sin 2\pi\nu t \sin 2\pi\mu t].$$

By the formulas for products on p. 85,

$$\sin 2\pi\nu t \sin 2\pi\mu t = \tfrac{1}{2}\cos 2\pi(\mu - \nu)t - \tfrac{1}{2}\cos 2\pi(\mu + \nu)t,$$

hence

$$i = A_0[\sin 2\pi\mu t + \tfrac{1}{2}m \cos 2\pi(\mu - \nu)t - \tfrac{1}{2}m \cos 2\pi(\mu + \nu)t].$$

Thus the radio signal broadcasting a pure musical note of frequency ν is the sum of three pure signals:

$A_0 \sin 2\pi\mu t$	at the carrier frequency μ
$\tfrac{1}{2}A_0 m \cos 2\pi(\mu - \nu)t$	the **left sideband** at frequency $\mu - \nu$
$\tfrac{1}{2}A_0 m \cos 2\pi(\mu + \nu)t$	the **right sideband** at frequency $\mu + \nu$.

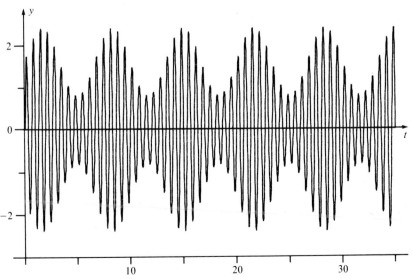

Fig. 4 Modulated AM signal $y = A(t) \sin 2\pi\mu t$ where $A(t) = A_0(1 + m \sin 2\pi\nu t)$. (In this computer-graphics example, $A_0 = 1.6$, $m = 0.5$, $\mu = 1.5$, and $\nu = 0.15$.)

The U.S. Federal Communications Commission requires each AM station to broadcast within a bandwidth of 10 khz centered at its assigned frequency μ. This means

$$\mu - 5000 < \mu - \nu \quad \text{and} \quad \mu + \nu < \mu + 5000,$$

hence, $0 \le \nu < 5000$ Hz. For all speech and most music this range of audio frequencies is adequate. (The highest piano note is about 4186 Hz.) However, the music transmitted is not HiFi; overtones are filtered out because of the bandwidth restrictions.

The graph of a (greatly distorted) modulated radio signal is shown in Fig. 4. You would hardly guess the decomposition of this graph into the carrier signal and the two sidebands.

REVIEW EXERCISES

For θ in the fourth quadrant and sec $\theta = 3$, find

1 $\sin \theta$

2 $\cot \theta$.

Prove the identity

3 $\tan^2 \theta - \sin^2 \theta = \sin^2 \theta \tan^2 \theta$

4 $\sin \theta \cos \theta = \dfrac{1 - \cos^2 \theta}{\tan \theta}$

5 $\dfrac{1}{\sec \theta + \tan \theta} = \dfrac{1 - \sin \theta}{\cos \theta}$

6 $\sec^4 \theta + \tan^4 \theta = 1 + 2 \sec^2 \theta \tan^2 \theta$

7 $\cos \tfrac{1}{2}\theta \cos \tfrac{1}{4}\theta = \dfrac{\sin \theta}{4 \sin \tfrac{1}{4}\theta}$

8 $\cos \tfrac{1}{2}\theta \cos \tfrac{1}{4}\theta \cos \tfrac{1}{8}\theta = \dfrac{\sin \theta}{8 \sin \tfrac{1}{8}\theta}$.

9 Use $105° = 60° + 45°$ to find $\sin 105°$.

10 Use $165° = 135° + 30°$ to find $\tan 165°$.

11 Express $-5 \cos \theta + 12 \sin \theta$ in the form $a \cos(\theta - \theta_0)$.

12 Express $\sin 5\theta \cos 3\theta$ as a sum of sines.

13 Express $\cos \theta - \cos 3\theta$ as a product of a constant and two sines.

14 Prove $\dfrac{\cos(\alpha - \beta)}{\cos \alpha \sin \beta} = \tan \alpha + \cot \beta$.

5 SOLVING RIGHT TRIANGLES

1 INTRODUCTION

We now begin a two-chapter unit on applications of the trigonometric functions to practical geometric problems. There are two steps in the solution of such a problem. First is the set-up: the analysis of the problem and its translation into an equation or systematic computation. Second is solving the equation, if there is one, and doing the computation to reach the answer, usually a number. We'll get to the set-up part later. Right now we are going to look at how we compute, and how accurate our computations should be.

In geometric problems from real life, we measure certain quantities with measuring tools that have certain limits of accuracy. We then use trigonometry (or other mathematical methods) to compute further quantities, which may not be accessible to direct measurement. Clearly, these computed quantities cannot be found to a higher degree of accuracy than that of the original measured quantities. This idea will become clearer as we proceed.

Accuracy

What does it mean that a measurement is accurate to 3 decimal places? It means that the error made is at most $\pm 5 \times 10^{-4}$.

For example, consider a chemist's analytical balance that weighs anything up to 150 grams with one-milligram accuracy. Since $1 \text{ mg} = 0.001 \text{ g}$, the read-out is accurate to 3 decimal places. If the read-out is 0.493 g, then the sample actually weighs between 0.4925 and 0.4935 g. If the read-out is 104.228 g, then the sample actually weighs between 104.2275 and 104.2285 g.

We say that 0.493 has 3 significant figures and 104.228 has 6 significant figures. In general, if a number is written with a decimal point, its number of **significant fig-**

ures is the number of digits from the left-most non-zero digit to the right-most digit.

Examples

NUMBER	SIG. FIGS.	NUMBER	SIG. FIGS.
12.8	3	0.04	1
12.80	4	1.336	4
1500.0	5	4.38×10^{-6}	3
1.5×10^3	2	3.1416	5
10^9	1	3.14159	6

The number of significant digits indicates how accurate the data is. For example, compare the readings 0.493 and 104.228. Both imply a maximum error of $\pm 5 \times 10^{-4}$. For the first (3 significant figures), this amounts to about 5 parts in 5000. But for the second (6 significant figures), it amounts to about 5 parts in 1,000,000. Thus more significant figures imply greater relative accuracy.

Note that 12.80 implies greater accuracy than 12.8. For 12.80 indicates a maximum error of $\pm 5 \times 10^{-3}$, whereas 12.8 indicates a maximum error of 5×10^{-2}.

Don't confuse the number of decimal places of accuracy with the number of significant figures. A weighing of 104.228 is accurate to 3 decimal places but has 6 significant figures.

Round-off

Suppose we have a 5-place read-out, but we only require 2-place accuracy. Then we must **round off** each 5-place reading to 2 places (and this means decimal places).

Examples

NUMBER	ROUNDED-OFF NUMBER
0.48265	0.48
0.48701	0.49
0.49013	0.49
0.49501	0.50
0.49500	?

The last entry is a problem. Do we round off to 0.49 or 0.50? The convention varies, but we shall adopt the rule "make the last digit even". So we round off to 0.50 because 0 is even and 9 is odd.

Rules for Rounding Off

(1) If the discarded portion is less than 5000 $\cdots$, then drop it.

(2) If the discarded portion is greater than 5000 $\cdots$, then drop it and add 1 to the last digit kept.

(3) If the discarded portion is exactly 5000 $\cdots$, then drop it; if the last digit kept is even, do nothing; if the last digit kept is odd, add 1 to it.

Each time we round off, we introduce an error. But we feel these rules are fair and hope that in a series of calculations with round-off, the errors will more or less average out, not pile up.

Example Round off 9.86507 to 0, 1, 2, 3, and 4 decimal places.
   Solution: 10, 9.9, 9.87, 9.865, 9.8651.

Example Round off 9.865 to 2 decimal places.
   Solution: 9.86.

Remark: Note the different results:

The one-step round-off is more accurate than the two-step procedure.

There is a lesson to be learned here: if you want 4 places it is better to use a 4-place table rather than round off from a 5-place table. The reason is that the entries in the tables are already rounded off by the table-makers; you may lose accuracy in rounding off again. For example, in 4-place and 5-place common log tables we find

   $\log 1.19 \approx 0.0755$ and $\log 1.19 \approx 0.07555$.

If we round off the second entry to 4 places we get 0.0756. But a 6-place table shows $\log 1.19 \approx 0.075547$ so that 0.0755 is more accurate.

Algebraic Priorities

Since we shall have to carry out some fairly complex computations, we should take a look at how our hand-held calculators work. Let's first review some relevant algebra.

In algebra, multiplications and divisions have higher priority than additions and subtractions. Thus the expression

   $2 \times 3 + 4 \times 5$

means

   $(2 \times 3) + (4 \times 5) = 6 + 20 = 26,$

because the multiplications are done before the addition. If we want

   $(2 + 3) \times (4 + 5)$ $[= 5 \times 9 = 45]$

we can *not* write

   $2 + 3 \times 4 + 5,$

because the rule of algebraic priority interprets this expression as

   $2 + (3 \times 4) + 5 = 2 + 12 + 5 = 19.$

Parentheses

Sometimes parentheses are essential to avoid ambiguity. For example, if you want to subtract from x the number that is 1 more than y, you must write

$$x - (y + 1) \quad \text{not} \quad x - y + 1.$$

These expressions are different because

$$x - (y + 1) = x - y - 1.$$

It is good policy to insert parentheses in case of doubt. Usually no harm can be done, but often lots of harm can be done by omitting them.

There are two rules for evaluating expressions with parentheses. The first says that you must complete all calculations inside of parentheses before doing anything outside. Thus in

$$(3 + 7) \cdot (5 - 2)$$

you must complete the addition $3 + 7$ and the subtraction $5 - 2$ before doing the multiplication.

The second rule says that in case of nested parentheses (parentheses within parentheses (like this)), you must complete the innermost one (or ones) first. Then if more nested parentheses remain, repeat the process. For instance,

$$14 - ((25 - 2) - (7 + (9 - 1)))$$
$$= 14 - (23 - (7 + 8))$$
$$= 14 - (23 - 15) = 14 - 8 = 6.$$

All the parentheses in this calculation make the first line hard to read. For greater ease of reading we often substitute brackets [ ] and braces { } for some of the parentheses. For example

$$14 - ((25 - 2) - (7 + (9 - 1))) = 14 - \{(25 - 2) - [7 + (9 - 1)]\}.$$

Calculator Priorities

A *left-to-right* calculator has its own priorities. It performs operations as they are keyed in, a procedure which sometimes agrees with the conventions of algebra and sometimes does not. For example, to compute $3 \cdot 9 + 1$, you key in

$$3 \;\boxed{\times}\; 9 \;\boxed{+}\; 1 \;\boxed{=}\;.$$

The calculator does the computations from left to right, that is, it computes

$$(3 \times 9) + 1 = 28,$$

which agrees with algebra. But suppose you use the algebraic equality

$$3 \cdot 9 + 1 = 1 + 3 \cdot 9$$

and, instead, key

$$1 \;\boxed{+}\; 3 \;\boxed{\times}\; 9 \;\boxed{=}$$

into a left-to-right calculator. The calculator computes

$$(1 + 3) \times 9 = 36,$$

not the expected answer, 28.

On the other hand, a calculator with *algebraic logic* (also called *algebraic operating system*) follows the priorities of algebra. It completes multiplications and divisions before additions and subtractions. Either key sequence

$$3 \;\boxed{\times}\; 9 \;\boxed{+}\; 1 \;\boxed{=} \quad \text{or} \quad 1 \;\boxed{+}\; 3 \;\boxed{\times}\; 9 \;\boxed{=}$$

results in 28.

Exercises

Round off to 2 decimal places

1 0.4444 0.3128 0.1075 0.2555   **2** 6.411 10.91 2.0041 3.0095.

Round off to 3 decimal places.

3 0.0005 0.00049 16.2445 3.7855   **4** 1.8125 3.14159265 0.9997 0.99946.

5 (a) Compute $1.255 + 0.395 + 2.116 + 1.336$, then round off to 2 places.
    (b) Round off each term first, then add. Compare the answers. Which answer is more accurate?

6 (cont.) Do the same for $0.255 + 0.365 + 0.166 + 0.823$.

Round off to 3 significant figures

7 1046.0 55.521 10.05   **8** 9.095 9.094 9.0949.

An inexperienced technician uses a voltmeter with 2 significant figure readings. Rewrite accurately his data

9 0.4   **10** 12.0   **11** 2.3   **12** 9.

Find the result of the given sequence of calculator steps assuming (a) left-to-right logic (b) algebraic logic

13 $6 \;\boxed{-}\; 2 \;\boxed{\times}\; 3 \;\boxed{=}$ **14** $6 \;\boxed{\times}\; 2 \;\boxed{-}\; 3 \;\boxed{=}$

15 $8 \;\boxed{\div}\; 4 \;\boxed{+}\; 6 \;\boxed{\div}\; 2 \;\boxed{=}$ **16** $8 \;\boxed{+}\; 4 \;\boxed{\div}\; 6 \;\boxed{+}\; 2 \;\boxed{=}$

17 $a \;\boxed{\times}\; b \;\boxed{-}\; c \;\boxed{\div}\; d \;\boxed{+}\; e \;\boxed{=}$ **18** $a \;\boxed{-}\; b \;\boxed{\times}\; c \;\boxed{+}\; d \;\boxed{\div}\; e \;\boxed{=}$

Using only $\boxed{+}, \boxed{-}, \boxed{\times}, \boxed{\div}, \boxed{=}$ find a sequence of calculator steps to compute the given quantity assuming (a) left-to-right logic (b) algebraic logic

19 $\dfrac{x + 3}{y}$ **20** $5(x - y)$ **21** $\dfrac{x}{yz}$

22 $\dfrac{a + b}{c} - d$ **23** $\tfrac{1}{2}\left(\dfrac{a}{b} + c\right)$ **24** $a\left(b + \dfrac{c}{d} + e\right).$

Compute

25 $1 - (2 - (3 - 4))$ **26** $6 + 5(4 + 3(2 + 1))$

27 $(2 + 3(7 + 4))((17 - 12)(13 - 8) + 2)$

28 $((6 - 4)(7 - 2) + (1 + 2)(6 - 3))(20 - 5(10 - 7))$

29 $((18 - 3)/5 + 5)/4 + ((18 - 3)/5 - 5)/2$

30 $1 + (1 \div (1 + (1 \div (1 + (1 \div (1 + 1)))))).$

2 USE OF CALCULATORS

Many models of hand-held calculators are available, each with its own special features. We cannot deal with all models and features here. Rather, we shall discuss some features common to all scientific calculators and some basic techniques of using them.

For the purposes of this course, and even more advanced courses, a fairly inexpensive scientific calculator will serve you well. It should display 5 to 8 significant figures; more are really not necessary. It should have the following types of keys:

(a) $\boxed{x^2}$ $\boxed{\sqrt{}}$ $\boxed{1/x}$ $\boxed{y^x}$

(b) $\boxed{\log}$ $\boxed{\ln}$ $\boxed{10^x}$ $\boxed{e^x}$

(c) One memory with

$\quad\quad$ $\boxed{\text{STO}}$ (also called $\boxed{\text{Min}}$ $\boxed{x{\to}M}$ etc.)

$\quad\quad$ $\boxed{\text{RCL}}$ (also called $\boxed{\text{MR}}$ $\boxed{\text{RM}}$ etc.)

$\quad\quad$ $\boxed{\text{M}+}$ (also called $\boxed{\text{SUM}}$ etc.)

(d) $\boxed{(}$ $\boxed{)}$ (to at least one level, more if possible)

(e) $\boxed{\sin}$ $\boxed{\cos}$ $\boxed{\tan}$

(f) $\boxed{\text{INV}}$ (also called $\boxed{\text{ARC}}$ $\boxed{\text{F}}$ $\boxed{\text{2nd}}$ etc.)

(g) $\boxed{\pi}$

As pointed out in Section 4, you don't really need all four keys of type (b). The key sequence $\boxed{\text{INV}}$ $\boxed{\log}$ and $\boxed{\text{INV}}$ $\boxed{\ln}$ are equivalent to $\boxed{10^x}$ and $\boxed{e^x}$ respectively.

Lots of other keys are available on various models. For example, there are statistical keys, $\boxed{e}$, $\boxed{x^{1/y}}$, $\boxed{n!}$, $\boxed{\circ\,'\,''}$, $\boxed{{\to}r\theta}$, and others, but we don't advise paying extra for them. Parentheses $\boxed{(}$ and $\boxed{)}$ are now standard on most scientific calculators, but many perfectly good older models lack them.

Since the main source of errors is hitting the wrong key, an uncluttered keyboard is preferable. Also you might like a calculator with a liquid crystal display. You can't read it in the dark, but the batteries last 500–2000 hours.

A final suggestion: Whatever calculator you buy, read the instructions carefully and do enough numerical experiments to familiarize yourself with its capabilities and peculiarities.

Left-to-Right Logic

This is the system usually found in the least expensive calculators. In our examples, we usually assume left-to-right logic unless noted otherwise.

On any calculators, the keys

$\quad\quad$ $\boxed{+}$, $\boxed{-}$, $\boxed{\times}$, $\boxed{\div}$, $\boxed{y^x}$

are called **binary** keys because they operate on *two* numbers. In left-to-right logic, each binary key completes all previous pending operations.

Examples

3 $\boxed{+}$ 4 $\boxed{\times}$ 2 $\boxed{=}$ produces $(3 + 4) \times 2 = 14$

1 0 $\boxed{-}$ 3 $\boxed{y^x}$ 2 $\boxed{=}$ produces $(10 - 3)^2 = 49$.

Each binary key acts as if it were preceded by $\boxed{=}$. For example the calculation

3 $\boxed{+}$ 4 $\boxed{\times}$ 2 $\boxed{=}$ is equivalent to 3 $\boxed{+}$ 4 $\boxed{=}$ $\boxed{\times}$ 2 $\boxed{=}$.

The keys

$\boxed{x^2}$ $\boxed{\sqrt{}}$ $\boxed{10^x}$ $\boxed{\log}$ $\boxed{\sin}$ etc.

are called **unary** keys because they operate on a single number. For example

x $\boxed{\sqrt{}}$ produces $\sqrt{x}$, x $\boxed{\sin}$ produces $\sin x$, etc.

The action of a unary key is completed *before* pending operations.

Examples

7 $\boxed{+}$ 9 $\boxed{\sqrt{}}$ $\boxed{=}$ produces $7 + \sqrt{9} = 10$

5 $\boxed{x^2}$ $\boxed{-}$ 3 $\boxed{x^2}$ $\boxed{=}$ $\boxed{\sqrt{}}$ produces $\sqrt{5^2 - 3^2} = 4$.

If you forget $\boxed{=}$ in either example, the answer will be 3. Why?

EXAMPLE 1 Assuming left-to-right logic and $\boxed{\deg}$ mode, find the calculation performed by the key sequence

(a) 6 $\boxed{+}$ 5 $\boxed{-}$ 2 $\boxed{\times}$ 3 $\boxed{\div}$ 7 $\boxed{=}$

(b) 5 $\boxed{x^2}$ $\boxed{x^2}$ $\boxed{-}$ 1 6 $\boxed{\sqrt{}}$ $\boxed{=}$ $\boxed{\cos}$.

Solution (a) Move from left to right performing each operation on the previous total. The result is

$$\frac{(6 + 5 - 2) \times 3}{7}.$$

(b) Similarly, proceeding from left to right we obtain

$$(5^2)^2 - \sqrt{16}$$

at the next to last step. The last step yields the cosine of this number of degrees.

Answer (a) $\dfrac{(6 + 5 - 2) \times 3}{7}$ (b) $\cos[(5^2)^2 - \sqrt{16}]^\circ$.

Algebraic Logic

In this system, some operations have higher priority than others. The usual order of priority is

highest $\longrightarrow$ 1 unary operations

2 $\boxed{y^x}$

$$3 \;\boxed{\times}\; \boxed{\div}$$
$$4 \;\boxed{+}\; \boxed{-}$$
$$\text{lowest} \longrightarrow 5 \;\boxed{=}.$$

An operation of higher priority is always completed before any pending previous operations of lower priority.

Examples

$3 \;\boxed{+}\; 4 \;\boxed{\times}\; 2 \;\boxed{=}\;$ produces $3 + (4 \times 2) = 11$

$1\;0 \;\boxed{-}\; 3 \;\boxed{y^x}\; 2 \;\boxed{=}\;$ produces $10 - 3^2 = 1$

$3 \;\boxed{\times}\; 5 \;\boxed{-}\; 2\;4 \;\boxed{\div}\; 8 \;\boxed{=}\;$ produces $(3 \times 5) - (24 \div 8) = 12$.

Calculators with algebraic logic often have parentheses. Many have parentheses within parentheses. All work on one basic rule: closing an open parenthesis automatically completes all operations pending since the last $\boxed{(}$.

Examples

$\boxed{(}\; 4 \;\boxed{+}\; 3 \;\boxed{\times}\; 2 \;\boxed{)}\;$ produces $4 + (3 \times 2) = 10$

$\boxed{(}\; 4 \;\boxed{+}\; 3 \;\boxed{)}\; \boxed{\times}\; \boxed{(}\; 5 \;\boxed{-}\; 3 \;\boxed{)}\;$ produces $5 - 3 = 2$

$\boxed{(}\; 4 \;\boxed{+}\; 3 \;\boxed{)}\; \boxed{\times}\; \boxed{(}\; 5 \;\boxed{-}\; 3 \;\boxed{)}\; \boxed{=}\;$ produces

$$(4 + 3) \times (5 - 3) = 14$$

Note that there is no $\boxed{=}$ in the first example; closing the parentheses completes all operations inside. In the second example, closing the last parentheses simply completes the operation inside that set of parentheses, $5 - 3 = 2$. If you want the product $(4 + 3) \times (5 - 3)$, you must complete the $\boxed{\times}$ operation by hitting $\boxed{=}$ at the end (third example).

> *EXAMPLE 2*   Assuming algebraic logic, find the calculation set up by the key sequence
>
>   (a) $4 \;\boxed{-}\; 5 \;\boxed{\div}\; 6 \;\boxed{+}\; 7 \;\boxed{\times}\; 8 \;\boxed{=}$
>
>   (b) $\boxed{(}\; 5 \;\boxed{\times}\; \boxed{(}\; 3 \;\boxed{-}\; 1 \;\boxed{)}\; \boxed{y^x}\; 4 \;\boxed{+}\; 1 \;\boxed{)}\;$.
>
> **Solution**   (a)   Since $\boxed{\div}$ and $\boxed{\times}$ have priority over $\boxed{-}$ and $\boxed{+}$, the sequence yields
>
> $$4 - (5 \div 6) + (7 \times 8).$$
>
> (b)   Closing the outer parentheses completes the sequence inside. The operation $\boxed{y^x}$ has highest priority, then $\boxed{\times}$, and last $\boxed{+}$. The result is
>
> $$[5 \times (3 - 1)^4] + 1.$$
>
> *Answer*   (a) $4 - (5 \div 6) + (7 \times 8)$ (b) $[5 \times (3 - 1)^4] + 1$

Warning   Be very careful using $\boxed{=}$. On some models, $\boxed{=}$ will not only complete all pending operations within ( ), but also all pending operations before the (. For example, the logical set-up for calculating

$$(21 - 5)\sqrt{59 + 22} \qquad (=144)$$

would seem to be

$\boxed{(}$ 2 1 $\boxed{-}$ 5 $\boxed{)}$ $\boxed{\times}$ $\boxed{(}$ 5 9 $\boxed{+}$ 2 2 $\boxed{=}$ $\boxed{\sqrt{}}$ $\boxed{)}$ $\boxed{=}$.

However, on a TI-58 we obtain 36. The joker is the first $\boxed{=}$. We thought it would complete only the addition $59 + 22$. It did, but it also completed the pending multiplication, giving $16 \cdot 81 = 1296$. The next step gave $\sqrt{1296} = 36$. Then nothing remained pending, so $\boxed{)}$ and $\boxed{=}$ had no effect.

Reverse Polish Logic

At first these calculators seem strange; they have no $\boxed{=}$ key! Yet the reverse Polish logic is highly efficient. In fact, we believe it is easier to use an RPL calculator than other types. When you are ready for an expensive calculator, consider a Hewlett-Packard; they are the Cadillacs among calculators.

This system uses an "operational stack," which consists of the displayed number plus other numbers stacked "above" the display. Pressing $\boxed{\text{ENTER} \uparrow}$ moves the whole stack up one space. Pressing a binary key combines the bottom two numbers (the displayed number and the one right "above" it) and moves the rest of the stack down one space.

Examples

3 $\boxed{\text{ENTER} \uparrow}$ 4 $\boxed{+}$ produces $3 + 4 = 7$.

5 $\boxed{\text{ENTER} \uparrow}$ 2 $\boxed{\text{ENTER} \uparrow}$ 7 $\boxed{+}$ $\boxed{\times}$ produces $(7 + 2) \times 5 = 45$.

Store and Recall

The key $\boxed{\text{STO}}$ stores the displayed number in the memory (without changing the display). The key $\boxed{\text{RCL}}$ recalls the stored number to the display (without changing the contents of the memory).

These features make for efficient computing. You can store a number in the memory and bring it back whenever needed. For example, $\boxed{+}$ $\boxed{\text{RCL}}$ $\boxed{=}$ adds the stored number to the display; $\boxed{y^x}$ $\boxed{\text{RCL}}$ $\boxed{=}$ raises the displayed number to the power of the stored number.

Examples   (left-to-right logic)

5 $\boxed{\text{STO}}$ $\boxed{x^2}$ $\boxed{+}$ $\boxed{\text{RCL}}$ $\boxed{=}$

produces $5^2 + 5 = 30$

8 3 $\boxed{+}$ 7 9 $\boxed{=}$ $\boxed{\text{STO}}$ 4 3 6 $\boxed{-}$ 8 7 $\boxed{\times}$ $\boxed{\text{RCL}}$ $\boxed{=}$

produces $(83 + 79) \times (436 - 87) = 56538$.

If the first $\boxed{=}$ is omitted in the second example, then only 79 (the displayed number) goes into the memory.

Keying in numbers is a major source of errors. When the same number occurs several times in a calculation, try to cut down the number of times you must key it in, hopefully to one time.

EXAMPLE 3   Let $a = 9.69699$. Compute $a(\sin a°)^{1/a} + \sqrt{a}$.

Solution   There is great danger of an error when keying in such a confusing number. The most obvious solution requires entering a four times. But that is poor technique and very risky. With efficient use of the memory, it's only necessary once.

Set $\boxed{\text{deg}}$ mode; then enter a and store it. Whenever you need a, just hit $\boxed{\text{RCL}}$. With this strategy, a reasonable key sequence is

$$9.6\,9\,6\,9\,9 \;\; \boxed{\text{STO}} \;\; \boxed{\sin} \;\; \boxed{y^x} \;\; \boxed{\text{RCL}} \;\; \boxed{1/x} \;\; \boxed{\times} \;\; \boxed{\text{RCL}} \;\; \boxed{+} \;\; \boxed{\text{RCL}} \;\; \boxed{\sqrt{}} \;\; \boxed{=}$$

Answer   11.18382

EXAMPLE 4   Let $a = 4.0767$. Compute $\dfrac{\sqrt{a}}{a^2 + 5}$.

Solution   Enter a and store it. Compute the denominator and take its reciprocal. Then multiply by $\sqrt{a}$:

$$4.0\,7\,6\,7 \;\; \boxed{\text{STO}} \;\; \boxed{x^2} \;\; \boxed{+} \;\; 5 \;\; \boxed{=} \;\; \boxed{1/x} \;\; \boxed{\times} \;\; \boxed{\text{RCL}} \;\; \boxed{\sqrt{}} \;\; \boxed{=}$$

One step shorter:

$$4.0\,7\,6\,7 \;\; \boxed{\text{STO}} \;\; \boxed{x^2} \;\; \boxed{+} \;\; 5 \;\; \boxed{\div} \;\; \boxed{\text{RCL}} \;\; \boxed{\sqrt{}} \;\; \boxed{=} \;\; \boxed{1/x}$$

Answer   0.09339187

Memory Plus

The key $\boxed{\text{M}+}$ adds the number in the display to the memory. Usually it *completes pending operations* before making this addition.

Example   (starting with memory clear)

$$2 \;\; \boxed{\times} \;\; 3 \;\; \boxed{\text{M}+} \;\; 4 \;\; \boxed{\times} \;\; 5 \;\; \boxed{\text{M}+} \;\; 6 \;\; \boxed{\times} \;\; 7 \;\; \boxed{\text{M}+} \;\; \boxed{\text{RCL}}$$

produces $(2 \times 3) + (4 \times 5) + (6 \times 7) = 68$.

Constant Factors

Suppose you need to multiply each of a whole slew of numbers by the same constant. Some calculators have a "constant factor" feature for doing this without reentering the factor each time. The sequence $c \;\boxed{\times}\; x \;\boxed{=}$ produces cx, as usual. But when you follow by $y \;\boxed{=}$, you get cy. Then $z \;\boxed{=}$ produces cz, etc. (Some models use a $\boxed{\text{K}}$ key to activate the constant factor feature.)

EXAMPLE 5   A weekend special at your record store announces a $27\frac{1}{2}\%$ discount on all marked prices. Allowing for a 4% sales tax, what is your actual cost on records and tapes at

$$\$1.98, \quad 2.37, \quad 3.19, \quad 3.89, \quad 4.89, \quad 5.57, \quad 7.29\,?$$

Round up each price to the nearest cent.

Solution First key in

$$1 \boxed{-} . 2 7 5 \boxed{\times} 1 . 0 4 \boxed{\times} .$$

This is your actual discount factor, taking tax into account. It is set up in the calculator as a constant factor. Now continue

$$1 . 9 8 \boxed{=} 2 . 3 7 \boxed{=} 3 . 1 9 \boxed{=} 3 . 8 9 \boxed{=} 4 . 8 9 \boxed{=}$$

$$5 . 5 7 \boxed{=} 7 . 2 9 \boxed{=} .$$

After each $\boxed{=}$ round up to get the corresponding cost.

Answer $1.50, 1.79, 2.41, 2.94, 3.69, 4.20, 5.50

Accuracy and Hidden Digits

Most calculators work internally with one to three more significant digits than are displayed. Usually the number displayed is a round-off of the internal number. Not always, however; in some models the final digits are truncated (chopped off) rather than rounded. Thus the internal number

$$\cdots 68 \qquad \text{may appear as} \qquad \cdots 6.$$

On some models you can see the hidden final digits by using a little trick. Let's illustrate the trick for $\sqrt{2}$ on three calculators:

$$2 \boxed{\sqrt{}} \begin{cases} 1.4142135 & \text{Casio fx-2000} \\ 1.4142135 & \text{Sharp EL-5805} \\ 1.414213562 & \text{Texas Instr. TI-59.} \end{cases}$$

We multiply by 1000, then subtract 1414. This deletes the first 4 significant digits, making space in the display for the hidden digits. The results:

$$0.213562 \qquad \text{fx-2000}$$
$$0.21357 \qquad \text{EL-5805}$$
$$0.21356273 \qquad \text{TI-59.}$$

Thus we get the following estimates for $\sqrt{2}$:

1.41421 3562	fx-2000
1.41421 357	EL-5805
1.41421 35623 73	TI-59.

Exercises

Find the result of the calculation in left-to-right logic

1 $a \boxed{+} b \boxed{\sqrt{}} \boxed{+} c \boxed{\sqrt{}} \boxed{=}$

2 $a \boxed{+} b \boxed{\sqrt{}} \boxed{+} c \boxed{=} \boxed{\sqrt{}}$

3 $a \boxed{\times} b \boxed{1/x} + c \boxed{1/x} \boxed{=}$

4 $a \boxed{\times} b \boxed{=} \boxed{1/x} \boxed{+} c \boxed{=} \boxed{1/x}$

5 $a \boxed{\times} b \boxed{\div} c \boxed{+} a \boxed{\div} b \boxed{\times} c \boxed{=}$

6 a $\boxed{-}$ b $\boxed{-}$ c $\boxed{\times}$ 3 $\boxed{+}$ a $\boxed{\times}$ b $\boxed{=}$

7 a $\boxed{+}$ b $\boxed{10^x}$ $\boxed{+}$ c $\boxed{y^x}$ 2 $\boxed{-}$ a $\boxed{=}$

8 a $\boxed{+}$ 1 $\boxed{y^x}$ b $\boxed{-}$ 1 $\boxed{\div}$ a $\boxed{=}$ $\boxed{1/x}$ $\boxed{\times}$ c $\boxed{=}$

9 a $\boxed{+}$ 1 $\boxed{=}$ $\boxed{1/x}$ $\boxed{+}$ 1 $\boxed{=}$ $\boxed{1/x}$ $\boxed{+}$ 1 $\boxed{=}$ $\boxed{1/x}$

10 a $\boxed{1/x}$ $\boxed{y^x}$ a $\boxed{+}$ a $\boxed{y^x}$ a $\boxed{1/x}$ $\boxed{+}$ a $\boxed{1/x}$ $\boxed{=}$ $\boxed{1/x}$

Find the result of the calculation in algebraic logic

11 a $\boxed{\times}$ b $\boxed{\div}$ c $\boxed{+}$ a $\boxed{\div}$ b $\boxed{\times}$ c $\boxed{=}$

12 a $\boxed{-}$ b $\boxed{-}$ c $\boxed{\times}$ 3 $\boxed{+}$ a $\boxed{\times}$ b $\boxed{=}$

13 a $\boxed{+}$ b $\boxed{y^x}$ c $\boxed{\times}$ d $\boxed{=}$

14 a $\boxed{\times}$ b $\boxed{y^x}$ c $\boxed{\times}$ d $\boxed{+}$ a $\boxed{y^x}$ b $\boxed{1/x}$ $\boxed{=}$

15 $\boxed{(}$ $\boxed{(}$ a $\boxed{-}$ b $\boxed{)}$ $\boxed{\times}$ c $\boxed{-}$ $\boxed{(}$ b $\boxed{-}$ c $\boxed{)}$ $\boxed{\times}$ a $\boxed{)}$

16 $\boxed{(}$ a $\boxed{\times}$ $\boxed{(}$ a $\boxed{+}$ b $\boxed{\times}$ c $\boxed{)}$ $\boxed{-}$ c $\boxed{)}$ $\boxed{\times}$ a $\boxed{=}$

17 a $\boxed{+}$ b $\boxed{y^x}$ $\boxed{(}$ a $\boxed{-}$ b $\boxed{)}$ $\boxed{=}$ $\boxed{\times}$ $\boxed{(}$ a $\boxed{-}$ b $\boxed{)}$ $\boxed{=}$

18 $\boxed{(}$ a $\boxed{-}$ b $\boxed{)}$ $\boxed{\div}$ a $\boxed{+}$ b $\boxed{\div}$ $\boxed{(}$ a $\boxed{+}$ b $\boxed{)}$ $\boxed{=}$

19 a $\boxed{+}$ $\boxed{1/x}$ $\boxed{\times}$ b $\boxed{=}$

20 (cont.) a $\boxed{+}$ $\boxed{1/x}$ $\boxed{\times}$ b $\boxed{+}$ $\boxed{1/x}$ $\boxed{\times}$ c $\boxed{=}$

Set up a routine on a left-to-right calculator with memory, keying a only once

21 $a \log(3 + a)$ **22** $\sqrt{1 + a^2} + a^4$ **23** $a - a^{1/a}$

24 $a^{(1+a)}$ **25** $\dfrac{1 + a}{1 - a}$ **26** $\dfrac{3 + a}{2 + a^2}$

27 $(\sqrt[4]{1 + 5a^4} - a)^a - a$ **28** $(1 + 2a)^{\sqrt{1+a}}$

Set up a routine on an algebraic calculator with memory and parentheses, keying a only once

29 $\dfrac{2 + a}{3 - a}$ **30** $\dfrac{a^2 + 4}{a^3 + 7}$

31 $(a^{1/a} + 1)[(1/a)^a + 1]$ **32** $a^{a^a} + a^a + a$

33 $3a^4 - 2a^3 + 5/a$ **34** $\dfrac{10}{3a^2 - 1} + (a^5 - 2) \times 10^a$.

Calculate

35 $\dfrac{2a + 1}{(a - 3)^5}$ for $a = 4.1891$

36 $[1 + \tan(4a + 3)]^a$ for $a = 2.5887$

37 $\sqrt{\dfrac{a}{a^2 + 6}}$ for $a = 0.6077$

38 $\sqrt{a + \sqrt{a + \sqrt{a}}}$ for $a = 3.9399$.

Try to find one additional hidden digit

39 $\sqrt{3}$ **40** $(1.56)^{2.81}$.

41 Suppose a is in the display and b is in the memory. Find a routine that interchanges a and b without reentering either (and without use of a memory exchange key if you happen to have one).

42 (cont.) With the same data, end up with a in the display and b^2 in the memory.

3   RIGHT TRIANGLES

Now we begin our work on applications of the trigonometric functions to problems of practical geometry. In the fields that use these applications: astronomy, navigation, surveying, civil engineering, machine shop, architecture, construction trades, mechanical engineering, etc., angles are always measured in degrees, not in radians. So we shall work in degrees almost exclusively.

Remark   Recall that the complete circle is divided into 360 degrees. This division is very old; perhaps it originated in the ancient world from the mistaken belief that the Earth's orbit was a circle traced in a year of 360 days, so one degree represented one day's arc.

Trigonometry

The word **trigonometry** comes from Greek and means the measurement of triangles. The general problem of trigonometry is this: A triangle has three sides and three angles. Given some of these six quantities, measure (compute) the others. We discuss this problem for right triangles here, and for oblique triangles in Chapter 6.

Recall that the trigonometric functions of an acute angle can be expressed as ratios of the sides of a right triangle (Fig. 1). Using these ratios, we can solve problems about right triangles: given two sides, or given one side and one acute angle, find the other sides and angles. ("Side" is short for side length, and "angle" for angle measure.)

$$\sin \theta = \frac{\text{opp}}{\text{hyp}} \qquad \csc \theta = \frac{\text{hyp}}{\text{opp}}$$

$$\cos \theta = \frac{\text{adj}}{\text{hyp}} \qquad \sec \theta = \frac{\text{hyp}}{\text{adj}}$$

$$\tan \theta = \frac{\text{opp}}{\text{adj}} \qquad \cot \theta = \frac{\text{adj}}{\text{opp}}$$

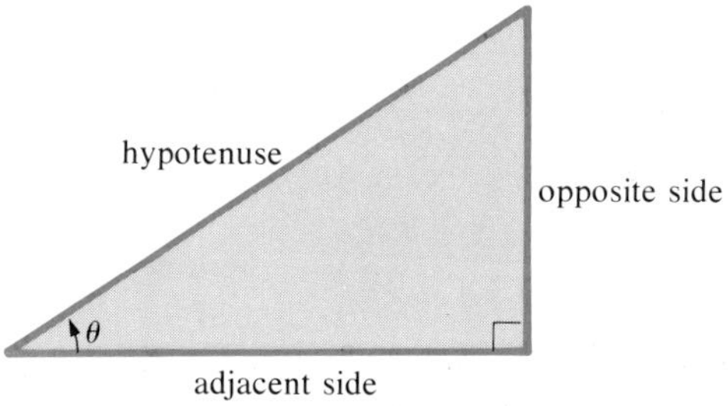

hypotenuse

opposite side

θ

adjacent side

Fig. 1

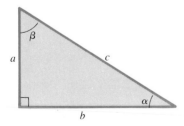

Fig. 2

EXAMPLE 1   The sides and angles of a right triangle are labeled as in Figure 2.

(a)   Given $a = 5.12$ and $c = 8.97$, find b, α, and β.

(b)   Given $a = 1.483$ and $\beta = 53.92°$, find b, c, and α.

Solution   (a)   First find b by the Pythagorean Theorem:

$$b = \sqrt{c^2 - a^2} = \sqrt{(8.97)^2 - (5.12)^2} \approx 7.37.$$

Next find α:

$$\sin \alpha = \frac{a}{c} \qquad \alpha = \text{arc} \sin \frac{a}{c} = \text{arc} \sin \frac{5.12}{8.97} \approx 34.8°$$

Finally, note that $\alpha + \beta = 90°$. Therefore

$$\beta = 90° - \alpha \approx 90° - 34.8° \approx 55.2°.$$

(b)   First find b:

$$\tan \beta = \frac{b}{a} \qquad b = a \tan \beta = 1.483 \tan 53.92° \approx 2.035.$$

Next find c:

$$\cos \beta = \frac{a}{c} \qquad c = \frac{a}{\cos \beta} = \frac{1.483}{\cos 53.92°} \approx 2.518$$

Finally,

$$\alpha = 90° - \beta = 90° - 53.92° = 36.08°.$$

Answer   (a) $b \approx 7.37$ $\qquad$ $\alpha \approx 34.8°$ $\qquad$ $\beta \approx 55.2°$

$\qquad\quad$ (b) $b \approx 2.035$ $\qquad$ $c \approx 2.518$ $\qquad$ $\alpha = 36.08°$

Remark   Accuracy. The answer to a problem cannot be more accurate than the data. For example, given that the two legs of a right triangle are 8.43 and 9.16, our calculator shows the hypotenuse to be $\sqrt{8.43^2 + 9.16^2} \approx 12.44871479$. But the extra decimal places are unjustified. The only reasonable answer consistent with the accuracy of the data is 12.45. We'll say more about measurement and accuracy in the next chapter.

Degrees, Minutes, and Seconds

For modern computation, it is convenient to express fractions of degrees in decimal parts, such as $31.742°$. However, there is a long tradition of using sixtieths of a degree and sixtieths of a sixtieth of a degree as fractional units. This tradition goes back to the ancient Sumerians (*ca.* 3000 B.C.), who introduced the sexagesimal (base 60) system of computation. This system was taken over by the Babylonians and then the Greek astronomers. For better or for worse, today most devices that measure fractions of degrees are calibrated in these sixtieth units.

> **Degrees, Minutes, and Seconds**   An angle of one **degree** ($1°$) is $\frac{1}{360}$ of a complete circle angle.
>
> An angle of one **minute** ($1'$) is $\frac{1}{60}$ of one degree.
>
> An angle of one **second** ($1''$) is $\frac{1}{60}$ of one minute.

We write $D°M'S''$ for an angle of D degrees plus M minutes plus S seconds. Thus

$$10°15'30'' = (10 + \tfrac{1}{60} \cdot 15 + (\tfrac{1}{60})^2 \cdot 30)° \approx 10.258333°.$$

In general,

$$D°M'S'' = (D + \tfrac{1}{60}M + (\tfrac{1}{60})^2 S)°.$$

In order to convert $D°M'S''$ to decimal degrees, we express the right-hand side as

$$(S/60 + M)/60 + D.$$

This form suggests a (left-to-right) calculator routine:

$$S \boxed{\div} \, 6 \, 0 \, \boxed{+} \, M \, \boxed{\div} \, 6 \, 0 \, \boxed{+} \, D \, \boxed{=}.$$

Example (set $\boxed{\text{deg}}$ mode) 35°22'41''

$$4 \, 1 \, \boxed{\div} \, 6 \, 0 \, \boxed{+} \, 2 \, 2 \, \boxed{\div} \, 6 \, 0 \, \boxed{+} \, 3 \, 5 \, \boxed{=} \qquad 35.37806$$

To calculate a trig function of an angle given as $D°M'S''$, convert to decimal degrees first, then press the appropriate trig key.

Example (set $\boxed{\text{deg}}$ mode) sin 12°17'23''

$$2 \, 3 \, \boxed{\div} \, 6 \, 0 \, \boxed{+} \, 1 \, 7 \, \boxed{\div} \, 6 \, 0 \, \boxed{+} \, 1 \, 2 \, \boxed{=} \, \boxed{\text{sin}} \qquad 0.212855$$

Now let's look at the reverse problem: given an angle in decimal degrees, express it in degrees, minutes, and seconds.

EXAMPLE 2 Convert 35.37806° to the form $D°M'S''$.

Solution The problem is to find D, M, and S so that

$$35.37806 = D + \tfrac{1}{60}M + (\tfrac{1}{60})^2 S.$$

Obviously $D = 35$. To find M, subtract 35 and then multiply by 60:

$$0.37806 = \tfrac{1}{60}M + (\tfrac{1}{60})^2 S$$
$$22.6836 = M + \tfrac{1}{60}S.$$

Hence $M = 22$. To find S, we subtract 22 and then multiply by 60:

$$0.6836 = \tfrac{1}{60}S \qquad S = 41.016.$$

Answer 35°22'41''

EXAMPLE 3 Express in degrees, minutes, and seconds the acute angle θ satisfying

$$\cos \theta = 0.343434.$$

Solution First calculate (in $\boxed{\text{deg}}$ mode) the desired angle in decimal degrees

$$. \, 3 \, 4 \, 3 \, 4 \, 3 \, 4 \, \boxed{\text{INV}} \, \boxed{\text{cos}} \qquad 69.913769$$

To convert to $D°M'S''$, continue the sequence of key strokes:

$$\boxed{-} \, 6 \, 9 \, \boxed{\times} \, 6 \, 0 \, \boxed{=} \qquad 54.82614$$

Finally,

$$\boxed{-} \, 5 \, 4 \, \boxed{\times} \, 6 \, 0 \, \boxed{=} \qquad 49.5684$$

Answer 69°54'49.6''

Remark   Many calculators have keys for converting DMS degrees to decimal degrees and vice versa. Their use varies from instrument to instrument, so you must read the instruction manual for details.

On one calculator, you enter $123°43'27.5''$ as

    1 2 3 . 4 3 2 7 5

and press $\boxed{\rightarrow \text{DEG}}$ for the decimal form $123.7243°$. To reverse this, you press $\boxed{\rightarrow \text{D.MS}}$ and the display reads 123.43275, which is interpreted as $123°43'27.5''$.

Negative angles   The notation $-D°M'S''$ is understood to mean $-(D°M'S'')$, or

$$-D - \tfrac{1}{60}M - (\tfrac{1}{60})^2 S \ \text{deg}.$$

For instance

$$-12°17'24'' = -12.2900°.$$

Exercises

In the following exercises you will be given two parts of the right triangle of Fig. 3. Find the remaining sides and angles.

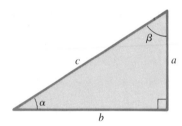

Fig. 3

1 $a = 12$ $c = 13$	**2** $a = 6$ $c = 13$
3 $a = 107.3$ $c = 175.5$	**4** $a = 24.03$ $b = 58.81$
5 $b = 15$ $c = 22$	**6** $b = 8.0$ $c = 12.5$
7 $b = 45.80$ $c = 66.75$	**8** $b = 237.6$ $c = 289.7$
9 $a = 4$ $b = 5$	**10** $a = 16$ $b = 9$
11 $a = 27.12$ $b = 9.315$	**12** $a = 18.06$ $c = 143.1$
13 $c = 30.0$ $\alpha = 14.5°$	**14** $c = 30.0$ $\alpha = 71.4°$
15 $a = 15.0$ $\alpha = 25.6°$	**16** $a = 9.50$ $\alpha = 40.8°$
17 $c = 235.5$ $\beta = 34.2°$	**18** $c = 1293$ $\beta = 68.8°$
19 $b = 56.03$ $\beta = 43.9°$	**20** $b = 526.2$ $\beta = 33.7°$
21 $c = 1.02570$ $\alpha = 12.43°$	**22** $a = 2.00143$ $\beta = 51.25°$
23 $a = 3.11701$ $\alpha = 5.08°$	**24** $b = 138.212$ $\beta = 72.21°$

Express in decimal degrees

25 $1''$	**26** $1'$	**27** $59''$	**28** $59'$
29 $12'12''$	**30** $6'3''$	**31** $\tfrac{1}{2}°$	**32** $\tfrac{1}{10}°$
33 $17°3'39''$	**34** $59°17'39''$	**35** $73°0'23''$	**36** $89°59'52''$
37 $71°57'51''$	**38** $1°13'47''$	**39** $157'$	**40** $10693''.$

Express in degrees, minutes, and seconds

41 $19.1082°$	**42** $57.1131°$	**43** $72.9499°$	**44** $61.7431°$
45 $0.0200°$	**46** $137.1933°$	**47** $0.000478°$	**48** $3.00015°$

Express in the form $D°M'S''$ the acute θ satisfying

49 $\sin \theta = 0.35162$ **50** $\sin \theta = 0.82141$

51 $\cos \theta = 0.35162$ **52** $\cos \theta = 0.82141$

53 $\tan \theta = 3.10482$ **54** $\tan \theta = 5.00000$

55 $\sin \theta = 4.84814 \times 10^{-6}$ **56** $\tan \theta = 2.90888 \times 10^{-4}$.

Compute

57 $\sin 15°12'19''$ **58** $\sin 9'13''$ **59** $\cos 73°51'39''$

60 $\cos 5°0'17''$ **61** $\tan 53°9'48''$ **62** $\tan 1''$

63 $\cos 89°59'59''$ **64** $\tan 89°59'59''$ **65** $\sin(223°34'12'')$

66 $\sin(349°12'0'')$ **67** $\cos(163°57'55'')$ **68** $\tan(300°12'23'')$.

Compute

69 $\sin 2' - 2\sin 1'$ **70** $\sin 30°1'' - \sin 30°$

71 $\sin 88°1'' - \sin 88°$ **72** $\sin 89°50'1'' - \sin 89°50'$

73 $\tan 1°1'' - \tan 1°$ **74** $\tan 89°1'' - \tan 89°$.

A hand of an analog clock sweeps an angle of $D°M'S''$. Find the elapsed time if the hand is

75 the hour hand   **76** the minute hand   **77** the second hand.

78 It is between 2 PM and 3 PM, and the minute hand is exactly 193° clockwise from the hour hand. What time is it?

4 APPLICATIONS

We begin with the type of geometric problem that is typical of applications of right triangle trigonometry.

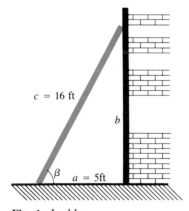

Fig. 1 Ladder

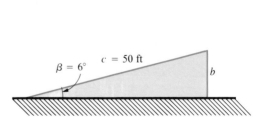

Fig. 2 Ramp (angle exaggerated)

EXAMPLE 1 A 16-ft ladder leans against a wall, with its base 5 ft from the wall. Find the angle the ladder makes with the ground and find how far up the wall the ladder extends.

Solution Draw a figure (Fig. 1). We want β and b; we are given $a = 5$ ft and $c = 16$ ft. Let us assume these measurements are accurate to one-half inch, and write $a = 5.0$, $c = 16.0$. To find β, we use

$$\cos \beta = \frac{a}{c}, \quad \text{so} \quad \beta = \text{arc cos} \frac{a}{c} = \text{arc cos} \frac{5.0}{16.0} \approx 71.8°.$$

By the Pythagorean theorem,

$$b = \sqrt{c^2 - a^2} = \sqrt{16^2 - 5^2} \approx 15.2 \text{ ft}$$

Answer 71.8°   15.2 ft

Query   Are we justified in writing 71.8°, or should we round to 72°? Try varying 5.0 and 16.0 by about $\frac{1}{2}$ inch, say by ± 0.05, and see what happens to β.

EXAMPLE 2   A 50-ft ramp is inclined at an angle of 6° with the ground. How high does it rise above the ground?

Solution   We seek b in Fig. 2, previous page; we are given $c = 50.0$ ft and $\beta = 6.0°$. Since

$$\sin \beta = \frac{b}{c}, \quad \text{we have} \quad b = c \sin \beta = 50 \sin 6° \approx 5.2 \text{ ft}.$$

Answer   5.2 ft

Multi-step Problems

EXAMPLE 3   Estimate θ in Fig. 3a.

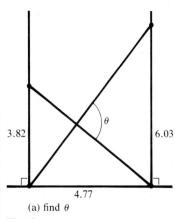

| (a) find θ | (b) auxiliary angles α, β, γ |

Fig. 3

Solution   Label the auxiliary angles α, β, γ as shown in Fig. 3b. Angles α and β can be found since each is in a right triangle whose legs are given. The idea is to express θ in terms of α and β. Now

$$\begin{aligned} \theta + \gamma &= 180° \quad \text{(supplementary angles)} \\ \underline{(\alpha + \beta) + \gamma} &= \underline{180°} \quad \text{(sum of angles of a triangle)} \end{aligned}$$

so $\theta = \alpha + \beta$ (subtracting).

From the figure

$$\tan \alpha = \frac{3.82}{4.77} \qquad\qquad \tan \beta = \frac{6.03}{4.77}$$

$$\alpha = \text{arc tan} \frac{3.82}{4.77} \approx 38.7° \qquad \beta = \text{arc tan} \frac{6.03}{4.77} \approx 51.7°$$

$$\theta = \alpha + \beta \approx 90.4°$$

Answer 90.4°

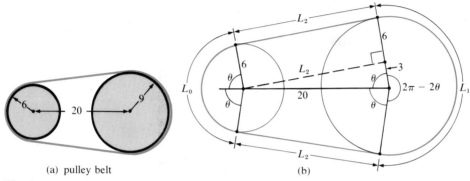

(a) pulley belt   (b)

Fig. 4

EXAMPLE 4 Two pulleys of radii 6.00 in. and 9.00 in. are connected by a belt (Fig. 4a). The distance between their centers is 20.00 in. Find the length of the belt.

Solution The belt consists of a small circular arc of length L_0, a large arc of length L_1, and two straight segments, each of length L_2. Thus the total length of the belt is

$$L = L_0 + L_1 + 2L_2.$$

Each straight piece is a common tangent to the two circles. Draw radii to the two upper points of tangency (Fig. 4b). These radii are parallel, since each is perpendicular to the common tangent. Therefore they determine equal central angles θ as shown.

Now draw an auxiliary line, through the center of the smaller circle and parallel to the upper common tangent. It forms a rectangle and a right triangle. From the right triangle,

$$L_2 = \sqrt{20^2 - 3^2} = \sqrt{391} \approx 19.77$$

and

$$\cos \theta = \tfrac{3}{20} = 0.15 \qquad \theta = \text{arc cos } 0.15 \approx 1.42 \text{ rad}.$$

The smaller circular arc is subtended by the central angle 2θ. Therefore its length (radius times central angle in *radians*) is

$$L_0 = 6 \cdot 2\theta = 12\theta \approx 17.04$$

The larger arc is subtended by the central angle $2\pi - 2\theta$, so its length is

$$L_1 = 9 \cdot (2\pi - 2\theta) \approx 30.99 \,.$$

Finally

$$L = L_0 + L_1 + 2L_2 \approx 17.04 + 30.99 + 2 \cdot 19.77 = 87.57$$

Answer 87.57 in.

EXAMPLE 5   A surveyor measures (Fig. 5) from A to B, using two intermediate stations S and T in the same vertical plane as A and B. Estimate the horizontal and vertical distances, h and v, from A to B.

Solution   The horizontal distance h from A to B is the sum of three horizontal distances: from A to S, from S to T, and from T to B. You must imagine that three right triangles with horizontal and vertical legs have been added to the figure. Then

$$h = a \cos \alpha + b \cos \beta + c \cos \gamma$$

$$= 27.72 \cos 5°12' + 34.39 \cos 3°43' + 23.41 \cos 7°35'$$

$$\approx 27.606 + 34.318 + 23.205 \approx 85.13 \text{ m} \,.$$

The vertical distance v from A to B consists of a negative contribution (downward from A to S) followed by two positive contributions (upward from S to T to B). Thus

$$v = -a \sin \alpha + b \sin \beta + c \sin \gamma$$

$$= -27.72 \sin 5°12' + 34.39 \sin 3°43' + 23.41 \sin 7°35'$$

$$\approx -2.512 + 2.229 + 3.089 \approx 2.81 \text{ m}$$

Answer $h \approx 85.13$ m $v \approx 2.81$ m

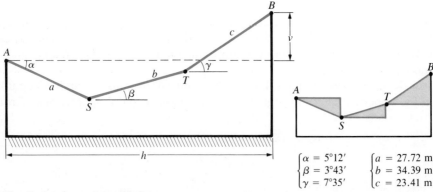

$$\begin{cases} \alpha = 5°12' \\ \beta = 3°43' \\ \gamma = 7°35' \end{cases} \qquad \begin{cases} a = 27.72 \text{ m} \\ b = 34.39 \text{ m} \\ c = 23.41 \text{ m} \end{cases}$$

Fig. 5 (angles exaggerated)

Remark   The computation goes briskly if you use $\boxed{\rightarrow \text{DEG}}$ and $\boxed{\text{M}+}$. For h, with slight variations for different calculators, it should be

5 . 1 2 $\boxed{\rightarrow \text{DEG}}$ $\boxed{\text{COS}}$ $\boxed{\times}$ 2 7 . 7 2 $\boxed{\text{M}+}$

3.43 →DEG COS × 34.39 M+

7.35 →DEG COS × 23.41 M+ RCL

 85.13

Be sure to clear the memory before you start calculating v.

Exercises

1 When the elevation of the Sun is 60.5°, a flagpole casts a 44-ft shadow. Find the height of the flagpole (Fig. 6).

2 (cont.) At what elevation will the shadow be twice as long?

3 A 15-ft ladder leans against a wall making a 70° angle with the ground. At what height does its upper end touch the wall?

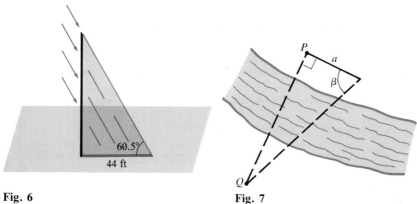

Fig. 6 $\qquad\qquad\qquad$ **Fig. 7**

4 You have a tape rule and a protractor and you wish to estimate the distance from P to an inaccessible point Q. See Fig. 7. You lay out a right triangle with leg $a = 50$ ft and measure $\beta = 67°$. What is the distance?

5 A road makes an angle of 4.5° with the horizontal. I start at sea level and drive exactly 5 km. What is my elevation above sea level?

6 I observe that the angle of elevation of a certain peak is 20°. I drive 2.0 km closer along a level road and find that now the angle is 32°. What is the height of the peak?

7 During the first few seconds of its flight, a rocket rises straight up. Five seconds after launch, an observer 1.5 km away notes that its angle of elevation is 7.5°. Four seconds later the angle is 58.1°. How far did the rocket rise during these four seconds?

8 A plane flying at an altitude of 300 m passes directly overhead. Two seconds later its angle of elevation is 41°. Compute its speed.

9 In Fig. 8 find ac/b^2. $\qquad\qquad$ 10 In Fig. 9 find $x + y$.

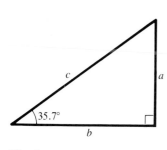

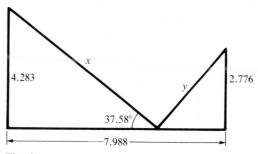

Fig. 8 $\qquad\qquad\qquad$ **Fig. 9**

11 Show that the length of the belt in Fig. 10 is

$$L = 2\sqrt{D^2 - (R_1 - R_0)^2} + 2\pi R_1 - 2(R_1 - R_0) \arccos\left(\frac{R_1 - R_0}{D}\right).$$

12 (cont.) Verify this formula for $R_0 = 0$.

13 (cont.) Verify this formula for $R_0 = R_1$.

14* (cont.) Suppose $R_0 = 3.00$, $R_1 = 6.00$, and $L = 60.00$. Estimate D numerically to two places.

15 Find the length of the belt in Fig. 11 if $R_0 = R_1$.

16* (cont.) Find the length if $R_0 < R_1$.

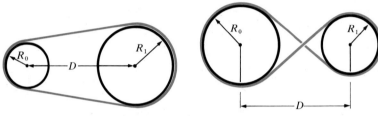

Fig. 10 **Fig. 11**

Find the horizontal and vertical distances from A to B.

	17 (Fig. 12)	18 (Fig. 12)	19 (Fig. 12)	20 (Fig. 12)
a	17.83 m	29.51 m	61.3 m	104.7 m
b	32.91 m	12.22 m	52.6 m	281.2 m
α	9.014°	4.112°	7.52°	3.104°
β	4.771°	2.790°	3.46°	1.892°

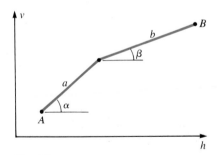

Fig. 12

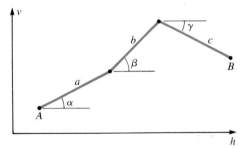

Fig. 13

	21 (Fig. 13)	22 (Fig. 13)	23 (Fig. 13)	24 (Fig. 13)
a	10.2 m	58.39 m	21.69 m	1056 m
b	8.6 m	41.11 m	31.82 m	1932 m
c	3.4 m	19.38 m	31.24 m	2158 m
α	5.1°	2.49°	3.52°	6.603°
β	6.6°	7.17°	6.14°	2.417°
γ	4.2°	3.94°	1.43°	3.528°

25 Fourteen pennies fit exactly in a ring around the base of a certain soda can. The diameter of each penny is 1.9 cm. Find the diameter of the can.

26 Thirteen nickels fit exactly in a ring around the base of a certain peach can of diameter 6.74 cm. Find the diameter of a nickel.

27 Seven of the pennies of Ex. 25 fit exactly in a ring *inside* a circular hoop. Find the diameter of the hoop.

28 Ten poker chips fit exactly in a ring *inside* a circular hoop of diameter 14.83 cm. Find the diameter of a chip.

5 FURTHER APPLICATIONS

Isosceles Triangles

An isosceles triangle (Fig. 1a) has equal base angles α, apex angle β, base b, and equal slant sides a. By drawing the altitude (Fig. 1b) of length h, we cut the isosceles triangle into two congruent right triangles. Several relations are clear from the figure:

$$\frac{\frac{1}{2}b}{a} = \cos \alpha = \sin \tfrac{1}{2}\beta \qquad h^2 + (\tfrac{1}{2}b)^2 = a^2$$

$$2\alpha + \beta = 180° \qquad h = a \sin \alpha.$$

With these formulas we can express various sides or angles in terms of others.

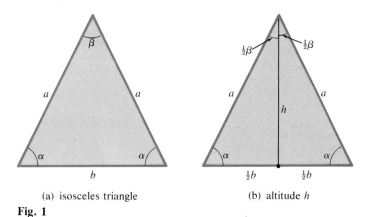

(a) isosceles triangle   (b) altitude h

Fig. 1

EXAMPLE 1   The tent in Fig. 2a, next page, has a rectangular floor, and its peak is centered. Find the base angles α and β of the front and side triangular flaps respectively, and their common slant height s.

Solution   The front flap is an isosceles triangle (Fig. 2b) with base $a = 7.0$ ft. We could find α and s from this triangle if we knew its altitude h_1. But h_1 is the hypotenuse (Fig. 2c) of a right triangle with legs h and $\tfrac{1}{2}b$, so

$$h_1{}^2 = h^2 + (\tfrac{1}{2}b)^2 = (3.7)^2 + (2.6)^2 = 20.45 \qquad h_1 \approx 4.5222$$

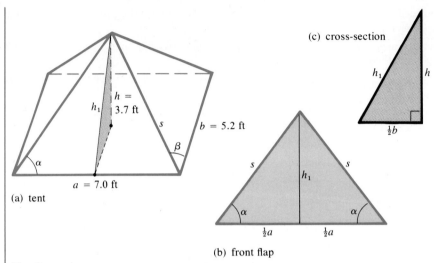

(c) cross-section

(a) tent

(b) front flap

Fig. 2

Now, returning to Fig. 2b, we have

$$\tan \alpha = \frac{h_1}{\frac{1}{2}a} \approx \frac{4.5222}{3.5} \approx 1.2921 \qquad \alpha \approx \text{arc tan } 1.2921 \approx 52.26°$$

and

$$s^2 = h_1^2 + (\tfrac{1}{2}a)^2 = 20.45 + (3.5)^2 = 32.7 \qquad s \approx 5.7184$$

We treat the right side flap similarly:

$$h_2^2 = h^2 + (\tfrac{1}{2}a)^2 = (3.7)^2 + (3.5)^2 = 25.94 \qquad h_2 \approx 5.0931$$

and

$$\tan \beta = \frac{h_1}{\frac{1}{2}b} \approx \frac{5.0931}{2.6} \approx 1.9589 \qquad \beta \approx \text{arc tan } 1.9589 \approx 62.96°$$

There is an easier way to find β. Once we have found s, we can use

$$\cos \beta = \frac{\frac{1}{2}b}{s} \approx \frac{2.6}{5.7184} \approx 0.45467 \qquad \beta \approx \text{arc cos } 0.45467 \approx 62.96°$$

This is a good check!

Answer $\alpha \approx 52°$ $\beta \approx 63°$ $s = 5.7$ ft

Remark We rounded off the answer to two significant figures, precisely what the data (and common sense) of the problem justifies. For the purpose of illustration, we carried 4 or 5 significant figures through the work. Actually, with a calculator, you never should do *any* intermediate rounding; only round your final answer.

For instance, if you want $\sqrt{5.7^2 + 3.8^2}$ to two significant figures, then

$$\sqrt{5.7^2 + 3.8^2} = \sqrt{32.49 + 14.44} = \sqrt{46.93} = 6.850547 \cdots \approx 6.9$$

is correct. However, rounding to two significant figures at each step gives

$$\sqrt{5.7^2 + 3.8^2} \approx \sqrt{32 + 14} = \sqrt{46} = 6.7823 \approx 6.8$$

Area

The area of a triangle with base b and height h equals $\frac{1}{2}bh$. From this fact follows a formula for the area of a triangle in terms of two sides and their included angle:

$$A = \tfrac{1}{2}ab \sin \gamma.$$

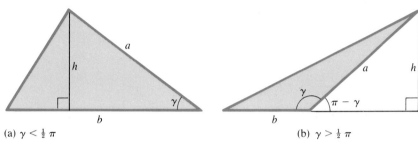

(a) $\gamma < \frac{1}{2}\pi$ (b) $\gamma > \frac{1}{2}\pi$

Fig. 3 Proof of the area formula

This is certainly true if $\gamma = 90°$. Otherwise, think of b as the base of the triangle and drop a perpendicular on b from the opposite vertex (Fig. 3). Its length h (the altitude of the triangle) is

$$h = a \sin \gamma = a \sin(\pi - \gamma).$$

Therefore

$$A = \tfrac{1}{2}bh = \tfrac{1}{2}ba \sin \gamma.$$

EXAMPLE 2 Express the area of an isosceles triangle in terms of its base b and slant side a.

Solution In Fig. 1b, we have

$$a^2 = h^2 + (\tfrac{1}{2}b)^2 \qquad h = \sqrt{a^2 - \tfrac{1}{4}b^2}.$$

Therefore

$$A = \tfrac{1}{2}bh = \tfrac{1}{2}b\sqrt{a^2 - \tfrac{1}{4}b^2} = \tfrac{1}{4}b\sqrt{4a^2 - b^2}.$$

Answer $A = \tfrac{1}{4}b\sqrt{4a^2 - b^2}$.

Exercises

1 A room measures $12 \times 15 \times 8$ ft. Compute the angles that the diagonal makes with each of the edges.

2 In Fig. 4, next page, express the chord length L in terms of the radius r and the central angle θ.

3 (cont.) Find the perimeter of an equilateral triangle inscribed in a circle of radius r.

4 (cont.) Find the perimeter of a regular hexagon inscribed in a circle of radius r.

5 (cont.) Find the perimeter of a regular octagon inscribed in a circle of radius r.

6 Find the height of the tent in Fig. 5, next page. The floor is rectangular and the peak centered.

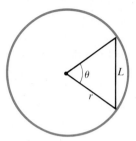

Fig. 4

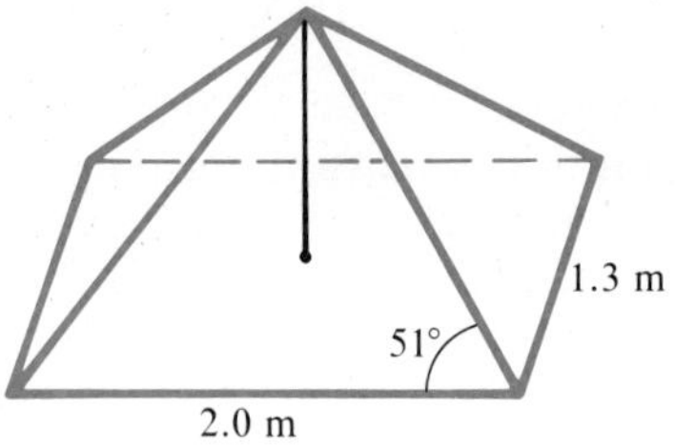

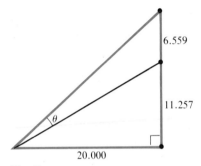

Fig. 5

7 Express the arc length s in terms of a and y (Fig. 6).

8 Estimate θ in Fig. 7 to three places.

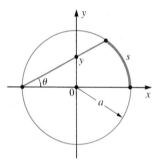

Fig. 6

Fig. 7

9 Express the area of an isosceles triangle in terms of its base b and its apex angle β.

10 Express the area of an isosceles triangle in terms of its slant side a and its apex angle β.

11 (cont.) Find the area of a regular octagon inscribed in a circle of radius r.

12 (cont.) Find the area of a regular dodecagon (12 sides) inscribed in a circle of radius r.

13 Find the apex angle of the isosceles triangle whose equal sides are 10 ft and area is 25 ft².

14 An isosceles triangle has base 1 and apex angle 20°. For what apex angle would its area be twice as much?

15 The pyramid (Fig. 8) with square base a has height h. Express its edge length b and apex angle in terms of a and h.

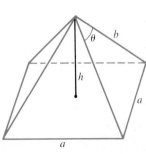

Fig. 8

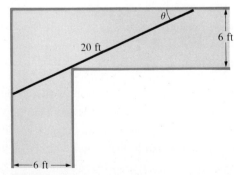

Fig. 9

16* The 20-ft pipe will never turn the corner in the 6-ft hall (Fig. 9). Approximate the angle at which it gets stuck.

The **sine bar** (Fig. 10) is a precision tool used in laboratories and machine shops to measure angles. It consists of two cylinders of equal radii, placed parallel, with their axes 25.0000 cm apart. They are connected to a flat surface, parallel to the plane that passes through their axes. In practice, the cylinders rest on highly accurate gauge blocks on a horizontal table (Fig. 11). The sine bar is then tilted to the horizontal through an angle θ.

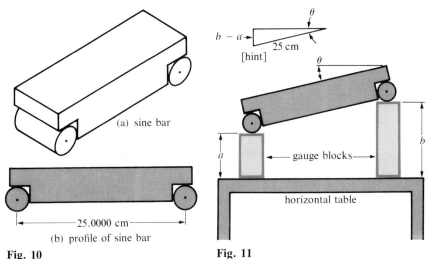

(a) sine bar

(b) profile of sine bar

Fig. 10   **Fig. 11**

Find θ in $° ' ''$

	17	18	19	20
a	1.0010 cm	2.0000 cm	2.0000 cm	1.0000 cm
b	3.3705 cm	6.0050 cm	12.5120 cm	7.3596 cm

Find b in cm

	21	22	23	24
a	1.0042 cm	2.0000 cm	1.5000 cm	1.3872 cm
θ	3°12′52″	5°0′17″	1°52′43″	25°5′21″.

The remaining exercises are related to practical surveying. The figures are stripped of instruments, terrain, and surveyors, giving only the essentials of the math.

cross hairs in surveyor's
telescope

25 (Fig. 12) Find the angle θ in °'" that gives 100 power magnification.

26 (Fig. 13) Find h if α = 1°17'29"

27 (Fig. 13) Find α if h = 37.804 m

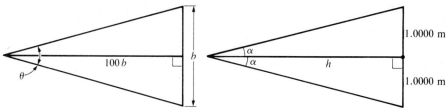

Fig. 12 (Isosceles triangle)   **Fig. 13**

28 (Fig. 14) Find h and v if θ = 13°17'5" and α = 48'19".

29 (Fig. 14) Find θ and h if v = 19.762 m and α = 38'17".

30 (Fig. 14) Find θ and v if h = 82.78 m and α = 32'54".

31 (Fig. 15) Find v and h if d = 1.4820 m and θ = 16°2'27".

32* (Fig. 15) Find v and θ if h = 100.00 m and d = 2.0000 m.

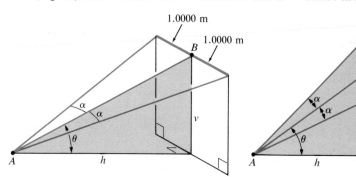

Fig. 14 (Horizontal subtense bar;   **Fig. 15** (Vertical subtense bar)
vertical plane shaded) α = 17' 11"

REVIEW EXERCISES

1 Compute 1 ×️ 2 +️ 3 ×️ 4 −️ 5 ×️ 6 =️

(a) in left-to-right logic   (b) in algebraic logic.

2 Estimate $\sqrt[4]{1} + \sqrt[3]{2^4 + 3^5}$ on a calculator.

Use a calculator to estimate to 6 significant figures

3 $\dfrac{3}{\sqrt{6} - \dfrac{2}{\sqrt{5} - 1}}$

4 $\dfrac{5 \tan a}{7 + \sin a} + \sqrt[4]{a}$ where a = 0.414392

Find the remaining sides and angles of the right triangle (Fig. 1)

5 a = 12.97 α = 21.14°   **6** c = 9.38614 β = 53°12'26"

7 The rope (Fig. 2) goes from A to C via the top of the pole at B. How long is the rope?

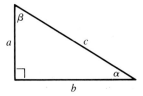

Fig. 1

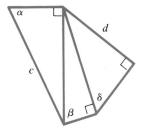

Fig. 2

8 Find the perimeter and area of an isosceles triangle with base 17.62 cm and base angle 13.13°.

9 In Fig. 3, given are $a = 16.34$, $\alpha = 35.21°$, $\beta = 25.48°$. Find b.

10 In Fig. 4, given are $c = 3.108$, $d = 1.500$, $\alpha = 63.82°$, $\beta = 71.43°$. Find δ.

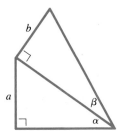

Fig. 3   **Fig. 4**

$\boldsymbol{6}$ SOLVING OBLIQUE TRIANGLES

1 LAW OF COSINES

We now take up oblique triangles, that is, non-right triangles. In this section we develop and apply a formula called the law of cosines, and in the next section another formula called the law of sines. Together these are adequate for all numerical problems concerning oblique triangles.

Generally we shall label the six "parts" of a triangle as in Fig. 1. Each side has a Latin label and its opposite angle has the corresponding Greek label.

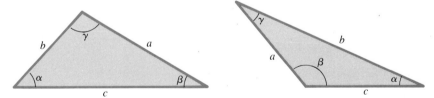

Fig. 1 The 6 parts of a triangle: 3 sides and 3 angles

Now we can state the law of cosines.

> **Law of Cosines**
>
> $$c^2 = a^2 + b^2 - 2ab \cos \gamma.$$

(The law of cosines is a generalization of the Pythagorean Theorem. If γ is a right angle, then $\cos \gamma = 0$, and the formula in the preceding box says that $c^2 = a^2 + b^2$.)

In words, the square of any side of a triangle equals the sum of the squares of the other two sides minus twice their product by the cosine of the included angle. Thus, for the triangle in Fig. 1, we also have

$$a^2 = b^2 + c^2 - 2bc \cos \alpha \quad \text{and} \quad b^2 = c^2 + a^2 - 2ca \cos \beta.$$

To prove the law of cosines, choose axes so that the angle γ is in standard position with side a along the positive x-axis (Fig. 2). The upper vertex of the triangle is at $(b \cos \gamma, b \sin \gamma)$. By the distance formula,

$$\begin{aligned}
c^2 &= (b \cos \gamma - a)^2 + (b \sin \gamma)^2 \\
&= (b^2 \cos^2 \gamma - 2ab \cos \gamma + a^2) + b^2 \sin^2 \gamma \\
&= b^2(\cos^2 \gamma + \sin^2 \gamma) + a^2 - 2ab \cos \gamma \\
&= a^2 + b^2 - 2ab \cos \gamma,
\end{aligned}$$

so the formula is proved.

The law of cosines is an effective tool when we are given either all three sides of a triangle or two sides and their included angle. Let us consider these two cases.

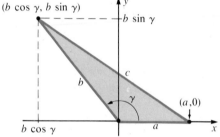

Fig. 2 Proof of the law of cosines   **Fig. 3** Given $a = 7.135$, $b = 6.002$, $c = 2.508$

Given: Three Sides (SSS)

Given three sides of a triangle, its three angles can be found by the law of cosines.

EXAMPLE 1   Given (Fig. 3)

$$a = 7.135 \qquad b = 6.002 \qquad c = 2.508,$$

find α, β, and γ.

Solution   By the law of cosines

$$a^2 = b^2 + c^2 - 2bc \cos \alpha.$$

Therefore

$$\cos \alpha = \frac{b^2 + c^2 - a^2}{2bc} = \frac{(6.002)^2 + (2.508)^2 - (7.135)^2}{(2)(6.002)(2.508)} \approx -0.2855,$$

$$\alpha \approx \arccos(-0.2855) \approx 106.59°$$

Similarly

$$\cos \beta = \frac{c^2 + a^2 - b^2}{2ca} = \frac{(2.508)^2 + (7.135)^2 - (6.002)^2}{(2)(2.508)(7.135)} \approx 0.5916$$

$$\beta \approx \text{arc} \cos (0.5916) \approx 53.73°$$

$$\cos \gamma = \frac{a^2 + b^2 - c^2}{2ab} = \frac{(7.135)^2 + (6.002)^2 - (2.508)^2}{(2)(7.135)(6.002)} \approx 0.9415$$

$$\gamma \approx \text{arc} \cos (0.9415) \approx 19.69°.$$

Check $106.59 + 53.73 + 19.69 = 180.01$

This is convincing since $\alpha + \beta + \gamma$ should be 180.00.

Answer $\alpha \approx 106.59°$ $\beta \approx 53.73°$ $\gamma \approx 19.69°$

EXAMPLE 2   Given

$$a = 4.102 \qquad b = 3.997 \qquad c = 8.165,$$

find α, β, and γ.

Solution   By the law of cosines

$$\cos \alpha = \frac{b^2 + c^2 - a^2}{2bc} = \frac{(3.997)^2 + (8.165)^2 - (4.102)^2}{(2)(3.997)(8.165)} \approx 1.008.$$

This is impossible, since $-1 \le \cos \alpha \le 1$, so there is no such triangle.
  Why? Because the sides of a triangle must satisfy the triangle inequalities

$$a < b + c \qquad b < c + a \qquad c < a + b$$

In this case

$$a + b = 4.102 + 3.997 = 8.099 < 8.165 = c.$$

Answer   There is no such triangle.

EXAMPLE 3   A pilot intends to fly from A to B, a straight line distance of 200.0
miles. His flight plan calls for a straight leg of 97.0 miles from A to a navigation
station C, then another straight leg of 108.0 miles from C to B. By what angle
should he change course at C?

Solution   The desired angle is θ in Fig. 4, next page. But $\theta = 180 - \gamma$, and we
can find γ by the law of cosines:

$$\cos \gamma = \frac{108^2 + 97^2 - 200^2}{(2)(108)(97)} \approx -0.9034$$

$$\gamma \approx \text{arc} \cos (-0.9034) \approx 154.6°$$

Therefore

$$\theta = 180 - \gamma \approx 180 - 154.6 = 25.4°$$

Answer   25.4°

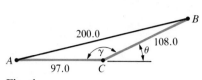

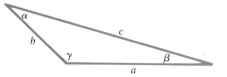

Fig. 4

Fig. 5 Given $a = 14.722$, $b = 8.624$, $\gamma = 137.41°$

Given: Two Sides and Their Included Angle (SAS)

EXAMPLE 4   Given in Fig. 5:

$$a = 14.722 \qquad b = 8.624 \qquad \gamma = 137.41°,$$

find c, α, and β.

Solution   By the law of cosines,

$$c^2 = a^2 + b^2 - 2ab \cos \gamma$$

$$= (14.722)^2 + (8.624)^2 - (2)(14.722)(8.624)(\cos 137.41°)$$

$$\approx 478.054$$

$$c \approx 21.864$$

Now we have the three sides, so we can proceed as in Example 1 to find α and β:

$$\cos \alpha = \frac{b^2 + c^2 - a^2}{2bc} = \frac{(8.624)^2 + (21.864)^2 - (14.722)^2}{(2)(8.624)(21.864)} \approx 0.8901$$

$$\alpha \approx \text{arc cos } 0.8901 \approx 27.11°$$

$$\cos \beta = \frac{c^2 + a^2 - b^2}{2ca} = \frac{(21.864)^2 + (14.722)^2 - (8.624)^2}{(2)(21.864)(14.722)} \approx 0.9637$$

$$\beta \approx \text{arc cos } 0.9637 \approx 15.48°$$

Check $\alpha + \beta + \gamma \approx 27.11 + 15.48 + 137.41 = 180°$

Answer $c \approx 21.864 \qquad \alpha \approx 27.11° \qquad \beta \approx 15.48°$

EXAMPLE 5   Two planes at the same altitude are approaching an airport. Radar shows plane A at 18.0 miles due east of the tower and plane B at 22.5 miles from the tower on a line 10° north of east. How far apart are the planes?

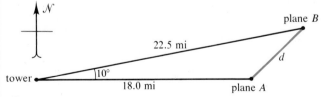

Fig. 6

Solution   Draw Fig. 6. To find the distance d between the planes, use the law of cosines:

$$d^2 = (18.0)^2 + (22.5)^2 - (2)(18.0)(22.5)(\cos 10°) \approx 32.56$$

$$d \approx 5.7$$

Answer   5.7 miles

Further applications of the law of cosines are given in Section 3.

Exercises

Find α, β, and γ to 2 places

1 $a = 2.00$ $b = 3.00$ $c = 4.00$ **2** $a = 7.00$ $b = 6.00$ $c = 5.00$

3 $a = 2.80$ $b = 1.90$ $c = 3.60$ **4** $a = 10.20$ $b = 13.60$ $c = 8.300$

5 $a = 1.043$ $b = 0.986$ $c = 1.527$ **6** $a = 200.8$ $b = 157.7$ $c = 166.5$

7 $a = 16.38$ $b = 24.43$ $c = 9.02$ **8** $a = 7.382$ $b = 7.889$ $c = 14.001$

9 $a = 3048$ $b = 3692$ $c = 6480$ **10** $a = 0.039$ $b = 0.846$ $c = 0.862$

11 $a = 0.416$ $b = 0.752$ $c = 0.329$ **12** $a = 20.97$ $b = 16.82$ $c = 3.970$.

Find the other side and two angles

13 $a = 5.000$ $b = 3.000$ $\gamma = 39.00°$ **14** $a = 4.000$ $b = 7.000$ $\gamma = 126.00°$

15 $a = 2.788$ $b = 9.303$ $\gamma = 11.14°$ **16** $a = 7.980$ $b = 8.003$ $\gamma = 9.368°$

17 $b = 130.2$ $c = 16.24$ $\alpha = 152.37°$ **18** $b = 0.01348$ $c = 0.5172$ $\alpha = 83.47°$

19 $b = 30348$ $c = 52802$ $\alpha = 39°17'12''$ **20** $b = 17.634$ $c = 62.115$ $\alpha = 108°46'53''$

21 $c = 99.44$ $a = 10.66$ $\beta = 3.06°$ **22** $c = 14.92$ $a = 17.89$ $\beta = 19.84°$

23 $c = 10,000$ $a = 10,003$ $\beta = 34''$ **24** $c = 100.0$ $a = 0.1362$ $\beta = 179°21'$.

25 The diagonals of a parallelogram are 10 cm and 15 cm, and they intersect at an angle of 28.7°. Compute the sides of the parallelogram.

26 A point A is 10.00 km due north of here. Another point B is 12.36 km from here on a line 17.4° east of north. What is the distance from A to B?

27 A plane travels 50 miles due east after take-off, then adjusts its course 10° northward and flies 100 miles. How far is it from the point of departure?

28 Two boats, one traveling 19 km/hr and the other 23 km/hr, leave a point on straight paths 65° from each other. How far apart are they after 2.5 minutes?

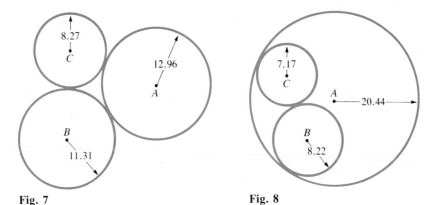

Fig. 7   Fig. 8

29 The circles in Fig. 7 are mutually tangent. Find the angles of the triangle ABC.

30 The circles in Fig. 8 are mutually tangent. Find the angles of the triangle ABC.

31   Use the law of cosines to prove this assertion about any parallelogram: the sum of the squares of the two diagonals equals the sum of the squares of the four sides.

32   (cont.) How does it follow that the diagonals of a rhombus (parallelogram with equal sides) are perpendicular?

33   The centerfielder stands 340 ft from home plate in straight centerfield. How far is he from third base? (The distance between bases is 90 ft.)

34   The leftfielder stands 330 ft from home plate. A straight line from him to home plate bisects the line from second to third base. How far is he from second base? (The distance between bases is 90 ft.)

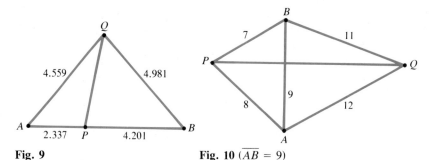

Fig. 9   **Fig. 10** ($\overline{AB} = 9$)

35*   Find $\overline{PQ}$ in Fig. 9. [*Hint*   Two steps are needed.]
36*   Find $\overline{PQ}$ in Fig. 10. [*Hint*   Three steps are needed.]

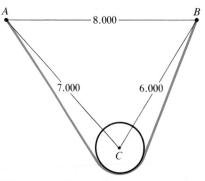 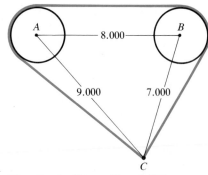

Fig. 11   (pulley radius = 1.000)   **Fig. 12**   (pulley radii = 1.500)

37   Find the length of the rope in Fig. 11. It is attached at A and at B, and taut around the pulley centered at C.

38*   Find the length of the rope in Fig. 12. The two pulleys centered at A and at B have equal radii 1.500, and the rope is pulled taut at C.

2   LAW OF SINES

As we saw in Section 1, given SSS (three sides) or SAS (two sides and their included angle), the remaining parts of a triangle can be found using the law of cosines. The cases SAA, ASA, and SSA cannot be handled by the law of cosines; they require the law of sines. (Note that AAA is simply not enough information.)

Law of Sines	$\dfrac{a}{\sin \alpha} = \dfrac{b}{\sin \beta} = \dfrac{c}{\sin \gamma}$

We'll postpone the proof until Section 3 and get right to its use in solving triangles.

Given: Two Angles and a Side (SAA or ASA)

Given any two angles of a triangle, we have the third angle automatically, because $\alpha + \beta + \gamma = 180°$. Therefore the cases SAA (two angles and an excluded side) and ASA (two angles and their included side) are really equivalent problems.

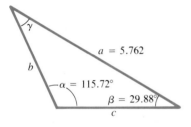

Fig. 1 (SAA)

EXAMPLE 1 Given

$$\alpha = 115.72° \qquad \beta = 29.88° \qquad a = 5.762,$$

find γ, b, and c.

Solution First,

$$\gamma = 180° - \alpha - \beta = 180° - 115.72° - 29.88° = 34.40°$$

To determine b we use part of the law of sines:

$$\frac{b}{\sin \beta} = \frac{a}{\sin \alpha}$$

$$b = \frac{a \sin \beta}{\sin \alpha} = \frac{(5.762) \sin 29.88°}{\sin 115.72°} \approx 3.186.$$

To determine c, we use the law of sines again:

$$\frac{c}{\sin \gamma} = \frac{a}{\sin \alpha}$$

$$c = \frac{a \sin \gamma}{\sin \alpha} = \frac{(5.762) \sin 34.40°}{\sin 115.72°} \approx 3.613.$$

Check $b^2 + c^2 - 2bc \cos \alpha$

$$\approx (3.186)^2 + (3.613)^2 - (2)(3.186)(3.613)(\cos 115.72°)$$

$$\approx 33.195$$

$$a^2 \approx (5.762)^2 \approx 33.201.$$

The law of cosines, $a^2 \approx b^2 + c^2 - 2bc \cos \alpha$, checks.

Answer $\gamma = 34.40°$ $b \approx 3.186$ $c \approx 3.613$

Given: Two Sides and an Excluded Angle (SSA)

Suppose sides a and c and angle γ are given. Then several possibilities can occur. Let us study them by drawing a few figures. To make life a little easier, we'll assume temporarily that γ is an acute angle.

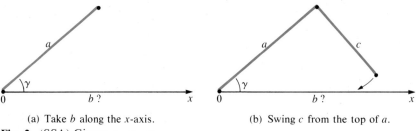

(a) Take b along the x-axis.   (b) Swing c from the top of a.
Fig. 2 (SSA) Given: a c γ

Think of side b as the base of the triangle lying on the x-axis with its left end-point at the origin (Fig. 2a). Then side a will make angle γ with the positive x-axis. Side c will dangle from the top end-point of a. Imagine it swinging like a pendulum, the lower end tracing an arc (Fig. 2b). This arc may meet the x-axis twice, may just touch it (tangent), or may miss the axis altogether.

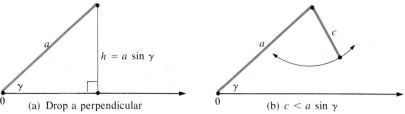

(a) Drop a perpendicular   (b) $c < a \sin \gamma$
Fig. 3

From Fig. 3a the distance h from the top end of side a to the x-axis is $a \sin \gamma$. This is a critical number, for if side c is too short, that is, if $c < a \sin \gamma$, then the pendulum will never reach the x-axis (Fig. 3b). Hence there is no triangle with the given parts a, c, and γ.

If $c = a \sin \gamma$, then the pendulum just reaches the x-axis. Hence a, c, γ determine one triangle, a right triangle (Fig. 4a).

If $c > a \sin \gamma$, then the pendulum is long enough to hit the x-axis twice, apparently forming two triangles (Fig. 4b) with the same parts a, c, and γ. But let's be careful: there really are *two* triangles formed *provided* c is not too large, that is, provided $c < a$. If $c \geq a$, the swinging pendulum hits the x-axis once on the right side of γ, and once (Fig. 4c) on the wrong (left) side; there is just one triangle with the given parts a, c, γ.

Remark The case $a \sin \gamma < c < a$ in which the same data determines two triangles is known as the **ambiguous case**.

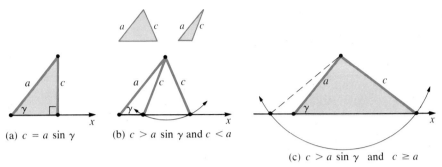

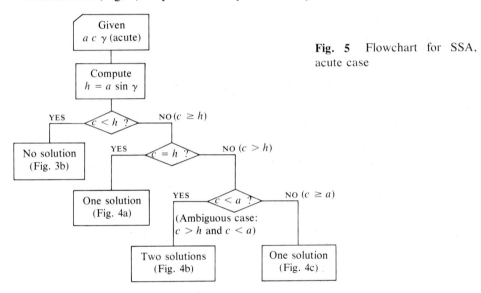

(a) $c = a \sin \gamma$

(b) $c > a \sin \gamma$ and $c < a$

(c) $c > a \sin \gamma$ and $c \geq a$

Fig. 4 $c \geq a \sin \gamma$.

A flowchart (Fig. 5) helps us work systematically.

Given
$a\ c\ \gamma$ (acute)

Compute
$h = a \sin \gamma$

$c < h$?

YES — No solution (Fig. 3b)

NO ($c \geq h$)

$c = h$?

YES — One solution (Fig. 4a)

NO ($c > h$)

$c < a$?

YES

(Ambiguous case: $c > h$ and $c < a$)

Two solutions (Fig. 4b)

NO ($c \geq a$)

One solution (Fig. 4c)

Fig. 5 Flowchart for SSA, acute case

EXAMPLE 2 Find α, β, b for a triangle with

(a) $a = 5$ $c = 2$ $\gamma = 30°$

(b) $a = 5$ $c = 2.5$ $\gamma = 30°$.

Solution (a) γ is acute, so we compare c with $h = a \sin \gamma$. Now $c = 2$ and

$$h = a \sin \gamma = 5 \sin 30° = 2.5 \quad \text{so} \quad c < h.$$

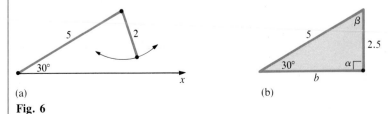

(a)

(b)

Fig. 6

There is no such triangle (Fig. 6a, previous page).

(b)   In this case $c = h$, so one triangle is formed, a right triangle (Fig. 6b). The right angle is α, and β is the complement of γ, that is, $\beta = 60°$. Finally,

$$b = 5 \cos \gamma = 5 \cos 30° = 2.5\sqrt{3} \approx 4.3301.$$

Answer   (a)   no such triangle

(b)   one possibility: $\alpha = 90°$ $\beta = 60°$ $b = 2.5\sqrt{3} \approx 4.3301$

EXAMPLE 3   Given a triangle with

$$a = 3.802 \qquad c = 2.766 \qquad \gamma = 41.09°$$

Find α, β, b.

Solution   Since γ is acute, we compute $h = a \sin \gamma$:

$$h = a \sin \gamma = (3.802) \sin 41.09° \approx 2.499.$$

Since $c = 2.766$ and $a = 3.802$,

$$c > h \quad \text{and also} \quad a > c.$$

This is the situation in Fig. 4b (ambiguous case). There are *two* triangles that fit the data. In each,

$$\frac{a}{\sin \alpha} = \frac{c}{\sin \gamma}$$

so

$$\sin \alpha = \frac{a \sin \gamma}{c} = \frac{(3.802) \sin 41.09°}{2.766} \approx 0.9034.$$

Expecting two answers, we are extra alert and realize that there are *two* angles α in the domain $0 < \alpha < 180°$ such that $\sin \alpha = 0.9034$. One is arc sin 0.9034 and the other is its supplement, $180° - $ arc sin 0.9034. Thus we have two cases.

Case 1 $\alpha \approx$ arc sin $0.9034 \approx 64.61°$

Then

$$\beta = 180° - \gamma - \alpha \approx 180° - 41.09° - 64.61° = 74.30°.$$

By the law of sines again

$$\frac{b}{\sin \beta} = \frac{c}{\sin \gamma} \qquad b = \frac{c \sin \beta}{\sin \gamma} = \frac{(2.766)(\sin 74.30°)}{\sin 41.09°} \approx 4.051.$$

See Fig. 7a.

Check   (law of sines)

$$\frac{a}{\sin \alpha} \approx \frac{3.802}{\sin 64.61°} \approx 4.208 \qquad \frac{b}{\sin \beta} \approx \frac{4.051}{\sin 74.30°} \approx 4.208$$

$$\frac{c}{\sin \gamma} \approx \frac{2.766}{\sin 41.09°} \approx 4.208$$

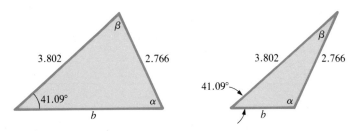

(a) $\alpha \approx 64.61°$ $\beta \approx 74.30°$ $b \approx 4.051$ (b) $\alpha \approx 115.39°$ $\beta \approx 23.52°$ $b \approx 1.679$

Fig. 7 Example of the ambiguous case

Check (law of cosines) Use the last angle found, β :

$$a^2 + c^2 - 2ac \cos \beta \approx (3.802)^2 + (2.766)^2$$
$$- (2)(3.802)(2.766)(\cos 74.30°) \approx 16.41$$

$$b^2 \approx (4.051)^2 \approx 16.41$$

Case 2 $\alpha \approx 180° - $ arc sin $0.9034 \approx 115.39°$

$$\beta = 180° - \gamma - \alpha \approx 180° - 41.09° - 115.39° = 23.52°.$$

As in Case 1,

$$b = \frac{c \sin \beta}{\sin \gamma} \approx \frac{(2.766)(\sin 23.52°)}{\sin 41.09°} \approx 1.679.$$

See Fig. 7b.

Check (law of cosines)

$$a^2 + c^2 - 2ac \cos \beta \approx (3.802)^2 + (2.766)^2$$
$$- (2)(3.802)(2.766)(\cos 23.52°) \approx 2.821$$

$$b^2 \approx (1.679)^2 \approx 2.819$$

Check (law of sines) Note that $c/\sin \gamma$ and $a/\sin \alpha$ are exactly the same as in Case 1, approximately 4.208. This time

$$\frac{b}{\sin \beta} \approx \frac{1.679}{\sin 23.52°} \approx 4.207.$$

Answer Two answers: $\begin{cases} \alpha_1 \approx 64.61° & \beta_1 \approx 74.30° & b_1 \approx 4.051 \\ \alpha_2 \approx 115.39° & \beta_2 \approx 23.52° & b_2 \approx 1.679 \end{cases}$

The ratio $c/\sin \gamma$ was used several times in the solution of Example 3. The computation could have been shortened by first computing and storing this ratio. Let's flowchart how the computation should be organized (Fig. 8, page 135), then try another problem.

EXAMPLE 4 Given a triangle with

$$a = 7002 \qquad c = 4468 \qquad \gamma = 19.91°$$

Find α, β, b.

Solution   Clearly $c < a$. Also $a \sin \gamma \approx 2384 < c$, so this is the ambiguous case. Compute

$$p = \frac{c}{\sin \gamma} = \frac{4468}{\sin 19.91°} \approx 13120 .$$

Then

$$\alpha_1 = \text{arc sin} \frac{a}{p} \approx \text{arc sin} \frac{7002}{13120} \approx 32.25°$$

$$\beta_1 = 180° - \gamma - \alpha_1 \approx 127.84° \qquad b_1 = p \sin \beta_1 \approx 10360$$

This completes the first case. Now

$$\alpha_2 = 180° - \alpha_1 \approx 147.75° \qquad \beta_2 = \alpha_1 - \gamma \approx 12.34°$$

$$b_2 = p \sin \beta_2 \approx 2804 ,$$

which completes the second case.

Check $b_1{}^2 \approx 1.073 \times 10^8 \qquad a^2 + c^2 - 2ac \cos \beta_1 \approx 1.074 \times 10^8$

$\qquad\qquad b_2{}^2 \approx 7.862 \times 10^6 \qquad a^2 + c^2 - 2ac \cos \beta_2 \approx 7.867 \times 10^6$

Answer   Two answers: $\begin{cases} \alpha_1 \approx 32.25° & \beta_1 \approx 127.84° & b_1 \approx 10360 \\ \alpha_2 \approx 147.75° & \beta_2 \approx 12.34° & b_2 \approx 2804 \end{cases}$

EXAMPLE 5   Given a triangle with

$$a = 75.22 \qquad c = 84.37 \qquad \gamma = 62.17°$$

Find α, β, b.

Solution   Since $c > a$, this is the case of Fig. 4c. No testing is required, only computing; there is one solution. Compute

$$p = \frac{c}{\sin \gamma} = \frac{84.37}{\sin 62.17°} \approx 95.40$$

By the law of sines,

$$\sin \alpha = a \frac{\sin \gamma}{c} = \frac{a}{p}$$

$$\alpha = \text{arc sin} \frac{a}{p} \approx \text{arc sin} \frac{75.22}{95.40} \approx 52.04°$$

$$\beta = 180° - \gamma - \alpha = 65.79°.$$

Again by the law of sines,

$$b = \frac{c}{\sin \gamma} \sin \beta = p \sin \beta \approx 95.40 \sin 65.79° \approx 87.01.$$

Check $b^2 \approx 7571 \qquad c^2 + a^2 - 2ac \cos \beta \approx 7571$

Answer $\alpha \approx 52.04° \qquad \beta \approx 65.79° \qquad b \approx 87.01$

Fig. 8 Organized computation for the ambiguous case

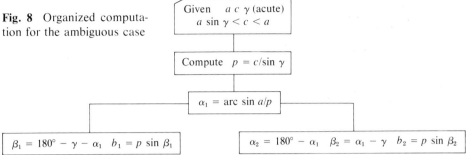

Given $a\ c\ \gamma$ (acute)
  $a \sin \gamma < c < a$

Compute $p = c/\sin \gamma$

$\alpha_1 = \text{arc sin } a/p$

$\beta_1 = 180° - \gamma - \alpha_1$ $b_1 = p \sin \beta_1$

$\alpha_2 = 180° - \alpha_1$ $\beta_2 = \alpha_1 - \gamma$ $b_2 = p \sin \beta_2$

Explanation   From the law of sines

$$\frac{a}{\sin \alpha} = \frac{c}{\sin \gamma} = p \quad \text{follows} \quad \sin \alpha = \frac{a}{p};$$

hence the third box in the flowchart. Similarly from

$$\frac{b}{\sin \beta} = \frac{c}{\sin \gamma} = p \quad \text{follows} \quad b_1 = p \sin \beta_1 \quad \text{and} \quad b_2 = p \sin \beta_2.$$

Finally, $\beta_2 = 180° - \gamma - \alpha_2 = 180° - \gamma - (180° - \alpha_1) = \alpha_1 - \gamma$.

SSA: The Obtuse Case

We have discussed the problem SSA assuming that γ is an acute angle. We know how to handle the triangle if γ is a right angle. Suppose γ is an obtuse angle, $90° < \gamma < 180°$. If $c < a \sin \gamma$, then just as before the pendulum will not reach the x-axis (Fig. 9a) and there is no solution. If $a \le c < a \sin \gamma$, the pendulum will meet the x-axis twice (Fig. 9b), but on the wrong side of γ. In this case too, there is no solution. If $c > a$, the pendulum will meet the x-axis once on the right side of γ, hence there is one solution (Fig. 9c). We shall not give a worked example illustrating Fig. 9c; step-by-step, the calculation is exactly like that in Example 5.

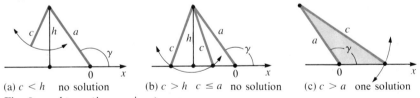

(a) $c < h$ no solution   (b) $c > h$ $c \le a$ no solution   (c) $c > a$ one solution
Fig. 9 γ obtuse ($h = a \sin \gamma$)

Remark 1   In the worked examples of SSA, the given sides were always a and c, and the given angle was always γ. Of course, two sides and an excluded angle might be given with different names, such as b, c, α, etc. Therefore you must know the method rather than formulas involving particular letters (names). Perhaps it is best to remember the law of sines in words: *in any triangle, the three quotients* (side) $\div$ sin(opp. angle) *are equal.* Always sketch the triangle(s)!

Remark 2   In the examples we have generally shown all numbers rounded to four significant figures. Of course, when doing this work on a calculator, you should not do any intermediate rounding. Only your answers should be rounded—to the accuracy of the data (or as you are instructed).

Exercises

Find the remaining angle and sides; check

1 $\alpha = 45.38°$ $\beta = 62.45°$ $a = 3.077$	**2** $\alpha = 71.32°$ $\beta = 21.01°$ $a = 5.000$
3 $\alpha = 103.05°$ $\beta = 13.11°$ $c = 12.97$	**4** $\alpha = 39.00°$ $\beta = 93.04°$ $c = 701.8$
5 $\beta = 24.17°$ $\gamma = 19.92°$ $b = 0.6523$	**6** $\beta = 138.22°$ $\gamma = 27.72°$ $b = 0.1628$
7 $\beta = 24.17°$ $\gamma = 19.92°$ $a = 0.6523$	**8** $\beta = 138.22°$ $\gamma = 27.72°$ $a = 0.1628$
9 $\beta = 24.17°$ $\gamma = 19.92°$ $c = 0.6523$	**10** $\beta = 138.22°$ $\gamma = 27.72°$ $c = 0.1628$.

Find the remaining parts; give all solutions and check

11 $a = 3.602$ $c = 3.403$ $\gamma = 75.64°$	**12** $a = 1042$ $c = 445.82$ $\gamma = 25.88°$
13 $a = 14.23$ $c = 21.38$ $\alpha = 47.08°$	**14** $a = 89.22$ $b = 103.4$ $\alpha = 63.30°$
15 $a = 3\sqrt{2}$ $c = 3$ $\gamma = 45°$	**16** $a = 7$ $b = 3$ $\beta = \text{arc sin } \frac{3}{7}$
17 $a = 4.990$ $c = 3.861$ $\gamma = 31.24°$	**18** $a = 12.63$ $c = 9.241$ $\gamma = 43.24°$
19 $a = 3111$ $c = 3048$ $\gamma = 73.94°$	**20** $a = 406.2$ $c = 378.7$ $\gamma = 63.78°$
21 $b = 0.6742$ $c = 0.5109$ $\gamma = 27.31°$	**22** $b = 10.43$ $c = 9.687$ $\gamma = 57.44°$
23 $a = 41.53$ $b = 47.22$ $\alpha = 54.99°$	**24** $a = 6.903$ $b = 8.211$ $\alpha = 31.31°$
25 $a = 40830$ $b = 20780$ $\beta = 18.35°$	**26** $a = 766.2$ $b = 683.1$ $\beta = 23.94°$
27 $a = 15.21$ $c = 38.97$ $\gamma = 83.12°$	**28** $a = 6.129$ $c = 9.204$ $\gamma = 52.51°$
29 $a = 524.6$ $b = 719.3$ $\beta = 9.38°$	**30** $a = 6048$ $b = 9121$ $\beta = 43.15°$
31 $a = 67.11$ $c = 72.38$ $\gamma = 138.61°$	**32** $a = 7.801$ $c = 9.316$ $\gamma = 100.41°$
33 $a = 3048$ $c = 2176$ $\alpha = 98.17°$	**34** $a = 607.0$ $c = 431.3$ $\alpha = 161.23°$
35 $a = 7.213$ $b = 7.014$ $\alpha = 118.30°$	**36** $b = 52.61$ $c = 61.52$ $\gamma = 126.34°$.

3 APPLICATIONS

In this section, we discuss some applications of the law of sines and some further applications of the law of cosines.

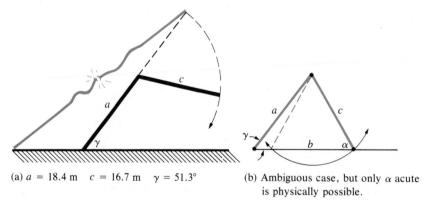

(a) $a = 18.4$ m $c = 16.7$ m $\gamma = 51.3°$

(b) Ambiguous case, but only α acute is physically possible.

Fig. 1 Falling crane

EXAMPLE 1   A cable supporting a large hinged crane snaps (Fig. 1a). How far from the base of the crane does the top hit the ground?

Solution   This is a case of SSA . Since

$$a \sin \gamma = 18.4 \sin 51.3° \approx 14.4 ,$$

we have

$$a \sin \gamma < c < a,$$

so it is the ambiguous case; there are two triangles that fit the data. Obviously we want the triangle with α acute (Fig. 1b). As usual, we compute

$$p = \frac{c}{\sin \gamma} = \frac{16.7}{\sin 51.3°} \approx 21.4$$

and use the law of sines:

$$\sin \alpha = \frac{a}{p} \approx \frac{18.4}{21.4} \approx 0.860 \qquad \alpha \approx 59.3°$$

Then

$$\beta = 180° - \gamma - \alpha \approx 69.4° \qquad b = p \sin \beta \approx 20.0.$$

Answer 20.0 m

EXAMPLE 2   A rangefinder (Fig. 2) sights a target at T. Find the distance d from A to T.

Solution   The angle at T in the big triangle is

$$\gamma = 180° - \alpha - \beta = 0.15°$$

By the law of sines

$$b = \frac{2 \sin \beta}{\sin \gamma} = \frac{2 \sin 100.26°}{\sin 0.15°} \approx 751.73$$

By the law of cosines

$$d^2 = b^2 + 1^2 - 2 \cdot 1 \cdot b \cos \alpha$$

$$\approx (751.73)^2 + 1 - 2(751.73) \cos 79.59° \approx 564827$$

$$d \approx 751.5$$

Check $a = \dfrac{2 \sin \alpha}{\sin \gamma} = \dfrac{2 \sin 79.59°}{\sin 0.15°} \approx 751.37$

$$d^2 \approx (751.37)^2 + 1 - 2(751.37) \cos 100.26° \approx 564826$$

$$d \approx 751.5$$

Answer $d \approx 751.5$ m

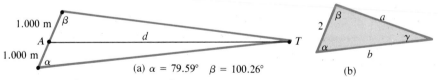

(a) $\alpha = 79.59°$ $\beta = 100.26°$ (b)

Fig. 2 Rangefinder (inaccurate figures)

Indirect Trigonometry Problems

Some geometric problems can be solved by trigonometry, but require several steps and, often, auxiliary constructions.

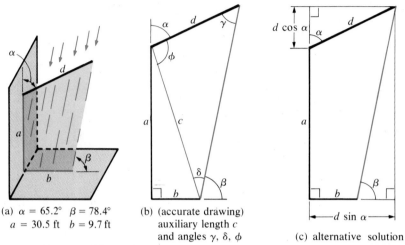

(a) $\alpha = 65.2°$ $\beta = 78.4°$
    $a = 30.5$ ft $b = 9.7$ ft

(b) (accurate drawing)
    auxiliary length c
    and angles γ, δ, ϕ

(c) alternative solution

Fig. 3 Shadow of flagstaff

EXAMPLE 3   A flagstaff on the south wall of a building casts a shadow at noon as shown in Fig. 3a. Find the length d of the flagstaff.

Solution   Since we would like d as the side of a triangle, we introduce c as in Fig. 3b, and mark the angles ϕ, δ, γ. From the shaded right triangle

$$c = \sqrt{a^2 + b^2} = \sqrt{(30.5)^2 + (9.7)^2} \approx 32.0$$

We could find d by the law of sines if we knew the angles δ and γ. Let's find them. First,

$$\beta + \delta + \arctan \frac{a}{b} = 180°$$

so

$$\delta = 180° - \beta - \arctan \frac{a}{b} = 180° - 78.4° - \arctan \frac{30.5}{9.7} \approx 29.2°$$

Next we find γ by first finding ϕ:

$$\phi = 180° - \alpha - \arctan \frac{b}{a} = 180° - 65.2° - \arctan \frac{9.7}{30.5} \approx 97.2°$$

Therefore, since $\delta + \phi + \gamma = 180°$,

$$\gamma = 180° - \delta - \phi \approx 180° - 29.2° - 97.2° = 53.6°.$$

Now we are ready for the law of sines:

$$\frac{d}{\sin \delta} = \frac{c}{\sin \gamma}$$

$$d = \frac{c \sin \delta}{\sin \gamma} \approx \frac{32.0 \sin 29.2°}{\sin 53.6°} \approx 19.4$$

Alternative Solution   Complete the rectangle (Fig. 3c). The horizontal projection of d is $d \sin \alpha$, and the vertical projection of d is $d \cos \alpha$. Thus the shaded right triangle has base $d \sin \alpha - b$ and height $d \cos \alpha + a$. Therefore

$$\frac{d \cos \alpha + a}{d \sin \alpha - b} = \tan \beta$$

Solve for d:

$$d = \frac{a + b \tan \beta}{\sin \alpha \tan \beta - \cos \alpha} = \frac{30.5 + 9.7 \tan 78.4°}{\sin 65.2° \tan 78.4° - \cos 65.2°} \approx 19.4$$

These are not the only ways to solve the problem!

Answer   19.4 ft

Area and the Law of Sines

On p. 117, we found a formula for the area of a triangle (Fig. 4) given SAS. The area is

$$A = \tfrac{1}{2}ab \sin \gamma.$$

Similarly

$$A = \tfrac{1}{2}bc \sin \alpha \quad \text{and} \quad A = \tfrac{1}{2}ac \sin \beta.$$

Area   The area of any triangle is

$$A = \tfrac{1}{2}ab \sin \gamma = \tfrac{1}{2}bc \sin \alpha = \tfrac{1}{2}ac \sin \beta.$$

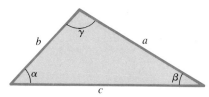
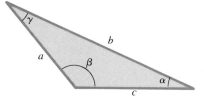

Fig. 4

A proof of the law of sines (postponed in Section 2 until now) drops out of these area formulas. For example, from

$$\tfrac{1}{2}bc \sin \alpha = \tfrac{1}{2}ac \sin \beta$$

follows

$$b \sin \alpha = a \sin \beta \qquad \frac{b}{\sin \beta} = \frac{a}{\sin \alpha}.$$

Similarly, from $\tfrac{1}{2}ab \sin \gamma = \tfrac{1}{2}ac \sin \beta$ follows

$$b \sin \gamma = c \sin \beta \qquad \frac{b}{\sin \beta} = \frac{c}{\sin \gamma}.$$

That's all there is to it!

The area formulas apply to the case SAS. Given SAA or SSS, however, we can compute the area of a triangle by combining the area formulas with the law of sines or the law of cosines.

EXAMPLE 4   Find the area of a triangle given

(a) $a = 5.107$ $\beta = 48.57°$ $\gamma = 26.19°$

(b) $a = 3.804$ $b = 7.112$ $c = 9.065$

Solution   (a)   The third angle is

$$\alpha = 180° - \beta - \gamma = 180° - 48.57° - 26.19° = 105.24°.$$

To apply one of the area formulas, we need another side, which we can find by the law of sines. For instance, let's find b:

$$\frac{b}{\sin \beta} = \frac{a}{\sin \alpha}$$

$$b = \frac{a \sin \beta}{\sin \alpha} = \frac{(5.107) \sin 48.57°}{\sin 105.24°} \approx 3.969.$$

Now it follows that

$$A = \tfrac{1}{2} ab \sin \gamma \approx \tfrac{1}{2}(5.107)(3.969) \sin 26.19° \approx 4.473.$$

(b)   To apply one of the area formulas, this time we need an angle (actually the sine of an angle). Let's find $\sin \gamma$. From the law of cosines,

$$\cos \gamma = \frac{a^2 + b^2 - c^2}{2ab} = \frac{(3.804)^2 + (7.112)^2 - (9.065)^2}{2(3.804)(7.112)} \approx -0.3165$$

$$\sin \gamma = \sqrt{1 - \cos^2 \gamma} \approx 0.9486.$$

Therefore

$$A = \tfrac{1}{2} ab \sin \gamma \approx \tfrac{1}{2}(3.804)(7.112)(0.9486) \approx 12.83$$

Answer   (a)   4.473   (b)   12.83

Heron's Formula

An ancient Greek geometer (probably Archimedes) found a beautiful formula for the area of a triangle in terms of its three sides.

> **Heron's Formula**   Set $s = \tfrac{1}{2}(a + b + c)$. Then
> $$A = \sqrt{s(s - a)(s - b)(s - c)}.$$

Let's sketch a proof. We start with $A = \tfrac{1}{2} ab \sin \gamma$ and square:

$$4A^2 = a^2 b^2 \sin^2 \gamma = a^2 b^2 (1 - \cos^2 \gamma) = a^2 b^2 - (ab \cos \gamma)^2$$

We multiply by 4:

$$16A^2 = 4a^2 b^2 - (2ab \cos \gamma)^2,$$

and use the law of cosines,

$$2ab \cos \gamma = a^2 + b^2 - c^2.$$

We find

$$16A^2 = 4a^2b^2 - (a^2 + b^2 - c^2)^2.$$

Factoring the right-hand side yields

$$16A^2 = (a + b + c)(a + b - c)(b + c - a)(c + a - b).$$

Now

$$a + b + c = 2s, \quad a + b - c = 2s - 2c, \quad \text{etc.,}$$

and we have

$$16A^2 = 16s(s - a)(s - b)(s - c),$$

from which Heron's formula follows.

Example $a = 3.804 \quad b = 7.112 \quad c = 9.065$

$$s = \tfrac{1}{2}(3.804 + 7.112 + 9.065) = 9.9905$$

$$s - a = 6.1865 \qquad s - b = 2.8785 \qquad s - c = 0.9255$$

$$s(s - a)(s - b)(s - c) \approx 164.655$$

$$A \approx \sqrt{164.655} \approx 12.832$$

Compare Example 4b.

Exercises

1 Two forest rangers 10.0 miles apart at points A and B observe a fire (Fig. 5). The ranger at A finds $\alpha = 52.6°$. The ranger at B finds $\beta = 31.5°$. How far is the fire from each ranger? From the line through A and B?

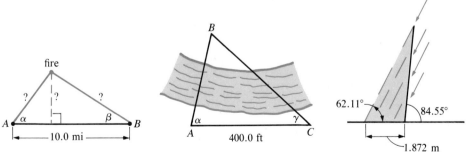

Fig. 5 $\alpha = 52.6°$ $\beta = 31.5°$ **Fig. 6** $\alpha = 77.3°$ $\gamma = 41.8°$ **Fig. 7**

2 Two points A and B are on opposite sides of a river (Fig. 6). A third point C is 400.0 ft from A. If $\alpha = 77.3°$ and $\gamma = 41.8°$, find the distance from A to B.

3 Find the length of the tilted pole (Fig. 7).

4 Find $\overline{CD}$ in Fig. 8, next page.

5 Find $\overline{CD}$ in Fig. 9, next page.

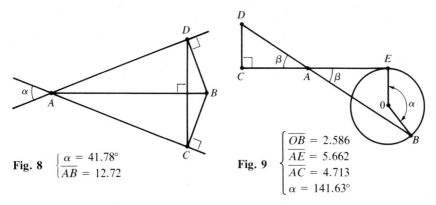

Fig. 8 $\begin{cases} \alpha = 41.78° \\ \overline{AB} = 12.72 \end{cases}$

Fig. 9 $\begin{cases} \overline{OB} = 2.586 \\ \overline{AE} = 5.662 \\ \overline{AC} = 4.713 \\ \alpha = 141.63° \end{cases}$

6   Refer to Example 2. Suppose $\alpha = 83.16°$ and $\beta = 96.73°$. Now what is the distance to the target?

7   A vertical telephone pole stands by the side of a road that slopes at an angle of $8.5°$ with the horizontal. When the angle of elevation of the sun is $61.3°$, the pole casts a shadow parallel to the road, downhill, and 13.2 ft long. How high is the pole?

8   (cont.) Suppose instead that the shadow is cast uphill. Then how high is the pole?

9   Find $\overline{BD}$ in Fig. 10.

10   Find $\overline{AB}$ in Fig. 11 to 4 significant figures.

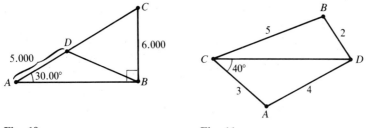

Fig. 10   **Fig. 11**

11   Find $\overline{AB}$ and $\overline{BC}$ in Fig. 12 to 5 significant figures.

12*   Find $\theta°$ in Fig. 13 to 4 significant figures.

13   Find $\overline{DC}$ and $\overline{EC}$ in Fig. 14 to 6 places.

14*   (cont.) Now find $\overline{DE}$, $\angle DEC$, and θ. Surprised?

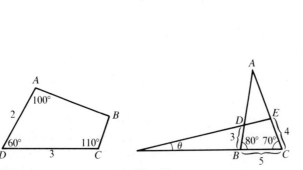

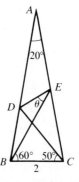

Fig. 12   **Fig. 13**   **Fig. 14** (*ABC* isosceles)

Find the area of the triangle, given

15	$b = 3$	$c = 5$	$\alpha = \frac{1}{6}\pi$	**16**	$b = 4$	$c = 1$	$\alpha = \frac{1}{4}\pi$
17	$c = 2.807$	$a = 1.442$	$\beta = 27.91°$	**18**	$c = 12.66$	$a = 17.41$	$\beta = 119.28°$
19	$a = 108.6$	$b = 349.7$	$\gamma = 156.34°$	**20**	$a = 7.803$	$b = 4.441$	$\gamma = 81.19°$
21	$a = 12.66$	$\beta = 15.29°$	$\gamma = 100.31°$	**22**	$a = 1069$	$\beta = 11.93°$	$\gamma = 16.01°$
23	$c = 5.131$	$\alpha = 63.88°$	$\beta = 41.70°$	**24**	$c = 15.31$	$\alpha = 63.88°$	$\beta = 41.70°$
25	$a = 16.52$	$b = 12.77$	$c = 19.51$	**26**	$a = 8.413$	$b = 2.718$	$c = 10.145$
27	$a = 506.1$	$b = 682.3$	$c = 415.5$	**28**	$a = 3004$	$b = 5172$	$c = 6131$
29	$a = 19.33$	$c = 17.61$	$\gamma = 55.22°$	**30**	$b.= 4.883$	$c = 6.107$	$\beta = 19.66°.$

Find the area by Heron's formula

31	$a = 11.26$	$b = 18.81$	$c = 23.44$	**32**	$a = 4.800$	$b = 7.343$	$c = 5.103$
33	$a = 703.1$	$b = 698.9$	$c = 721.5$	**34**	$a = 69.96$	$b = 39.74$	$c = 28.77$

35   The 8×8 square in Fig. 15a is cut into 4 pieces. The pieces are fitted together in Fig. 15b. But the square has area 64 and the rectangle 65. Explain this paradox.

36   Find the area to 3 places of triangle ABC in Fig. 16. Its vertices are on 3 vertical rods.

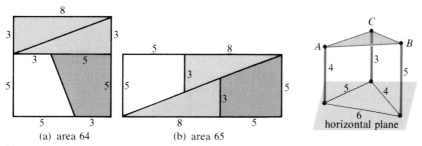

Fig. 15   (a) area 64   (b) area 65

Fig. 16   horizontal plane

4   MEASUREMENT AND ACCURACY [Optional]

We first describe briefly the tools currently used to measure angles and distances, and their accuracy. Then we look at some examples of how sensitive quantities computed by trigonometry are to small errors in the measured data.

Divided Circle

One of the most basic tools of angle measurement is the protractor, or equally divided circumference. If we assume that we can distinguish clearly rulings that are $\frac{1}{2}$ mm apart, then a divided circumference that distinguishes degrees must have diameter at least

$$\frac{360 \times 0.05}{\pi} \approx 5.73 \text{ cm} \approx 2\frac{1}{4} \text{ in}.$$

But if we want the division marks to represent minutes, then we need a diameter of 60 times this, that is, about 60×5.73 cm ≈ 3.44 m $\approx 11\frac{1}{4}$ ft. It is unrealistic to make an accurate divided circle this large.

Astronomical Measurements

Astronomers are very dependent on highly accurate angle measurements. These measurements are used indirectly to determine distances. Telescopes commonly have divided circles attached to read the azimuth (horizontal) angle and the elevation (vertical) angle of rotation. These divided circles are read with microscopes, and the usual accuracy is to 0.03″ with a maximum error of ±0.003″. The most accurate of these devices, used for elevation angles only, is called a **meridian circle.** It is generally about 1 m in diameter and can be read to 0.01″. The reading is indirect and painstaking to avoid errors.

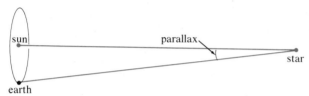

Fig. 1 Measurement of parallax

The distance to nearby stars is determined by measuring an angle called **parallax** (Fig. 1). It is the angle at the star subtended by the average radius of the Earth's orbit. For the *nearest* star, α-Centauri, the parallax is a mere 0.760″.

The mean radius of the Earth's orbit is called an **astronomical unit** (A.U.), and

$$1 \text{ A.U.} \approx 1.495979 \times 10^8 \text{ km .*}$$

Other important units of distance in astronomy are the **light year** (ly), where

$$1 \text{ ly} \approx 9.460530 \times 10^{12} \text{ km} \approx 6.32397 \times 10^4 \text{ A.U.}$$

and the **parsec.** One parsec (pc) is the distance from the Sun to a point in space with parallax 1″. Thus

$$1 \text{ pc} \approx \frac{1 \text{ A.U.}}{\tan 1''} \approx 3.085678 \times 10^{13} \text{ km} \approx 3.26163 \text{ ly .}$$

By measuring parallax, distances up to 30 parsecs can be estimated within a 10% error.

Distances to the Sun and the nearby planets can be measured more accurately by using reflected radar signals. The most accurate distance measurement in space is the reflection of a laser beam off a reflector on the Moon. This measurement of about 3.8×10^5 km is accurate to ±6 cm .

Machine Shop Measurements

Let us start with distance measurements. Here are three common measuring devices with their ranges and accuracies (for high quality models):

	range	error
Vernier caliper gauge	0–100 mm	±0.025 mm
Dial gauge	0–25 mm	±0.0025 mm
Micrometer gauge	0–25 mm	±0.004 mm

---

* If you are unfamiliar with scientific notation, please refer to page 266.

For very accurate distance measurements, **gauge blocks** are used. These are metal rectangular solids, two of whose faces are ground parallel at a measured distance to high accuracy. They must be used at 20 °C to avoid errors of thermal expansion. For "grade 0" gauge blocks the range and errors are

range	error
0—20 mm	$\pm 10^{-4}$ mm
. . . .	. . . .
80—100 mm	$\pm 2.5 \times 10^{-4}$ mm

To be able to measure any distance from 0 to 10 cm by increments of, say, 0.002 mm would seem to require $100/0.002 = 50{,}000$ gauge blocks! Actually one simply "wrings" together (stacks up) pieces of standard sizes to make desired heights. A standard 112-piece set consists of the following:

Lengths (in mm)	Number of pieces
1.0005	1
1.001 , 1.002 , 1.003 , $\cdots$, 1.009	9
1.01 , 1.02 , 1.03 , $\cdots$, 1.49	49
0.5 , 1.0 , 1.5 , 2.0 , $\cdots$, 24.5	49
25 , 50 , 75 , 100	4
	total: 112

For example,

$$34.3755 = 1.0005 + 1.005 + 1.37 + 6 + 25 .$$

Of course, there are many other devices used in precision machining. One device, called a coordinate cathetometer, combines a microscope and micrometer adjustments in two directions. It is used for two-dimensional measurements in a vertical plane. An instrument with a range of 75 mm $\times$ 75 mm is typically accurate to ± 0.001 mm.

Angles are measured in various ways. A precision protractor that is about 23 cm in diameter will have circumference ≈ 720 mm, so $\frac{1}{2}°$ will be indicated by division marks 1 mm apart. By adding a micrometer or vernier for fine adjustment, you can easily interpolate to $1'$ (or less) with about $\pm \frac{1}{10}'$ error.

For highly accurate angle measurements, a set of **angle gauge blocks** are used. Since they can be wrung together either to add or *to subtract* angles, very few are needed in a set. A standard 13 (thirteen!) piece set consists of

$$1° \quad 3° \quad 9° \quad 27° \quad 41° \quad 90° \quad 1' \quad 3' \quad 9' \quad 27' \quad 3'' \quad 9'' \quad 27''$$

The error is $\pm 0.5''$ for $1°{-}90°$ and $\pm 0.3''$ for $3''{-}27'$. Any angle can be "wrung" in increments of $3''$, with interpolation to $1\frac{1}{2}''$. See Fig. 2, next page, for an example of "wringing."

A 16-block set can produce all 324,000 angles from 0 to 90° in $1''$ increments:

$$1° \quad 3° \quad 5° \quad 15° \quad 30° \quad 45° \quad 1' \quad 3' \quad 5' \quad 20' \quad 30' \quad 1'' \quad 3'' \quad 5'' \quad 20'' \quad 30''$$

We mentioned the **sine bar** in the exercises on p. 119. Used with a set of (distance) gauge blocks, a 250 mm sine bar measures angles in the range 0°–30° with an accuracy of $\pm 1''$.

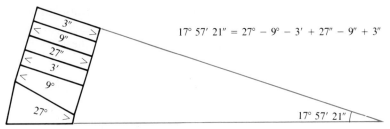

Fig. 2 Example of angle gauge blocks wrung together

Laboratory Distance Measurements

The ultimate distance measurement instrument is the **interferometer** with a laser light source. It measures the interference pattern of a split light beam, one half of which is reflected off the target. Accuracies of 3 parts in 10^8 are possible. That means that in a measurement of 1 m, the error is $\pm 3 \times 10^{-8}$ m.

Angle Measurement in Surveying

A surveyor's **transit** measures horizontal angles at $20''$ intervals; the error is $\pm 10''$. The (generally) more accurate **theodolite** measures both horizontal and vertical angles at $1''$ intervals, so the accuracy is about $\pm\frac{1}{2}''$. By averaging several readings, accuracies to $0.1''$ are possible. We shouldn't forget the mariner's **sextant,** which measures vertical angles with an accuracy of $\pm 12''$, usually from the horizon to a star (*sextant* because its scale is $\frac{1}{6}$ of a circle). These instruments are entirely optical.

There is a new generation of electronic instruments using infrared beams, laser beams, and microwave radar. They are not more accurate than the best theodolites, but are much easier to use as they have digital readouts and data recording possibilities.

Distance Measurement in Surveying

Most basic is the good old steel tape, in lengths of 30 to 300 m, calibrated in cm or mm. At standard temperature and tension, the tape is accurate to about 1 part in 10,000. But thermal expansion, sag, uneven ground, etc. contribute to a possible large error.

Surveyors commonly use the angle subtended by a measured length at the target to estimate distance. Exs. 26–32 on p. 120 illustrate the principles. For accurate distance measurements, the standard is a 2-meter **subtense bar,** so called because you measure the angle it subtends. The main property of a subtense bar is that its length is nearly constant over a wide range of temperatures. This is accomplished by a clever design using metals with different coefficients of thermal expansion. A good subtense bar has length 2 ± 0.003 m over the temperature range from $-10\,°C$ to $50\,°C$.

The new generation electronic measuring instruments are remarkably precise, easy to use, and expensive! We list in Table 1 some instrument types, their ranges and accuracies.

Table 1

Type	Beam source	Range	Accuracy
E.O.	Infrared	0–1.6 km	±3 mm
E.O.	Infrared	0–10 km	±5 mm ± 1 mm/km
E.O.	Laser	0–60 m	±10⁻⁵ mm
E.O.	Laser	0–45 km	±5 mm ± 1 mm/km
EDMI	Microwave	50–2000 m	±10 mm
EDMI	Microwave	2–30 km	Depends on weather

E.O. = electro-optical
EDMI = electronic distance measuring instrument

A model of the third instrument listed, also capable of accurate angle measurement, sells for about $50,000.

Examples of Sensitivity

EXAMPLE 1 A surveyor estimates a distance (Fig. 3) by a subtense measurement. He measures 1°13′15.6″ for the angle subtended by the subtense bar.

(a) Find his distance estimate for $h = \overline{AB}$.

(b) Suppose his angle measurement has an error of ±0.3″. How does this affect his distance estimate?

Solution (a) By the upper right triangle,

$$\frac{1}{h} = \tan\tfrac{1}{2}\theta \qquad h = \frac{1}{\tan\tfrac{1}{2}\theta} = \frac{1}{\tan 36′37.8″} \approx 93.847 \text{ m}.$$

(b) If the reading of θ is too large, then θ should be 1°13′15.6″ − 0.3″ = 1°13′15.3″, so

$$h \approx \frac{1}{\tan\tfrac{1}{2}(1°13′15.3″)} = \frac{1}{\tan 36′37.65″} \approx 93.853 \text{ m}.$$

If the reading of θ is too small, then θ = 1°13′15.6″ + 0.3″ = 1°13′15.9″, so

$$h \approx \frac{1}{\tan\tfrac{1}{2}(1°13′15.9″)} = \frac{1}{\tan 36′37.95″} \approx 93.841 \text{ m}.$$

Now

$$93.853 - 93.847 = 0.006 \text{ m} \qquad 93.847 - 93.841 = 0.006 \text{ m}.$$

Answer (a) 93.847 m   (b) by ±6 mm.

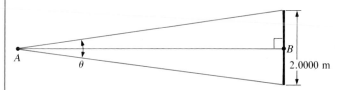

Fig. 3 $\theta = 1° 13′ 15.6″$

EXAMPLE 2 In the triangle the measured parts are

$a = 4$ $b = 6$ $\gamma = 35°$.

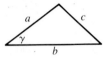

Fig. 4

(a) Compute c to 6 places.

(b) Suppose the measurements are accurate to 5 parts in 10,000. Find by how much c can differ from the value calculated in (a) in parts per 10,000.

Solution (a) By the law of cosines

$$c = \sqrt{4^2 + 6^2 - 2 \cdot 4 \cdot 6 \cdot \cos 35°} \approx 3.560997$$

(b) It is clear from the figure that c increases if a decreases, if b increases, or if γ increases. Therefore the worst case for an increase in c occurs for

$a = 4 \cdot (1 - 5 \times 10^{-4}) = 3.998$ $b = 6 \cdot (1 + 5 \times 10^{-4}) = 6.003$

$\gamma = 35 \cdot (1 + 5 \times 10^{-4}) = 35.0175°$.

In this case the law of cosines yields

$c = 3.564987$.

The relative increase from the previous c is

$$\frac{3.564987 - 3.560997}{3.560997} \approx 0.00112 < 0.0012$$

Similarly, the worst case for a decrease in c occurs for

$a = 4 \cdot (1 + 5 \times 10^{-4}) = 4.002$ $b = 6 \cdot (1 - 5 \times 10^{-4}) = 5.997$

$\gamma = 35 \cdot (1 - 5 \times 10^{-4}) = 34.9825°$.

The corresponding c is 3.557010 and the relative decrease is

$$\frac{3.560997 - 3.557010}{3.560997} \approx 0.00112 < 0.0012$$

Answer (a) 3.560997   (b) 12 parts in 10,000

Exercises

1 Find the diameter of a circle graduated in seconds, if the division points are $\frac{1}{2}$ mm apart.

2 An eye with good vision can distinguish points that subtend $1'$. From how far away can this eye distinguish two points the same distance from the eye and 1 cm apart? One meter apart?

3 How many A.U. are there in one parsec?

4 How many parsecs are there in one light year?

5 Show that $1/\alpha \approx d$, where α is the parallax in radians and d the distance in A.U. to a star.

6 Show that $1/\alpha \approx d$, where α is the parallax in seconds and d the distance in parsecs to a star.

Fill in the missing entries for the eight nearest stars to our Sun (use Ex. 6 and the conversion factors in the text).

	Star	Parallax (")	pc	ly	km
7	α-Centauri	0.760"			
8	Barnard's	0.552"			
9	Wolf 359		2.32		
10	Lalande 21185		2.49		
11	Sirius			8.65	
12	Luyten 726-8			8.94	
13	Ross 154				8.94×10^{13}
14	Ross 248				9.73×10^{13}

Show how to make up the given length from the standard 112-piece gauge block set (p. 145)

15 153.976 mm   **16** 81.043 mm

Show how to make up the given angle from the standard 13-piece set of angle gauge blocks (p. 145)

17 82°   **18** 70°

19 14°38′51″   **20** 60°20′30″

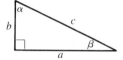

Fig. 5

Fig. 6

21 In Fig. 5, suppose $a = 4$ and $b = 3$. Find c. Now suppose a and b can be in error by one part in 1000. Find the maximum error in c.

22 In Fig. 5, suppose $a = 4$ and $\beta = 25°$. Find c. Now suppose β can be in error by $\pm 10′$ and a by ± 0.003. Find the maximum error in c.

23 In Fig. 6, suppose $a = 3$, $b = 4$, and $\gamma = 118°$. Find c. Now suppose a, b, and γ can be in error by one part in 1000. Find the worst possible error in c in parts per 1000.

24 In Fig. 6, suppose $a = 4$, $b = 6$, and $\gamma = 123° \pm 5′$. Find the possible error in β in minutes.

25 In Fig. 6, suppose $b = 10$, $\alpha = 21°$, and $\gamma = 105°$. Find the worst possible error in c due to $\frac{1}{10}$ of 1% errors in α, b, and γ.

26 In Fig. 6, suppose $\alpha = 24°$, $c = 7$, $a = 3$ and γ is obtuse as drawn. If α has a possible error of 2 parts in 1000, find the worst possible error in b in parts per 1000.

27 The gunsights on a rifle are 25 in. apart, and the rifle is aimed at a target 170 ft from the near sight. Suppose the far sight is $\frac{1}{10}$ in. out of line. How much error will this cause?

28 (cont.) Suppose instead the near sight is $\frac{1}{10}$ in. out of line. Now how much error?

REVIEW EXERCISES

1 The hour hand of a clock is 3.7 cm long and the minute hand 5.2 cm. At a certain instant the ends of the hands are exactly 8 cm apart. What is the angle between the hands at that instant?

2 An oblique triangle has

$$a = 4.308 \qquad b = 3.977 \qquad \gamma = 118.52°$$

Find the length of the altitude on side c.

3 (cont.) Find the length of the median to side c.

Find the remaining parts; give all solutions and check

4 $\alpha = 42.78°$ $\qquad\qquad$ $\beta = 97.71°$ $\qquad\qquad$ $c = 112.4$

5 $a = 3.466$ $\qquad\qquad$ $c = 4.991$ $\qquad\qquad$ $\alpha = 14.20°$

6 $b = 146.2$ $\qquad\qquad$ $c = 131.4$ $\qquad\qquad$ $\beta = 127.82°$

7 A flagpole is tilted 15.0° to the west. When the sun is due south and making a 38.4° angle with the horizon, the shadow of the flagpole has length 4.60 m. How long is the flagpole? [*Hint* Draw a careful picture; express the shadow length in terms of the pole length.]

8* (cont.) Solve the same problem if the tilt is not due west, but is 21.0° south of west.

9 In Figure 1, the measurements are $a = 3$ $\quad b = 4$ $\quad \gamma = 25°$, accurate to 1%. Find c and the possible error in c.

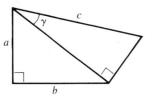

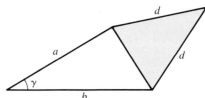

Fig. 1 $\qquad\qquad\qquad\qquad\qquad$ **Fig. 2**

10 In Figure 2, the measurements are $a = 5$ $\quad b = 6$ $\quad \gamma = 30°$ $\quad d = 4$, accurate to one part in 50. Find the shaded area, and its possible error. (You may assume that the triangle is truly isosceles; both measurements of d give equal values.)

7 INVERSE TRIGONOMETRIC FUNCTIONS

1 INVERSE FUNCTIONS

We wish to invert the relation $y = \sin \theta$ to express θ as a function of y, namely $\theta = \arcsin y$, and do the same for the other trigonometric functions. This requires a preliminary digression into the concept of inverse functions, the topic of this and the next section. We'll split the work so that this section will be descriptive of the theory of inverting functions, and Section 2 will contain practical work on examples and graphs.

The Concept of an Inverse Function

We start with an easy example, $f(x) = 2x$. Suppose x is given. The function f assigns to x the number $y = 2x$. Now suppose that not x, but y is given. Then to y corresponds $x = \frac{1}{2}y$. This correspondence is a function, $g(y) = \frac{1}{2}y$. It is obtained by solving $y = 2x$ for x in terms of y.

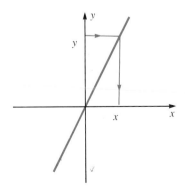

Fig. 1 The graph of $y = 2x$ assigns a y to each x. But it also assigns an x to each y!

.ok at the graph of $y = 2x$ in Fig. 1. To each x on the x-axis corresponds a
.que y on the y-axis. But seen from the y-axis, the graph assigns to each y a
unique x on the x-axis. Thus, related to the function $y = f(x)$ there is a function
$x = g(y)$. The line $y = 2x$ is the graph of either of these functions; which one
depends on your point of view.

The function f takes you from x to y. Then g takes you from this y back to
x. Thus

$$g[f(x)] = \tfrac{1}{2}f(x) = \tfrac{1}{2}(2x) = x.$$

Similarly, g takes you from y to x, then f takes you from x back to y:

$$f[g(y)] = 2g(y) = 2(\tfrac{1}{2}y) = y.$$

The actions of f and g neutralize each other. Apply one after the other and you
are back where you started. In the language of composite functions,

$$(g \circ f)(x) = x \qquad (f \circ g)(y) = y.$$

Inverse Function Suppose $f(x)$ and $g(y)$ are functions such that

$g[f(x)] = x$ for all x in the domain of f

$f[g(y)] = y$ for all y in the domain of g.

Then g is the **inverse function** of f and f is the **inverse function** of g.

For a black box interpretation, see Fig. 2.

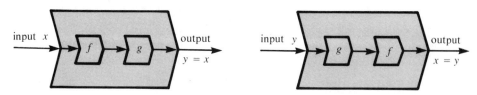

Fig. 2 Inverse functions: $\begin{cases} (g \circ f)(x) = x \\ (f \circ g)(y) = y \end{cases}$

Examples

(1) $f(x) = 2x - 5 \qquad g(y) = \tfrac{1}{2}(y + 5).$

Then $g[f(x)] = \tfrac{1}{2}[f(x) + 5] = \tfrac{1}{2}[(2x - 5) + 5] = x$

and $f[g(y)] = 2g(y) - 5 = 2 \cdot \tfrac{1}{2}(y + 5) - 5 = y.$

(2) $f(x) = x^3 \qquad g(y) = y^{1/3}.$

Then $g[f(x)] = [f(x)]^{1/3} = (x^3)^{1/3} = x$

and $f[g(y)] = [g(y)]^3 = (y^{1/3})^3 = y.$

In our examples so far, we have found the inverse function $x = g(y)$ by
solving $y = f(x)$ explicitly for x in terms of y. But difficulties can arise. We
may not be able to solve for x, as in the case of $y = x^7 + 3x$. Or, as in the case

of $y = x^2$, there may be more than one value of x for a given y, which is not allowed in the definition of a function. Thus, it may not be clear whether a given function has an inverse. We need a closer look at the question.

Domains and Ranges

We shall have to be very precise about the domains and ranges of our function. The **domain** of a function is the set of points on which it is defined. When specifying a function, we sometimes explicitly state its domain. Other times, we just give a formula for the function without mentioning a domain. Then it is understood that the domain is the set of all real numbers for which the formula makes sense. For instance, we can define $f_1(x) = \sqrt{x}$ on the domain $\{1 \leq x < 2\}$. Then f_1 is a function whose domain is $\{1 \leq x < 2\}$. We can also define f_2 by $f_2(x) = \sqrt{x}$ without giving a domain. Then it is understood that the domain of f_2 is the set of all non-negative real numbers, since that is where $\sqrt{x}$ makes sense. The functions f_1 and f_2 are *not* equal because they have different domains.

We haven't used the range of a function much, so let's recall that the **range** of f is the set of all values of f, that is, the set of all real numbers $f(x)$, where x is in the domain of f.

It is useful to have the notation

dom(f)	domain of f
range(f)	range of f.

Intervals

We'll be dealing with functions whose domains and ranges are intervals. Recall that an **interval** is the set of all real numbers between two given real numbers, possibly including one or both of the end points. Examples are

$$\{x \mid -2 < x < \sqrt{19}\} \quad \{x \mid -2 \leq x \leq \sqrt{19}\}$$
$$\{x \mid -2 \leq x < \sqrt{19}\} \quad \{x \mid -2 < x \leq \sqrt{19}\}.$$

We'll use the shorter set notation

$$\{-2 < x < \sqrt{19}\} \quad \{-2 \leq x \leq \sqrt{19}\} \quad \text{etc.}$$

We allow also infinite intervals where the "end points" may be $-\infty$ or $+\infty$. Examples are

$$\{-\infty < x < \infty\} \quad \{-\infty < x < 2\} \quad \{10 \leq x < \infty\}$$

which we abbreviate as

$$\{\text{all } x\} \quad \{x < 2\} \quad \{10 \leq x\}.$$

Betweenness Property   One property distinguishes intervals from all other sets on the line: if an interval contains two points a and b, then it contains all points between a and b.

This is proved in more advanced courses.

Existence of Inverse Functions

Examine the graphs in Fig. 3. Each of the functions shown has as domain an interval **I** and as range an interval **J** . In (a) and (b), it is clear that to each y in **J** corresponds a unique x in **I**. Hence an inverse function exists. But in (c), there is a y in **J** to which corresponds several points x in **I**. Hence no inverse function can exist.

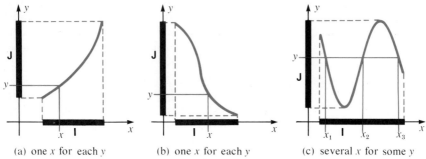

(a) one x for each y (b) one x for each y (c) several x for some y

Fig. 3   The domain **I** is the set on which the function is *defined*.
The range **J** is the set of *values* of the function.

With these examples in mind, we now consider functions $f(x)$ with three properties:

(1)   The domain of $f(x)$ is an interval **I**.

(2) $f(x)$ is **strictly increasing,** that is, whenever x_1 and x_2 are points of **I** and $x_1 < x_2$, then $f(x_1) < f(x_2)$.

(3)   the range of $f(x)$ is an interval **J**.

Fig. 4 shows examples of such functions. Note their domains and ranges.

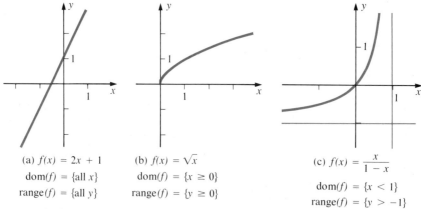

(a) $f(x) = 2x + 1$ (b) $f(x) = \sqrt{x}$ (c) $f(x) = \dfrac{x}{1-x}$
dom(f) = {all x}   dom(f) = {$x \geq 0$}
range(f) = {all y}   range(f) = {$y \geq 0$}   dom(f) = {$x < 1$}
                                                                range(f) = {$y > -1$}

Fig. 4   Strictly increasing functions

We also consider functions for which condition (2) is replaced by

(2′) $f(x)$ is **strictly decreasing,** that is, whenever x_1 and x_2 are points of **I** and $x_1 < x_2$, then $f(x_1) > f(x_2)$.

Examples of such functions are the negatives of the functions in Fig. 4. Further examples are shown in Fig. 5.

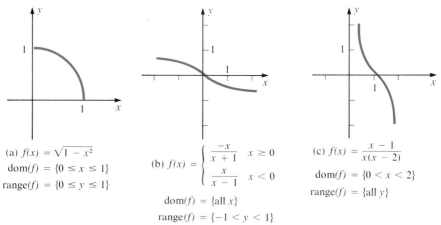

(a) $f(x) = \sqrt{1 - x^2}$
dom$(f) = \{0 \leq x \leq 1\}$
range$(f) = \{0 \leq y \leq 1\}$

(b) $f(x) = \begin{cases} \dfrac{-x}{x+1} & x \geq 0 \\[2mm] \dfrac{x}{x-1} & x < 0 \end{cases}$

dom$(f) = \{$all $x\}$
range$(f) = \{-1 < y < 1\}$

(c) $f(x) = \dfrac{x-1}{x(x-2)}$
dom$(f) = \{0 < x < 2\}$
range$(f) = \{$all $y\}$

Fig. 5   Strictly decreasing functions

We now come to the main point of this section. It is a theorem that guarantees the existence of an inverse function for any function that satisfies (1), (2) or (2′), and (3).

Existence of Inverse Functions

Let $f(x)$ be a strictly increasing function whose domain and range are intervals **I** and **J** respectively. Then there exists a unique inverse function $g(y)$ with domain **J** and range **I** such that

$$f[g(y)] = y \text{ for all } y \text{ in } \mathbf{J},$$

and

$$g[f(x)] = x \text{ for all } x \text{ in } \mathbf{I}.$$

The function $g(y)$ is strictly increasing.

The corresponding statements hold if $f(x)$ is strictly decreasing.

This theorem is proved in advanced courses in analysis. The proof depends on the Betweenness Property of intervals. Notice that the theorem is only an existence statement; it does not give any formula for the inverse function $g(y)$ or any method for computing it.

2 EXAMPLES AND GRAPHS

EXAMPLE 1 Let $f(x) = ax + b$ $(a > 0)$. Show that $f(x)$ has an inverse function $g(y)$. Compute $g(y)$ and give its range and domain.

Solution Obviously $f(x)$ is strictly increasing and dom$(f) = $ range(f)

= {all reals}. Therefore it has an inverse function $g(y)$ with dom(g) = range (g) = {all reals}. To compute $g(y)$, solve $y = ax + b$ for x:

$$x = \frac{1}{a}y - \frac{b}{a}, \qquad g(y) = \frac{1}{a}y - \frac{b}{a}.$$

Check

$$f[g(y)] = ag(y) + b = a\left(\frac{1}{a}y - \frac{b}{a}\right) + b = y$$

$$g[f(x)] = \frac{1}{a}f(x) - \frac{b}{a} = \frac{1}{a}(ax + b) - \frac{b}{a} = x.$$

Answer $g(y) = \frac{1}{a}y - \frac{b}{a}$ dom(g) = range(g) = {all reals}

EXAMPLE 2   Let $f(x) = x^2$ dom(f) = {$0 \leq x$}. Show that $f(x)$ has an inverse function $g(y)$. Compute $g(y)$ and give its range and domain.

Solution   If $0 \leq x_1 < x_2$, then $x_1^2 < x_2^2$. Hence $f(x)$ is strictly increasing. Its range is the interval {$0 \leq y$}. Therefore $f(x)$ has an inverse $g(y)$ with domain {$0 \leq y$} and range {$0 \leq x$}. Solve $y = x^2$ for x:

$$x = \sqrt{y}, \quad \text{that is,} \quad g(y) = \sqrt{y} \quad y \geq 0.$$

Check

$$f[g(y)] = [g(y)]^2 = (\sqrt{y})^2 = y$$

$$g[f(x)] = \sqrt{f(x)} = \sqrt{x^2} = x.$$

Answer $g(y) = \sqrt{y}$ dom(g) = range(g) = {reals ≥ 0}

Remark   The natural domain of $f(x) = x^2$ is {all x}. But on this domain, $f(x)$ has no inverse function. That's because $y = x^2$ implies $x = \pm\sqrt{y}$, two values of x for each positive y. To remove the difficulty, we restricted the domain of $f(x) = x^2$ to {$x \geq 0$}. Then $x = +\sqrt{y}$ only, so an inverse function exists. Geometrically speaking, we cut down the domain of $f(x)$ to the largest sub-domain on which the function is strictly increasing.

EXAMPLE 3   Let $f(x) = \frac{1}{x}$ dom(f) = {$x > 0$}. Show that $f(x)$ has an inverse function $g(y)$. Compute $g(y)$ and give its range and domain.

Solution   If $0 < x_1 < x_2$, then $1/x_1 > 1/x_2$ so $f(x)$ is strictly decreasing. Its range is {$0 < y$}. Hence $f(x)$ has an inverse function $g(y)$ with dom(g) = {$0 < y$} and range(g) = {$0 < x$}. Solve $y = 1/x$ for x:

$$x = g(y) = 1/y.$$

Check

$$f[g(y)] = \frac{1}{g(y)} = \frac{1}{1/y} = y \qquad g[f(x)] = \frac{1}{f(x)} = \frac{1}{1/x} = x.$$

Answer $g(y) = \frac{1}{y}$ dom(g) = range(g) = {positive reals}

EXAMPLE 4   Let $f(x) = \dfrac{x}{1-x}$ $\operatorname{dom}(f) = \{x < 1\}$. Show that $f(x)$ has an inverse function $g(y)$. Compute $g(y)$ and give its range and domain.

Solution   Rewrite $f(x)$:

$$f(x) = \frac{(x-1)+1}{1-x} = -1 + \frac{1}{1-x}.$$

As x increases on the interval $-\infty < x < 1$, then $1 - x$ decreases from ∞ to 0. Hence $1/(1-x)$ increases from 0 to ∞. As a result, $f(x)$ is a strictly increasing function and its values run from -1 to ∞. See Fig. 4c, page 154.

According to the theorem, $f(x)$ has an inverse function $g(y)$ with $\operatorname{dom}(g) = \{-1 < y\}$ and $\operatorname{range}(g) = \{x < 1\}$. To compute $g(y)$, solve $y = f(x)$ for x:

$$y = \frac{x}{1-x} \qquad y(1-x) = x \qquad y - yx = x.$$

Hence

$$y = yx + x = (y+1)x \qquad x = g(y) = \frac{y}{y+1}.$$

Check

$$f[g(y)] = \frac{g(y)}{1-g(y)} = \frac{\dfrac{y}{y+1}}{1 - \dfrac{y}{y+1}} = \frac{y}{(y+1)-y} = y$$

$$g[f(x)] = \frac{f(x)}{f(x)+1} = \frac{\dfrac{x}{1-x}}{\dfrac{x}{1-x}+1} = \frac{x}{x+(1-x)} = x.$$

Answer $g(y) = \dfrac{y}{y+1}$ $\operatorname{dom}(g) = \{-1 < y\}$ $\operatorname{range}(g) = \{x < 1\}$

In the examples above, $f(x)$ was simple enough that we could solve $y = f(x)$ for $x = g(y)$. But that is not always the case. For example, take $y = x^7 + 3x$. This function is strictly increasing, so our theorem guarantees the existence of an inverse function $g(y)$. But there is no formula for x in terms of y. Even approximating $g(1)$ isn't easy. It amounts to approximating the real solution of

$$x^7 + 3x = 1.$$

(Actually $x \approx 0.3331814032$.) We simply have to swallow the fact that $g(y)$ is a legitimate function but there is no algebraic formula for it.

Some Questions of Notation

The function f that assigns to each real number its square can be written

$$f(x) = x^2 \quad \text{or} \quad f(t) = t^2 \quad \text{or} \quad f(y) = y^2.$$

It doesn't matter what symbol represents the independent variable.

We have been writing $x = g(y)$ for the inverse of $f(x)$. But this makes us a little uncomfortable because we are used to x as the independent variable and y as the dependent variable. However, it doesn't matter what symbols we use for the variables; there is nothing to stop us from writing $y = g(x)$. For instance, we can write $y = \sqrt{x}$ as the inverse of $y = x^2$. Here $x \geq 0$ for both functions. But in general $y = f(x)$ and $y = g(x)$ may have different domains. Therefore, when we write $y = f(x)$, then x is in the domain of f; when we write $y = g(x)$, then x is in the domain of g.

Graphs

Suppose we have the graph of a strictly increasing (or decreasing) function $y = f(x)$. Then we can easily find the graph of its inverse function $y = g(x)$. We just flip the graph of $y = f(x)$ over the line $y = x$.

Here is why it works. Each point (x, y) on the graph of $y = f(x)$ satisfies $x = g(y)$. In other words, the graph is also the graph of $x = g(y)$. But we want the graph of $y = g(x)$, so we interchange x and y. Now interchanging the coordinates of a point simply reflects that point in the line (mirror) $y = x$. See Fig. 1a. Therefore the graph of $y = g(x)$ is the reflection of the graph of $x = g(y)$, that is, the reflection of the graph of $y = f(x)$. See Fig. 1b.

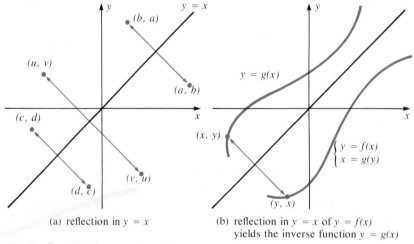

(a) reflection in $y = x$

(b) reflection in $y = x$ of $y = f(x)$ yields the inverse function $y = g(x)$

Fig. 1   Graph of inverse function

> **Graphs of Inverse Functions**   Let f and g be a pair of inverse functions. Then the graphs of $y = f(x)$ and $y = g(x)$ are mirror images of each other in the line $y = x$.

Some examples are shown in Fig. 2.

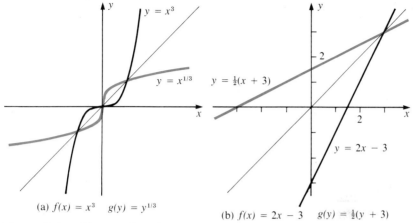

(a) $f(x) = x^3$ $g(y) = y^{1/3}$

(b) $f(x) = 2x - 3$ $g(y) = \frac{1}{2}(y + 3)$

Fig. 2 Pairs of inverse functions

Exercises

Explain why the function has an inverse, and find the domain of the inverse function $g(y)$

1 $f(x) = x^3$ $\{$all $x\}$ **2** $f(x) = -1/x$ $\{x < 0\}$

3 $f(x) = -1/x$ $\{x > 0\}$ **4** $f(x) = 1/x^2$ $\{x < 0\}$

5 $f(x) = x^2$ $\{3 < x < 4\}$ **6** $f(x) = 2x^2 - 5$ $\{3 < x < 4\}$

7 $f(x) = x^5 + x^3$ $\{$all $x\}$ **8** $f(x) = x^7 + 3x$ $\{$all $x\}$

9 $f(x) = 1/x^3$ $\{x > 0\}$ **10** $f(x) = 1/x$ $\{x < 0\}$

11 $f(x) = x^2$ $\{x \le 0\}$ **12** $f(x) = x^4 + x^2$ $\{x \le 0\}$.

Find the inverse function and its domain

13 $f(x) = \frac{1}{8}x^3$ $\{$all $x\}$ **14** $f(x) = 1/x$ $\{x < 0\}$

15 $f(x) = -1/x$ $\{x > 0\}$ **16** $f(x) = 4x^2$ $\{x \le 0\}$

17 $f(x) = 5/x^2$ $\{x > 0\}$ **18** $f(x) = 1 - 3x^2$ $\{x \ge 0\}$

19 $f(x) = \sqrt{x + 1}$ $\{x \ge -1\}$ **20** $f(x) = \sqrt[3]{x}$ $\{$all $x\}$

21 $f(x) = \sqrt{x^2 + 4}$ $\{x \ge 0\}$ **22** $f(x) = \sqrt{x^3 + 27}$ $\{x \ge -3\}$

23 $f(x) = \dfrac{1}{x^4 + 10}$ $\{x \le 0\}$ **24*** $f(x) = x + \dfrac{1}{x}$ $\{x \ge 1\}$

25 $f(x) = \dfrac{-1}{x^2}$ $\{x < 0\}$ **26*** $f(x) = \dfrac{x}{1 + |x|}$ $\{$all $x\}$.

Compute $g(4)$ where g is the inverse function of f

27 $f(x) = x^7 + 3x + 4$ $\{$all $x\}$ **28** $f(x) = x^3 - 23$ $\{$all $x\}$

29 $f(x) = \dfrac{x}{1 - x}$ $\{x < 1\}$ **30** $f(x) = x^4 + x^2$ $\{x \ge 0\}$.

On the same graph plot $y = f(x)$ and $y = g(x)$, where g is the inverse function of f

31 $f(x) = \frac{1}{2}x + 1$ **32** $f(x) = -3x + 2$

33 $f(x) = x^2$ $\{x \ge 0\}$ **34** $f(x) = 1/x$ $\{x > 0\}$

35 $f(x) = 1/x^2$ $\{x > 0\}$ **36** $f(x) = \sqrt{1 + x}$ $\{x \ge -1\}$

37 $f(x) = \begin{cases} x + 1 & 0 \le x \le 1 \\ x^2 + 1 & 1 \le x < \infty \end{cases}$
38 $f(x) = \begin{cases} x^2 & 0 \le x \le 1 \\ \sqrt{x} & 1 \le x < \infty. \end{cases}$

39 Suppose $g(y)$ is the inverse of $f(x)$. Find the domain and inverse of $f(-x)$.

40* Suppose $f_1(x)$ and $f_2(x)$ have inverses $g_1(x)$ and $g_2(x)$ respectively. Suppose the range of f_1 lies in the domain of f_2. Show that $f_2 \circ f_1$ has the inverse $g_1 \circ g_2$.

3   ARC SIN AND ARC COS

We now apply this theory to the functions sine and cosine.

Arc Sine

The function $y = \sin \theta$, defined for all real θ, has no inverse because it is neither strictly increasing nor strictly decreasing. In fact, if $-1 \le c \le 1$, then there are infinitely many values of θ for which $\sin \theta = c$. But the very point of an inverse is to provide a *unique* θ for each given c.

The situation is not hopeless, however. We can limit ourselves to a smaller domain where $\sin \theta$ is either strictly increasing or strictly decreasing. A natural choice is the interval $-\frac{1}{2}\pi \le \theta \le \frac{1}{2}\pi$. There $\sin \theta$ increases strictly and takes all values from -1 to 1. See Fig. 1.

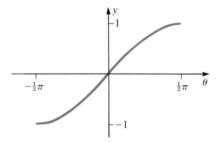

Fig. 1 $y = \sin \theta \qquad -\frac{1}{2}\pi \le \theta \le \frac{1}{2}\pi;$
y is strictly increasing.

Now we have all the ingredients for an inverse: the function $y = \sin \theta$, defined only on the interval $\mathbf{I} = \{-\frac{1}{2}\pi \le \theta \le \frac{1}{2}\pi\}$, is strictly increasing on $\mathbf{I}$, and has range the interval $\mathbf{J} = \{-1 \le y \le 1\}$. Hence there exists a unique inverse function called the **arc sine** and written

$\theta = \text{arc sin } y.$

Stated briefly, arc sin y is that unique real number for which $\sin \theta = y$ and $-\frac{1}{2}\pi \le \theta \le \frac{1}{2}\pi$. Figure 2 illustrates the idea geometrically.

Examples

(1) $\text{arc sin}(-1) = -\frac{1}{2}\pi \qquad \text{arc sin } 0 = 0 \qquad \text{arc sin } 1 = \frac{1}{2}\pi.$

(2) $\text{arc sin } \frac{1}{2} = \frac{1}{6}\pi$ because $\sin \frac{1}{6}\pi = \frac{1}{2}$ and $-\frac{1}{2}\pi \le \frac{1}{6}\pi \le \frac{1}{2}\pi.$

(3) $\text{arc sin } \frac{1}{2}\sqrt{2} = \frac{1}{4}\pi$ because $\sin \frac{1}{4}\pi = \frac{1}{2}\sqrt{2}$ and $-\frac{1}{2}\pi \le \frac{1}{4}\pi \le \frac{1}{2}\pi.$

(4) $\text{arc sin}(-\frac{1}{2}\sqrt{3}) = -\frac{1}{3}\pi$
      because $\sin(-\frac{1}{3}\pi) = -\frac{1}{2}\sqrt{3}$ and $-\frac{1}{2}\pi \le -\frac{1}{3}\pi \le \frac{1}{2}\pi.$

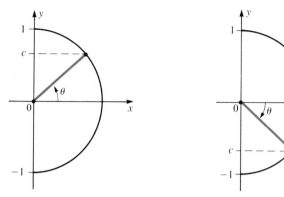

Fig. 2 To each c with $-1 \le c \le 1$, there corresponds a unique θ with $-\frac{1}{2}\pi \le \theta \le \frac{1}{2}\pi$ such that $\sin \theta = c$.

Arc Sine There is a unique function $\theta = \text{arc sin } y$ with domain $\{-1 \le y \le 1\}$ and satisfying

$$\sin(\text{arc sin } y) = y \qquad (-1 \le y \le 1)$$

$$\text{arc sin}(\sin \theta) = \theta \qquad (-\tfrac{1}{2}\pi \le \theta \le \tfrac{1}{2}\pi).$$

The function arc sin y is strictly increasing.*

We obtain the graph of $\theta = \text{arc sin } y$ from the graph of $y = \sin \theta$ shown in Fig. 1. Since

$$y = \sin \theta \quad \text{and} \quad \theta = \text{arc sin } y$$

are equivalent statements (provided $-\frac{1}{2}\pi \le \theta \le \frac{1}{2}\pi$), Figure 1 is the graph of $\theta = \text{arc sin } y$. However, the independent variable is on the vertical axis and the dependent variable is on the horizontal axis. For a more conventional graph, we interchange the axes and reflect the graph in the line $y = \theta$. The result (Fig. 3) is the graph of $\theta = \text{arc sin } y$.

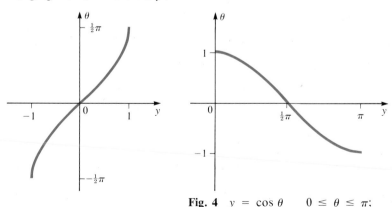

Fig. 3 $\theta = \text{arc sin } y$

Fig. 4 $y = \cos \theta \qquad 0 \le \theta \le \pi$; y is strictly decreasing.

---

* The notations arcsin y and $\sin^{-1} y$ are also used for arc sin y.

Arc Cosine

Like $\sin \theta$, the function $\cos \theta$ defined for all real θ has no inverse. But we can limit $\cos \theta$ to a smaller domain. A natural choice is the interval $0 \le \theta \le \pi$. There $\cos \theta$ decreases strictly and takes all values from 1 to -1. See Fig. 4, previous page.

We are now set up for an inverse: the function $y = \cos \theta$, defined only on the interval $\mathbf{I} = \{0 \le \theta \le \pi\}$, is strictly decreasing on $\mathbf{I}$, and has range the interval $\mathbf{J} = \{-1 \le y \le 1\}$. Hence there exists a unique inverse function called the **arc cosine** and written

$$\theta = \text{arc cos } y.$$

Note that arc cos y is that unique real number θ for which $\cos \theta = y$ and $0 \le \theta \le \pi$.

Examples

(1)   arc cos $1 = 0$ because $\cos 0 = 1$ and $0 \le 0 \le \pi$.

(2)   arc cos $0 = \frac{1}{2}\pi$ because $\cos \frac{1}{2}\pi = 0$ and $0 \le \frac{1}{2}\pi \le \pi$.

(3)   arc cos $(-1) = \pi$ because $\cos \pi = -1$ and $0 \le \pi \le \pi$.

(4)   arc cos $\frac{1}{2} = \frac{1}{3}\pi$ because $\cos \frac{1}{3}\pi = \frac{1}{2}$ and $0 \le \frac{1}{3}\pi \le \pi$.

(5)   arc cos $(-\frac{1}{2}\sqrt{2}) = \frac{3}{4}\pi$ because $\cos \frac{3}{4}\pi = -\frac{1}{2}\sqrt{2}$ and $0 \le \frac{3}{4}\pi \le \pi$.

Arc Cosine   There is a unique function $\theta = \text{arc cos } y$ with domain $-1 \le y \le 1$ and satisfying

$$\cos(\text{arc cos } y) = y \qquad (-1 \le y \le 1)$$
$$\text{arc cos}(\cos \theta) = \theta \qquad (0 \le \theta \le \pi).$$

The function arc cos y is strictly decreasing.

We obtain the graph of $\theta = \text{arc cos } y$ from Fig. 4 by interchanging the axes and reflecting the graph in the line $y = \theta$. See Fig. 5.

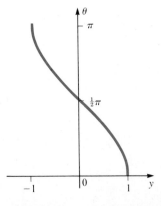

Fig. 5 $\theta = \text{arc cos } y$

Notation   Since we are used to calling the horizontal axis the x-axis, let us replace y by x and write $\theta = \text{arc sin } x$ and $\theta = \text{arc cos } x$.

Relation Between Arc Sin and Arc Cos

From the examples above we have

$$\text{arc sin}(-1) + \text{arc cos}(-1) = -\tfrac{1}{2}\pi + \pi = \tfrac{1}{2}\pi$$

$$\text{arc sin } 0 + \text{arc cos } 0 = 0 + \tfrac{1}{2}\pi = \tfrac{1}{2}\pi$$

$$\text{arc sin } \tfrac{1}{2} + \text{arc cos } \tfrac{1}{2} = \tfrac{1}{6}\pi + \tfrac{1}{3}\pi = \tfrac{1}{2}\pi.$$

These relations suggest the following identity:

$$\text{arc sin } x + \text{arc cos } x = \tfrac{1}{2}\pi \quad (-1 \le x \le 1).$$

To prove the identity, let $\theta = \text{arc sin } x$. Then $x = \sin \theta$ and $-\tfrac{1}{2}\pi \le \theta \le \tfrac{1}{2}\pi$. It follows that $-\theta$ is in the same interval, $-\tfrac{1}{2}\pi \le -\theta \le \tfrac{1}{2}\pi$. Add $\tfrac{1}{2}\pi$ to the three terms in this (double) inequality:

$$0 \le \tfrac{1}{2}\pi - \theta \le \pi.$$

But

$$\cos(\tfrac{1}{2}\pi - \theta) = \sin \theta = x,$$

hence $\tfrac{1}{2}\pi - \theta = \text{arc cos } x$. Therefore

$$\text{arc sin } x + \text{arc cos } x = \theta + (\tfrac{1}{2}\pi - \theta) = \tfrac{1}{2}\pi.$$

The identity has a simple geometric interpretation in the case $0 \le x \le 1$. Figure 6 shows a right triangle in which $\sin \alpha = \cos \beta = x$. That means $\alpha = \text{arc sin } x$ and $\beta = \text{arc cos } x$. But $\alpha + \beta = \tfrac{1}{2}\pi$.

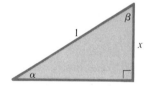

Fig. 6

EXAMPLE 1   Show that $\text{arc sin } x = \text{arc cos }\sqrt{1 - x^2}$ for $0 \le x \le 1$.

Solution   Set $\theta = \text{arc sin } x$. Then $0 \le \theta \le \tfrac{1}{2}\pi$ since $0 \le x \le 1$, and $x = \sin \theta$. Since θ is in the first quadrant, $\cos \theta \ge 0$. Therefore

$$\cos \theta = \sqrt{1 - \sin^2 \theta} = \sqrt{1 - x^2} \qquad \theta = \text{arc cos }\sqrt{1 - x^2}$$

But $\theta = \text{arc sin } x$, so $\text{arc sin } x = \text{arc cos }\sqrt{1 - x^2}$.

EXAMPLE 2   Find $\text{arc sin}(\cos 15°)$ exactly.

Solution $\cos 15° = \sin 75° = \sin \tfrac{5}{12}\pi$ and $-\tfrac{1}{2}\pi \le \tfrac{5}{12}\pi \le \tfrac{1}{2}\pi$, hence

$$\text{arc sin}(\cos 15°) = \tfrac{5}{12}\pi.$$

Answer $\tfrac{5}{12}\pi$

Exercises

Give exact values

1 $\arc\sin(-\frac{1}{2})$ **2** $\arc\sin(\frac{1}{2}\sqrt{3})$ **3** $\arc\sin(-\frac{1}{2}\sqrt{2})$
4 $\arc\cos(\frac{1}{2}\sqrt{2})$ **5** $\arc\cos(-\frac{1}{2})$ **6** $\arc\cos(\frac{1}{2}\sqrt{3})$
7 $\arc\cos(-\frac{1}{2}\sqrt{3})$ **8** $\arc\cos(\cos\frac{1}{7}\pi)$ **9** $\sin(\arc\sin 0.3)$
10 $\arc\sin(\sin 0.3)$ **11** $\cos[\arc\cos(-\frac{1}{3})]$ **12** $\arc\cos[\cos(-\frac{1}{3})]$.

Prove

13 $\arc\sin(-x) = -\arc\sin x$ **14** $\arc\cos(-x) = \pi - \arc\cos x$
15 $180\cdot\arc\cos(\cos\theta°) = \pi\cdot\theta$ **16** $\cos(\arc\sin x) = \sqrt{1-x^2}$
for $0 \le \theta \le 180$
17 $\arc\sin(\cos\theta) = \frac{1}{2}\pi - \theta$ **18** $2\arc\cos x = \arc\cos(2x^2 - 1)$
for $0 \le \theta \le \pi$ for $0 \le x \le 1$.

Find exactly

19 $\arc\cos(\sin 35°)$ **20** $\cos[\arc\sin(-0.7)]$.

21 By Example 1, the functions $\arc\sin x$ and $\arc\cos\sqrt{1-x^2}$ are equal for $0 \le x \le 1$. Find their relation for $-1 \le x \le 0$.

22 According to Ex. 18, the functions $2\arc\cos x$ and $\arc\cos(2x^2 - 1)$ are equal for $0 \le x \le 1$. Find their relation for $-1 \le x \le 0$.

Use trigonometric identities to find exactly

23 $\sin(2\arc\sin\frac{1}{3})$ **24** $\cos(2\arc\sin\frac{1}{3})$
25 $\cos(\arc\sin\frac{3}{5} + \arc\cos\frac{5}{13})$ **26** $\sin[\arc\sin\frac{4}{5} - \arc\cos(-\frac{7}{25})]$.

4 ARC TANGENT

The graph of $\tan\theta$ shows that $\tan\theta$ is strictly increasing for $-\frac{1}{2}\pi < \theta < \frac{1}{2}\pi$ and takes on every possible real value (Fig. 1). Therefore an inverse function exists.

> **Arc Tangent** There is a unique function $\theta = \arc\tan x$ with domain $\{\text{all } x\}$ and satisfying
>
> $\tan(\arc\tan x) = x$ $(-\infty < x < \infty)$
>
> $\arc\tan(\tan\theta) = \theta$ $(-\frac{1}{2}\pi < \theta < \frac{1}{2}\pi)$.
>
> The function $\arc\tan x$ is strictly increasing.

Note that $\arc\tan x$ is that unique real number θ for which $\tan\theta = x$ and $-\frac{1}{2}\pi < \theta < \frac{1}{2}\pi$.

Examples

(1) $\arc\tan 1 = \frac{1}{4}\pi$ since $\tan\frac{1}{4}\pi = 1$ and $-\frac{1}{2}\pi < \frac{1}{4}\pi < \frac{1}{2}\pi$

(2) $\arc\tan(-1) = -\frac{1}{4}\pi$ since $\tan(-\frac{1}{4}\pi) = -1$ and $-\frac{1}{2}\pi < -\frac{1}{4}\pi < \frac{1}{2}\pi$

(3) $\arc\tan(-\sqrt{3}) = -\frac{1}{3}\pi$ since $\tan(-\frac{1}{3}\pi) = -\sqrt{3}$ and $-\frac{1}{2}\pi < -\frac{1}{3}\pi < \frac{1}{2}\pi$

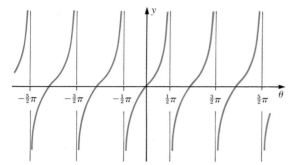

Fig. 1 $y = \tan \theta$

We obtain the graph of $\theta = \text{arc tan } x$ by reflecting the graph of $x = \tan \theta$. See Fig. 2. The lines $\theta = \frac{1}{2}\pi$ and $\theta = -\frac{1}{2}\pi$ are horizontal asymptotes* of the graph. That checks: for example, as $x \longrightarrow \infty$, the first quadrant angle whose tangent is x approaches $\frac{1}{2}\pi$.

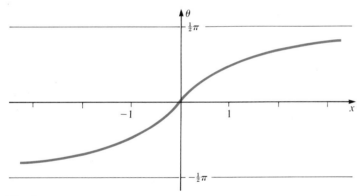

Fig. 2 $\theta = \text{arc tan } x$

EXAMPLE 1 Prove $\text{arc sin } x = \text{arc tan } \dfrac{x}{\sqrt{1 - x^2}}$ $(-1 < x < 1)$.

Solution This identity is a disguised version of a statement that looks easier: if θ is in the first or fourth quadrants and if $\sin \theta = x$, then $\tan \theta = x/\sqrt{1 - x^2}$.

Let $\theta = \text{arc sin } x$, so $\sin \theta = x$. Since θ is in the first or fourth quadrant, $\cos \theta \geq 0$, hence

$$\tan \theta = \frac{\sin \theta}{\cos \theta} = \frac{\sin \theta}{\sqrt{1 - \sin^2 \theta}} = \frac{x}{\sqrt{1 - x^2}}.$$

Therefore $\text{arc tan } \dfrac{x}{\sqrt{1 - x^2}} = \theta = \text{arc sin } x$.

EXAMPLE 2 Find exactly

(a) $\tan[\text{arc cos}(-\frac{1}{2})]$ (b) $\sin[\text{arc tan}(-\frac{3}{4})]$.

---

* A horizontal line $\theta = c$ is a **horizontal asymptote** of a graph $\theta = f(x)$ if $f(x) \longrightarrow c$ as $x \longrightarrow -\infty$ or as $x \longrightarrow \infty$, or both.

Solution

(a)   arc cos $(-\frac{1}{2}) = \frac{2}{3}\pi$ because $\cos\frac{2}{3}\pi = -\frac{1}{2}$ and $0 \le \frac{2}{3}\pi \le \pi$. Therefore

$\tan[\arc\cos(-\frac{1}{2})] = \tan\frac{2}{3}\pi = -\sqrt{3}$.

(b)   Let $\theta = $ arc $\tan(-\frac{3}{4})$. Then $-\frac{1}{2}\pi < \theta < \frac{1}{2}\pi$ and $\tan\theta = -\frac{3}{4} < 0$, so $-\frac{1}{2}\pi < \theta < 0$. Put θ in standard position; its terminal side must be in the fourth quadrant, and $(x,y) = (4,-3)$ is a point on this terminal side. Clearly $r^2 = 4^2 + (-3)^2 = 25$, so $r = 5$ and

$$\sin[\arc\tan(-\tfrac{3}{4})] = \sin\theta = \frac{y}{r} = -\frac{3}{5}.$$

Answer   (a) $-\sqrt{3}$ (b) $-\frac{3}{5}$

Calculating Inverse Trig Functions

Your calculator has a key

| INV | or | ARC | or | F |

It is used in combination with | sin |, | cos |, and | tan | to estimate the corresponding inverse functions. (These are sometimes written $\sin^{-1}$, $\cos^{-1}$, etc.)

Examples   (First set | rad | mode.)

(1)   arc sin 0.628   . 6 2 8 | INV | | sin | *0.678981*

(2)   arc cos (-0.922) . 9 2 2 | +/− | | INV | | cos | *2.74401*

(3)   arc sin (1.485) 1 . 4 8 5 | INV | | sin | **ERROR**

(First set | deg | mode.)

(4)   arc tan 0.505   . 5 0 5 | INV | | tan | *26.7938*

(5)   arc tan (-3.190) 3 . 1 9 | +/− | | INV | | tan | *−72.5949*

(6)   arc sin $\frac{3}{17}$ 3 | ÷ | 1 7 | = | | INV | | sin | *10.1642*

Exercises

Give exact values

1   arc tan 0   **2**   arc tan $(\frac{1}{3}\sqrt{3})$ **3**   arc tan $(-\frac{1}{3}\sqrt{3})$
4   arc tan $(\sqrt{3})$ **5**   tan (arc tan 35)   **6**   arc tan (tan 1)
7   arc tan$[\tan(-1)]$ **8**   arc tan 5 + arc tan 0.2 .

Prove

9   arc tan $(-x) = -$arc tan x **10**   arc tan x + arc tan $\dfrac{1}{x} = \frac{1}{2}\pi$
                                                                                          for $x > 0$

11   cot (arc tan x) = $\dfrac{1}{x}$ **12**   arc tan $x = $ arc sin $\dfrac{x}{\sqrt{1+x^2}}$
       for $x \ne 0$

13 arc tan(cot θ) = $\frac{1}{2}\pi$ − θ
for $0 < \theta < \pi$

14 2 arc tan x = arc tan $\dfrac{2x}{1 - x^2}$
for $-1 < x < 1$

15 arc tan $\frac{1}{2}$ + arc tan $\frac{1}{3}$ = $\frac{1}{4}\pi$

16 arc tan $\frac{1}{3}$ + arc tan $\frac{1}{7}$ = arc tan $\frac{1}{2}$.

Find exactly

17 cos[arc tan(−$\frac{1}{3}$)]

18 tan[arc sin(−$\frac{2}{3}$)].

Use trigonometric identities to find exactly

19 tan(2 arc tan 5)

20 sin(2 arc tan 5)

21 tan(2 arc cos $\frac{1}{3}$)

22 tan(arc sin $\frac{12}{13}$ − arc sin $\frac{5}{13}$).

23 Show that x = cot θ on the domain $0 < \theta < \pi$ has an inverse function θ = arc cot x, with domain {all x}, and strictly decreasing.

24 (cont.) Graph θ = arc cot x.

25 (cont.) Prove arc tan x + arc cot x = $\frac{1}{2}\pi$.

26 (cont.) Give a sequence of calculator steps for estimating arc cot x.

27 Graph x = sec θ for $0 \le \theta \le \pi$, $\theta \ne \frac{1}{2}\pi$. Show that sec θ restricted to this domain has an inverse function, arc sec x.

28 (cont.) Show that arc sec x is defined for $|x| \ge 1$, and sketch its graph.

29 Graph x = csc θ for $-\frac{1}{2}\pi \le \theta \le \frac{1}{2}\pi$, $\theta \ne 0$. Show that csc θ restricted to this domain has an inverse function, arc csc x.

30 (cont.) Show that arc csc x is defined for $|x| \ge 1$, and sketch its graph.

Estimate on a calculator; answer in radians to 4 places

31 arc sin $\frac{1}{4}$

32 arc cos $\frac{1}{6}$

33 arc tan $\frac{10}{3}$

34 arc tan(−10)

35 arc sin(−$\frac{9}{11}$)

36 arc cos(−$\frac{13}{29}$).

Estimate on a calculator; answer in degrees to 2 places

37 arc sin 0.1203

38 arc cos 0.8177

39 arc tan(−294.7)

40 arc cos(0.1212)

41 arc tan 1.352 + arc tan $\dfrac{1}{1.352}$

42 arc sin(−0.944) + arc cos(−0.944).

43 Tabulate x, arc sin x, and $x + \frac{1}{6}x^3$ for x = 0.1, 0.05, 0.01. Conclusion?

44 Tabulate x, arc tan x, and $x - \frac{1}{3}x^3$ for the same values as in Ex. 43. Conclusion?

45 There are 3 squares in Fig. 3. Show that $\alpha + \beta = \gamma$. [*Hint* See Ex. 15.]

46 P bisects one leg of the 45-45-90 triangle in Fig. 4. Find tan β. [*Hint* See Ex. 15.]

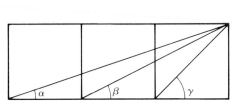

Fig. 3

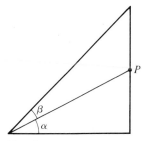

Fig. 4

REVIEW EXERCISES

Find the inverse function; give its domain and range

1 $y = \begin{cases} 3x \\ 2x \end{cases}$ for $\begin{cases} x \leq 0 \\ x \geq 0 \end{cases}$

2 $y = x^4 + 2x^2$ for $x \geq 0$

Find exactly

3 arc sin$(\cos \frac{5}{7}\pi)$

4 $\sin[\text{arc cos}(-0.3)]$.

Graph (for all x)

5 $y = \text{arc sin}(\sin x)$

6 $y = \text{arc cos}(\cos x)$.

7 The function $y = \sin\theta$ for $\frac{1}{2}\pi \leq \theta \leq \frac{3}{2}\pi$ is strictly decreasing. Express its inverse function $g(y)$ in terms of arc sin y.

8 The function $y = \tan\theta$ for $-\frac{3}{2}\pi < \theta < -\frac{1}{2}\pi$ is strictly increasing. Express its inverse function $g(y)$ in terms of arc tan y.

8 TRIGONOMETRIC EQUATIONS

1 PERIODIC EQUATIONS

An equation involving trigonometric functions is called a **trigonometric equation.** In this section we study equations that involve *only* trigonometric functions of θ such as $\sin \theta$, $\cos \theta$, $\cdots$ and perhaps also $\sin 2\theta$, $\cos 3\theta$, etc. Examples are

$$\sin 2\theta = \cos \theta \qquad \csc \theta = 3 \sin \theta \qquad \sec^2 \theta = 4 \tan \theta - 2$$

$$\cot 3\theta = 5 + \cos 2\theta$$

Whenever θ is a root of a trigonometric equation of the type considered here, so is $\theta + 2\pi n$ for each integer n. Therefore it suffices to find all the roots in a conveniently chosen interval of length 2π, such as $0 \leq \theta < 2\pi$ or $-\pi < \theta \leq \pi$.

We shall always choose the interval

$$-\pi < \theta \leq \pi$$

for the simple and practical reason that the values of arc $\sin x$, arc $\cos x$, and arc $\tan x$ all fall into this interval.

Thus when you are asked for *all* solutions of a trigonometric equation, you must first find *all* solutions $\theta_1, \theta_2, \cdots, \theta_r$ in the interval $-\pi < \theta \leq \pi$, and then write

$$\theta_1 + 2\pi n \qquad \theta_2 + 2\pi n \cdots \theta_r + 2\pi n \qquad (n \text{ an integer})$$

for *all solutions* of the equation. Remember that θ must appear *only* in the form $\sin m\theta$, $\cos m\theta$, $\cdots$, $\csc m\theta$, where m is an integer, but it can appear so more than once.

We begin our work with the simplest type of trigonometric equation.

EXAMPLE 1 Find all solutions in $-\pi < \theta \le \pi$ of

(a) $\sin \theta = \frac{1}{2}\sqrt{3}$ (b) $\cos \theta = 0.2598$ (c) $\tan \theta = 3.9810$.

Solution (a) $\sin \theta$ is positive in the first and second quadrants. There is a solution in each (Fig. 1a):

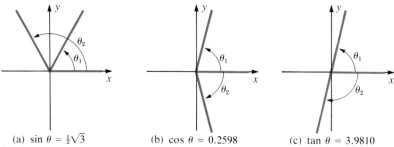

(a) $\sin \theta = \frac{1}{2}\sqrt{3}$ (b) $\cos \theta = 0.2598$ (c) $\tan \theta = 3.9810$

Fig. 1

$$\theta_1 = \arcsin(\tfrac{1}{2}\sqrt{3}) = \tfrac{1}{3}\pi \qquad \text{in the 1-st quadrant}$$

$$\theta_2 = \pi - \arcsin(\tfrac{1}{2}\sqrt{3}) = \tfrac{2}{3}\pi \qquad \text{in the 2-nd quadrant}.$$

(b) $\cos \theta$ is positive in the first and fourth quadrants. There is a solution in each (Fig. 1b):

$$\theta_1 = \arccos 0.2598 \approx 1.30798 \qquad \text{in the 1-st quadrant}$$

$$\theta_2 = -\arccos 0.2598 \approx -1.30798 \qquad \text{in the 4-th quadrant}.$$

(c) $\tan \theta$ is positive in the first and third quadrants. There is a solution in each (Fig. 1c):

$$\theta_1 = \arctan 3.9810 \approx 1.32470 \qquad \text{in the 1-st quadrant}$$

$$\theta_2 = \arctan 3.9810 - \pi \approx -1.81690 \qquad \text{in the 3-rd quadrant}$$

(We subtract (rather than add) π to make θ fall into the interval $-\pi < \theta \le \pi$.)

Answer (a) $\tfrac{1}{3}\pi$ $\tfrac{2}{3}\pi$

(b) $\theta_1 = \arccos 0.2598 \approx 1.30798$ $\theta_2 = -\theta_1 \approx -1.30798$

(c) $\theta_1 = \arctan 3.9810 \approx 1.32470$ $\theta_2 = \theta_1 - \pi \approx -1.81690$

In the example we were given positive values for sin, cos, and tan. We handle negative values of cos and tan similarly; sin must be treated a bit differently. Suppose we seek all θ in the interval $-\pi < \theta \le \pi$ such that

$$\sin \theta = a \quad \text{where} \quad -1 < a < 0.$$

The 4-th quadrant solution is $\theta_1 = \arcsin a$, and

$$-\tfrac{1}{2}\pi < \theta_1 < 0$$

The 3-rd quadrant solution θ_2 lies in the interval $-\pi < \theta < -\frac{1}{2}\pi$. It satisfies $\theta_2 + \theta_1 = -\pi$. See Fig. 2. Therefore

$$\theta_2 = -\theta_1 - \pi = -\arcsin a - \pi.$$

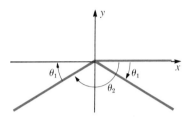

Fig. 2 $-1 < \sin \theta < 0$:
$$\theta_1 + \theta_2 = -\pi$$

EXAMPLE 2   Find all solutions in the interval $-\pi < \theta \le \pi$ of

(a) $\sin \theta = -0.3122$ (b) $\cos \theta = -\frac{1}{2}$ (c) $\tan \theta = -0.2419$

Solution   (a) $\theta_1 = \arcsin(-0.3122) \approx -0.31751$ (4-th quadrant)

$\qquad \theta_2 = -\arcsin(-0.3122) - \pi \approx -2.82408$ (3-rd quadrant)

(b) $\theta_1 = \arccos(-\frac{1}{2}) = \frac{2}{3}\pi$ (2-nd quadrant)

$\qquad \theta_2 = -\arccos(-\frac{1}{2}) = -\frac{2}{3}\pi$ (3-rd quadrant)

(c) $\theta_1 = \arctan(-0.2419) \approx -0.23734$ (4-th quadrant)

$\qquad \theta_2 = \arctan(-0.2419) + \pi \approx 2.90425$ (2-nd quadrant)

We add (rather than subtract) π to fall into the interval $\frac{1}{2}\pi < \theta < \pi$.

Answer

(a) $\theta_1 = \arcsin(-0.3122) \approx -0.31751$ $\theta_2 = -\theta_1 - \pi \approx -2.82408$

(b) $\theta_1 = \frac{2}{3}\pi$ $\theta_2 = -\frac{2}{3}\pi$

(c) $\theta_1 = \arctan(-0.2419) \approx -0.23734$ $\theta_2 = \theta_1 + \pi \approx 2.90425$

EXAMPLE 3   Solve $\sin 2\theta = \frac{1}{2}$.

Solution   Clearly, if $-\pi < 2\theta \le \pi$ and $\sin 2\theta = \frac{1}{2}$, then $2\theta = \frac{1}{6}\pi$ or $2\theta = \pi - \frac{1}{6}\pi = \frac{5}{6}\pi$. Thus *any* solution θ of $\sin 2\theta = \frac{1}{2}$ must satisfy

$$2\theta = \tfrac{1}{6}\pi + 2\pi n \quad \text{or} \quad 2\theta = \tfrac{5}{6}\pi + 2\pi n$$

that is,

$$\theta = \tfrac{1}{12}\pi + n\pi \quad \text{or} \quad \theta = \tfrac{5}{12}\pi + n\pi,$$

where n is an arbitrary integer. This is a correct answer as it stands; however, it is usually preferable to display all solutions θ in the interval $-\pi < \theta \le \pi$, and then add $2\pi n$ to each. They are (choose $n = 0$ and $n = -1$)

$$\tfrac{1}{12}\pi \qquad \tfrac{1}{12}\pi - \pi = -\tfrac{11}{12}\pi \qquad \tfrac{5}{12}\pi \qquad \tfrac{5}{12}\pi - \pi = -\tfrac{7}{12}\pi.$$

Answer $\tfrac{1}{12}\pi + 2\pi n \qquad -\tfrac{11}{12}\pi + 2\pi n \qquad \tfrac{5}{12}\pi + 2\pi n \qquad -\tfrac{7}{12}\pi + 2\pi n$

Now we can handle equations of the form $\sin m\theta = c$, etc., in which the value of a trigonometric function is known. This generally represents the final step in the solution of a (more general) trigonometric equation. Before that we try to use algebra technique and trigonometric identities to solve for a trigonometric function of θ, from which we find θ.

EXAMPLE 4   Find all solutions of $\sin 2\theta = \cos \theta$.

Solution   Use the double angle formula for $\sin 2\theta$:

$$2 \sin \theta \cos \theta = \cos \theta \qquad (2 \sin \theta - 1)\cos \theta = 0.$$

The product is zero only if one of the factors is zero, that is, only if

$$2 \sin \theta - 1 = 0 \quad \text{or} \quad \cos \theta = 0.$$

In the first case, $\sin \theta = \frac{1}{2}$. The solutions of this equation in the interval $-\pi < \theta \le \pi$ are $\frac{1}{6}\pi$ and $\frac{5}{6}\pi$.

In the second case, $\cos \theta = 0$, and the solutions of this equation in the interval $-\pi < \theta \le \pi$ are $\frac{1}{2}\pi$ and $-\frac{1}{2}\pi$.

Answer $\frac{1}{6}\pi + 2\pi n \qquad \frac{5}{6}\pi + 2\pi n \qquad \frac{1}{2}\pi + 2\pi n \qquad -\frac{1}{2}\pi + 2\pi n$

EXAMPLE 5   Solve $\sec^2 \theta = 4 \tan \theta - 2$ in the interval $-\pi < \theta \le \pi$.

Solution   Replace $\sec^2 \theta$ by $1 + \tan^2 \theta$. The result is a quadratic equation for $\tan \theta$:

$$1 + \tan^2 \theta = 4 \tan \theta - 2 \qquad \tan^2 \theta - 4 \tan \theta + 3 = 0.$$

The quadratic factors:

$$(\tan \theta - 1)(\tan \theta - 3) = 0 \qquad \text{hence} \quad \tan \theta = 1 \quad \text{or} \quad \tan \theta = 3.$$

If $\tan \theta = 1$, then $\theta = \frac{1}{4}\pi$ or $\frac{1}{4}\pi - \pi = -\frac{3}{4}\pi$. If $\tan \theta = 3$, then

$$\theta = \text{arc tan } 3 \approx 1.24905 \quad \text{or} \quad \text{arc tan } 3 - \pi \approx -1.89255.$$

Answer $\frac{1}{4}\pi \quad -\frac{3}{4}\pi \quad$ arc tan $3 \approx 1.24905 \quad$ arc tan $3 - \pi \approx -1.89255$

EXAMPLE 6   Solve $\tan \theta° = 2 \cos \theta°$ in the interval $-180° < \theta° \le 180°$.

Solution   Since $\tan \theta° = \sin \theta°/\cos \theta°$,

$$\frac{\sin \theta°}{\cos \theta°} = 2 \cos \theta° \qquad \sin \theta° = 2 \cos^2 \theta°.$$

Replace $\cos^2 \theta°$ by $1 - \sin^2 \theta°$; the result is a quadratic equation for $\sin \theta°$:

$$2 \sin^2 \theta° + \sin \theta° - 2 = 0.$$

Now solve for $\sin \theta°$ by the quadratic formula:

$$\sin \theta° = \frac{-1 \pm \sqrt{17}}{4} = \begin{cases} \frac{1}{4}(-1 + \sqrt{17}) \approx 0.78078 \\ \frac{1}{4}(-1 - \sqrt{17}) \approx -1.28. \end{cases}$$

Since $\sin \theta° < -1$ is impossible, only $\sin \theta° = \frac{1}{4}(\sqrt{17} - 1)$ yields solutions. In this case $\sin \theta° > 0$, so there is one solution in the first quadrant and one in the second:

$$\theta_1° = \text{arc sin} \frac{1}{4}(\sqrt{17} - 1) \approx 51.3317° \qquad \theta_2° = 180° - \theta_1° \approx 128.6683°$$

Answer   51.3317°   128.6683°

Exercises

Solve on the interval $-\pi < \theta \le \pi$, exactly if possible, otherwise to 4-place accuracy

1 $\sin \theta = \frac{1}{2}\sqrt{2}$	**2** $\sin \theta = -\frac{1}{2}\sqrt{2}$	**3** $\cos \theta = -0.8121$
4 $\cos \theta = 0.8121$	**5** $\tan \theta = \sqrt{3}$	**6** $\tan \theta = -\sqrt{3}$
7 $\sin 2\theta = \frac{1}{2}\sqrt{3}$	**8** $\sin 2\theta = -\frac{1}{2}\sqrt{3}$	**9** $\cos 3\theta = \frac{1}{2}\sqrt{2}$
10 $\cos 3\theta = -\frac{1}{2}\sqrt{2}$	**11** $\tan 4\theta = -\sqrt{3}$	**12** $\tan 4\theta = \sqrt{3}$
13 $\cos 3\theta = -0.6917$	**14** $\tan 2\theta = 5.000$	**15** $\sin 2\theta = 0.8162$
16 $\tan 3\theta = -0.8164$.		

Solve exactly on the interval $0 \le \theta° < 360°$

17 $\sin \theta° = -\frac{1}{2}$	**18** $\cos \theta° = -\frac{1}{2}$
19 $\tan \theta° = \frac{1}{3}\sqrt{3}$	**20** $\tan \theta° = -\frac{1}{3}\sqrt{3}$.

Solve on the interval $-\pi < \theta \le \pi$, exactly if possible, otherwise to 4-place accuracy

21 $\sin^2 \theta = \cos^2 \theta$	**22** $16 \sin^4 \theta = \cos^4 \theta$
23 $3 \sin^2 \theta + 7 \cos \theta = 5$	**24** $6 \cos^4 \theta - 5 \cos^2 \theta + 1 = 0$
25 $\cos 2\theta = 2 \cos \theta$	**26** $\sin 2\theta = \sin \theta$
27 $\cot 2\theta = 1 + \tan \theta$	**28** $\sec^2 \theta - 5 \tan \theta - 1 = 0$
29 $\sec \theta + \cos \theta = 4$	**30** $12 \sin \theta + 5 = 2 \csc \theta$
31 $\sin^2 \theta + \sin \theta = 3$	**32** $2 \sin \theta - 2 \cos \theta = 3$
33 $2 \cos^2 \theta + \sin \theta = 2$	**34** $2 \cos^3 \theta = \cos \theta$
35 $\tan \theta = \sin \theta$	**36** $\tan \theta = \cot \theta$
37 $\sin^6 \theta = \sin^2 \theta$	**38** $\sec^3 \theta - 2 \sec \theta = 0$
39 $\tan \theta = \cot^2 \theta$	**40** $\tan^2 \theta + \tan \theta = 0$
41 $2 \sin^2 \theta = -3 \cos \theta$	**42** $2 \sin \theta \cos \theta = \cos 2\theta$
43 $\cos \theta = \cos 2\theta$	**44** $\tan \theta = \tan 2\theta$
45 $\tan \theta + \cot \theta = \sec \theta \csc \theta$	**46** $\tan^2 \theta \sec^2 \theta = 9 + \tan^2 \theta$
47 $\cos \theta - \cos 3\theta = 0$	**48** $\cos \theta - \sqrt{3} \sin \theta = 1$
49 $\tan \theta + \sec \theta = 1$	**50** $\tan \theta = \sin 2\theta$
51 $\sin 3\theta + \sin 5\theta = 0$	**52** $\cos 3\theta + \cos 5\theta = 0$.

2   CALCULATOR TECHNIQUE [Optional]

In the next section we shall consider equations involving both trigonometric functions of θ and also θ in other forms, for example,

$$\sin \theta = \theta^2 \qquad \theta \tan \theta = 1 \qquad \theta + \sin \theta = 2.5.$$

Such equations may have no roots, a finite number of roots, or infinitely many. Periodicity doesn't apply to them in general. If θ_0 is a root, there is no reason whatsoever for $\theta_0 + 2\pi n$ to be a root also.

There are no algebraic methods for solving such equations. We can only hope to approximate a root by numerical methods. In this section we discuss a method that fits well on hand-held calculators. Part of the method uses interpolation, so we digress briefly to this topic.

Linear Interpolation

Interpolation is a way of "reading between the lines" for values of a function, often used when a function is only accessible from a table. Many useful functions are not built into calculators but do exist in tables, and many results of highly accurate measurements are tabulated in handbooks.

For example, suppose a portion of a table of values of a certain function $y = f(x)$ is

x	3.150	3.151	3.152	3.153
y	0.49831	0.49845	0.49859	0.49872

Suppose we want the value of y corresponding to $x = 3.1517$. From the table, the closest values are

$$f(3.151) \approx 0.49845 \qquad f(3.152) \approx 0.49859$$

We pretend that the function $f(x)$ is linear for $3.151 \le x \le 3.152$. If so, since 3.1517 is $\frac{7}{10}$ of the way from 3.151 to 3.152, then $f(3.1517)$ is $\frac{7}{10}$ of the way from 0.49845 to 0.49859. See Fig. 1. But

$$\tfrac{7}{10}(0.49859 - 0.49845) = \tfrac{7}{10}(0.00014) \approx 0.00010.$$

Therefore

$$f(3.1517) \approx 0.49845 + 0.00010 = 0.49855.$$

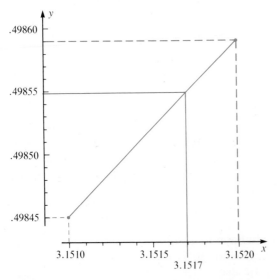

Fig. 1 Example of linear interpolation

Solutions of Equations

Complicated equations such as

$$x^4 = x + 1 \qquad \text{and} \qquad x^5 + 2x^3 + x^2 - 1 = 0$$

usually cannot be solved exactly. However, solutions can be approximated by numerical methods. These require lots of computation, so the calculator is a valuable aid. The method we use might be considered "educated" trial-and-error.

EXAMPLE 1   Estimate to 3 places the positive solution of $x^4 = x + 1$.

Solution   First graph $y = x^4$ and $y = x + 1$ on the same grid (Fig. 2).

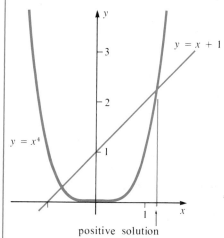

Fig. 2 Graphical estimate of the positive root of $x^4 = x + 1$

The figure shows that the graphs cross for x around 1.1 or 1.2. This gives a first, crude approximation to the solution. For more accurate information, let us tabulate x, x^4, and $x + 1$ by tenths, starting at 1.0. The calculator sequence $x \boxed{x^2} \boxed{x^2}$ gives x^4 easily.

x	1.0	1.1	1.2	1.3
x^4	1.0000	1.4641	2.0736	2.8561
$x + 1$	2.0	2.1	2.2	2.3

The table shows that $x^4 < x + 1$ for $x = 1.2$ and $x^4 > x + 1$ for $x = 1.3$. Hence the positive solution is trapped between 1.2 and 1.3. To pin it down tighter, we tabulate by hundredths, starting at 1.2, working towards 1.3 until $x^4 > x + 1$.

	x	1.20	1.21	1.22	1.23
(to 4 places)	x^4	2.0736	2.1436	2.2153	2.2889
	$x + 1$	2.20	2.21	2.22	2.23

We see that $x^4 < x + 1$ for $x = 1.22$ and $x^4 > x + 1$ for $x = 1.23$, so the solution is between 1.22 and 1.23. We continue:

x	1.220	1.221
x^4	2.2153	2.2226
$x + 1$	2.220	2.221

The solution is between 1.220 and 1.221. Which is closer? To decide, we just check the midpoint 1.2205:

$$(1.2205)^4 \approx 2.2190 < 1.2205 + 1 = 2.2205 \, .$$

Therefore, the solution is between 1.2205 and 1.2210, so it is closer to 1.221 than to 1.220.

Answer   1.221

Use of Linear Interpolation

Suppose Example 1 called for 6 places instead of 3. The way we were going, that would require much more work. At the next step, we would calculate x^4 and $x + 1$ for $x = 1.2201, 1.2202, 1.2203, \cdots$. We might have to go all the way to 1.2209 before $x^4 > x + 1$. Then we might need 9 calculations at steps of 0.00001, etc.

There is a way to cut down the work considerably. For convenience, we write the equation $x^4 = x + 1$ in the form

$$f(x) = 0, \quad \text{where} \quad f(x) = x^4 - x - 1 \, .$$

From the last table in the solution of Example 1, we find

$$f(1.220) \approx 2.2153 - 2.2200 = -0.0047 \, ,$$

$$f(1.221) \approx 2.2226 - 2.2210 = 0.0016 \, .$$

Since the second value is closer to 0, the solution should be closer to 1.221 than to 1.220. In fact, the value $f(1.220)$ is about 3 times as far from 0 as the value $f(1.221)$ is from 0. So we suspect that the solution is about 3 times as far from 1.220 as from 1.221. That would place it about $\frac{3}{4}$ of the way from 1.220 to 1.221, at 1.22075. Therefore, instead of testing 1.2201, 1.2202, $\cdots$, we jump right to 1.2207. If we feel especially confident, we could jump right to 1.22075 and start testing at intervals of 10^{-5}. That would pay off since, to 6 places, the solution turns out to be 1.220744.

This method worked so well that we should examine what is really going on. Basically we assumed that as x increased from 1.220 to 1.221, the values of $f(x)$ varied proportionally from $f(1.220)$ to $f(1.221)$. In other words, we assumed that $f(x)$ was *linear* and estimated the solution by linear interpolation (Fig. 3a).

The technique of linear interpolation is useful for approximating solutions of equations $f(x) = 0$ in general. We assume that the graph of $y = f(x)$ is a smooth curve. By graphical or numerical methods, we find a small interval $[a, b]$ such that $f(a)$ and $f(b)$ are of opposite sign. Then there is a solution of $f(x) = 0$ inside $[a, b]$. We replace $f(x)$ by a linear function $g(x)$ on $[a, b]$, then approximate the solution of $f(x) = 0$ by the solution of $g(x) = 0$. See Fig. 3b. This linear interpolation can be done graphically. With a piece of millimeter graph paper and a straight edge you can get accurate estimates.

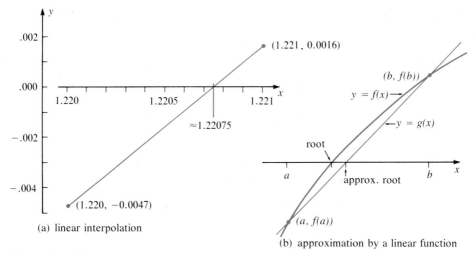

(a) linear interpolation

(b) approximation by a linear function

Fig. 3 Root location

The linear interpolation can also be done numerically on a calculator rather than graphically. Assume that $a < b$ and that $f(a)$ and $f(b)$ are of opposite sign. The linear function interpolating

$$(a, f(a)) \quad \text{and} \quad (b, f(b))$$

is

$$g(x) = \frac{f(b) - f(a)}{b - a}(x - a) + f(a).$$

The solution of $g(x) = 0$ is

$$x = a + \frac{-f(a)}{f(b) - f(a)}(b - a).$$

One can check that

$$x = a + \frac{|f(a)|}{|f(b)| + |f(a)|}(b - a),$$

which is easy to calculate. Actually only the multiplier of $b - a$ must be calculated, since $b - a = 10^{-n}$ at each step. For instance, in the graphical interpolation above,

$$a = 1.220 \qquad b - a = 0.001$$

and

$$\frac{|f(a)|}{|f(a)| + |f(b)|} \approx \frac{0.0047}{0.0047 + 0.0016} \approx 0.75$$

so the next approximate root is

$$1.220 + (0.75)(0.001) = 1.22075.$$

A shortcut is

4 7 $\boxed{\text{STO}}$ $\boxed{+}$ 1 6 $\boxed{\div}$ $\boxed{\text{RCL}}$ $\boxed{=}$ $\boxed{1/x}$ 0.75

Programmable Calculators

A programmable calculator can remember a sequence of instructions and execute the sequence repeatedly. This capability is useful whenever a function $f(x)$ must be evaluated for many values of x . You load into the program memory a sequence of steps that computes the function. Then if you enter any number x and press the run/stop ⌈R/S⌉ key, the calculator will compute $f(x)$.

But a programmable calculator can do more. It can make decisions (do one thing if the display is positive, another if it is negative) and it can loop (repeat a sequence of steps a prescribed number of times). These are tremendous capabilities. For example, they enable the entire process of searching for a solution of $f(x) = 0$ as in Example 1, interpolation and all, to be loaded into the program. (Any particular function $f(x)$ must be programmed and loaded as a subroutine.) Once this is set up, you tell the calculator where to start looking, and what accuracy you want. Then you just press ⌈R/S⌉ , sit back, and wait for the answer.

Exercises

x	0.500	0.501	0.502	0.503	0.504
$S(x)$	0.52110	0.52222	0.52335	0.52448	0.52561
$C(x)$	1.12763	1.12815	1.12867	1.12919	1.12972

Use the table to estimate

1 $S(0.5013)$ **2** $S(0.5031)$ **3** $S(0.5028)$
4 $C(0.5006)$ **5** $C(0.5019)$ **6** $C(0.5035)$.

Use the table to estimate x, where

7 $S(x) = 0.52300$ **8** $C(x) = 1.12800$.

Solve approximately to 6 places

9 $x^3 + x^2 = 1$ **10** $x = \dfrac{1}{2 + x^4}$

11 $x^4 = \dfrac{1}{x + 1}$ **12** $x^3 = \dfrac{1}{x + 1}$.

3   EQUATIONS INVOLVING θ [Optional]

Now we are ready for equations involving trig functions and θ in other ways. A good way to start is by drawing a graph. That should show the number of roots and their approximate locations.

EXAMPLE 1   Locate roughly the roots of

(a) $\sin \theta = \theta^2$ (b) $\theta \tan \theta = 1$.

Solution   (a)   Graph $y = \sin \theta$ and $y = \theta^2$ on the same grid (Fig. 1). The curves cross at two points, indicating two roots. One is $\theta = 0$ and the other is near $\theta = 1$.

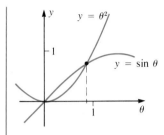

Fig. 1 Graphical solution of $\sin \theta = \theta^2$

Fig. 2 Graphical solution of $\theta \tan \theta = 1$

(b) Graph the even function $y = \theta \tan \theta$. One way is to start with a graph of $y = \tan \theta$ (Fig. 1, p. 52), superimpose the line $y = \theta$, and multiply corresponding values (Fig. 2). Even the roughest sketch shows infinitely many branches and the graph crossing the horizontal line $y = 1$ infinitely often. The intersections correspond to roots $\pm \theta_0, \pm \theta_1, \pm \theta_2, \cdots$ of $\theta \tan \theta = 1$, where

$$\tfrac{1}{4} \pi < \theta_0 < \tfrac{1}{2} \pi \qquad \pi < \theta_1 < \tfrac{3}{2} \pi \cdots n \pi < \theta_n < (n + \tfrac{1}{2}) \pi .$$

As we move to the right, the branches become steeper and steeper at $\theta = n \pi$, so we expect θ_n to be quite close to $n \pi$ if n is large.

The graphical method is a good start, but for more accurate approximations to roots, computational techniques are necessary.

EXAMPLE 2 Estimate to 6 places the positive root of $\sin \theta = \theta^2$.

Solution Set

$$f(\theta) = \theta^2 - \sin \theta .$$

We'll tabulate $f(\theta)$; for this we need a calculator routine. For example, in the $\boxed{\text{rad}}$ mode:

$\theta \ \boxed{\text{STO}} \ \boxed{x^2} \ \boxed{-} \ \boxed{\text{RCL}} \ \boxed{\sin} \ \boxed{=} \ .$

(One step shorter:

$\theta \ \boxed{x^2} \ \boxed{-} \ \boxed{\sqrt{}} \ \boxed{\sin} \ \boxed{=} \ .)$

We start at $\theta = 1.0$ and work backwards by tenths:

θ	.8	.9	1.0
$f(\theta)$	$-.077$	.027	.159

The root lies between 0.8 and 0.9, probably closer to 0.9. To estimate at what fraction of the way from 0.8 to 0.9 it lies, we interpolate:

$$\text{fractional part} = \frac{77}{77 + 27} \approx 0.7$$

Hence we now suspect that $\theta \approx 0.8 + 0.07 = 0.87$. We tabulate again, starting at 0.87 and moving by steps of 0.01:

θ	.86	.87	.88
$f(\theta)$		$-.00743$	.00366

The root lies in the interval $0.87 < \theta < 0.88$. We interpolate again.

$$\text{fractional part} = \frac{743}{743 + 366} \approx 0.67$$

Now we are in a tiny interval so we can safely move ahead *two* more places; we try $\theta \approx 0.8767$ and tabulate:

θ	.8766	.8767	.8768
$f(\theta)$		-2.92×10^{-5}	8.22×10^{-5}

One more interpolation:

$$\text{fractional part} = \frac{292}{292 + 822} \approx 0.26 \qquad \theta \approx 0.876726$$

Answer   0.876726

EXAMPLE 3   Estimate to 4 places the root of $\theta \tan \theta = 1$ near $\theta = -2\pi$.

Solution   Set $f(\theta) = \theta \tan \theta - 1$. Since $-2\pi \approx -6.28$ we start tabulating $f(\theta)$ at -6.3:

θ	-6.5	-6.4	-6.3
$f(\theta)$	.432	$-.249$	$-.894$

The root is in the interval $-6.5 < \theta < -6.4$. Interpolate:

$$\text{fractional part} = \frac{432}{432 + 249} \approx 0.6.$$

We must go $\frac{6}{10}$ of the way from -6.5 to -6.4, that is, to -6.44. (Since we are working with negative numbers, it would have been wiser to have computed the fractional part from -6.4 to -6.5:

$$\text{fractional part} = \frac{249}{249 + 432} \approx 0.4 \qquad \theta \approx -6.44.)$$

Tabulate values again:

θ	-6.45	-6.44	-6.43
$f(\theta)$		.0182	$-.0491$

$$\text{fractional part} = \frac{491}{491 + 182} \approx 0.73 \qquad \theta \approx -6.4373$$

Answer -6.4373

Iteration

The next example will illustrate an entirely different method of solution, called **iteration.**

EXAMPLE 4   Estimate to 4 places the root of $\cos \theta = \theta$.

Solution   First graph $y = \cos \theta$ and $y = \theta$ on the same grid (Fig. 3). Clearly $\cos \theta = \theta$ only for one value of θ, about 0.8.

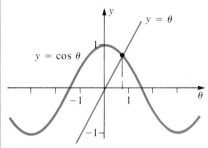

Fig. 3 $\cos \theta = \theta$

Fig. 4   Assume $f(\theta)$ increases *slowly* near θ_0. Set $\theta_{n+1} = f(\theta_n)$. Then $\theta_n \longrightarrow \theta_0$.

On your calculator (set [rad] mode), watch closely as you press

.8 [cos] [cos] [cos]

and continue until 4 places stabilize. After pressing [cos] 18 times the display is 0.739136, and after one more press, 0.739051. Since the values alternate up down up down, we conclude that all further values will round to 0.7391.

Answer   0.7391

This iteration method is remarkable. Unfortunately, it does not always work. The equation in Example 4 has the form $\theta = f(\theta)$. Near a root, computing $f(\theta)$ practically gives you back θ again. The theory of such equations says that if you start sufficiently close to a root θ_0 with a guess θ_1 and if $f(\theta)$ is changing sufficiently slowly near θ_0, then the sequence of iterations

$$\theta_2 = f(\theta_1) \qquad \theta_3 = f(\theta_2) \qquad \theta_4 = f(\theta_3)$$

will approach closer and closer to θ_0. Perhaps Fig. 4 helps show why.

EXAMPLE 5   Find the root of $\tan \theta = \theta$ near $\frac{3}{2}\pi$ to 5 places.

Solution   A graph (Fig. 5a, next page) of $y = \tan \theta$ and $y = \theta$ helps us see what is going on. The root is just a little less than $\frac{3}{2}\pi \approx 4.71$, so let us set [rad] mode, then start at 4.6 and iterate the tangent:

4.6 [tan] [tan] [tan] [tan] · · ·

The results are

8.86 -0.633 -0.734 -0.903 · · ·

Most unsatisfactory; we have drifted far away from the root near $\frac{3}{2}\pi$.

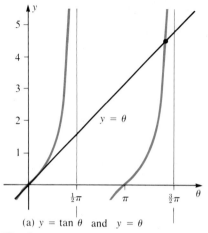
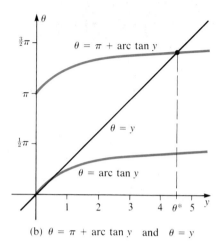

(a) $y = \tan \theta$ and $y = \theta$ (b) $\theta = \pi + \text{arc} \tan y$ and $\theta = y$

Fig. 5

The trouble is that the function $\tan \theta$ increases rapidly near the root (Fig. 5a). We can remedy this rapid growth by using the *inverse* function to the function $y = \tan \theta$ on the domain $\frac{1}{2}\pi < \theta < \frac{3}{2}\pi$. Where the function increases rapidly, its inverse function increases slowly. By Fig. 5b, the inverse function is

$$g(y) = \text{arc} \tan y + \pi \quad -\infty < y < \infty.$$

Clearly, if $\tan \theta_0 = \theta_0$, then

$$\theta_0 = g(\tan \theta_0) = g(\theta_0),$$

so the problem of solving $\tan \theta = \theta$ near $\frac{3}{2}\pi$ is equivalent to the problem of solving $g(y) = y$ near $\frac{3}{2}\pi$. We iterate, again in ⌐rad⌐ mode and starting at 4.6:

$$4.6 \boxed{g} \quad \boxed{g} \quad \boxed{g} \cdot \cdot \cdot \,,$$

where

$$\boxed{g} \text{ denotes* } \boxed{\text{INV}} \quad \boxed{\tan} \quad \boxed{+} \quad \boxed{\pi} \quad \boxed{=}.$$

The results are

4.4983   4.4936   4.49342   4.493410   4.493409   4.493409 · · ·

After only 6 iterations, 6 decimal places are fixed.

Answer   4.49341

Exercises

Estimate to 4 places the root of

1 $\theta + \sin \theta = 2.5$ **2** $\cos \theta = \sqrt{\theta}$

3 $\theta = 1 + \sin \theta$ **4** $\sin \theta = \theta^3 \quad \theta > 0$

---

* On a programmable calculator with "user definable" keys, this $\boxed{g}$ might indeed be a single key stroke.

5 $\theta \tan \theta = 0.5$ $0 < \theta < \frac{1}{2}\pi$

7 $\cos 2\theta = \frac{1}{6}\theta^{3/2}$ $\frac{1}{2}\pi < \theta < \pi$

9 $\sin \theta = 2/\theta$ $2\pi < \theta < \frac{5}{2}\pi$

6 $\cos \theta = e^{-\theta}$ $0 < \theta < \frac{1}{2}\pi$

8 $\sin \pi\theta^2 = 1 - \theta$ $0 < \theta < 1$

10 $\tan \theta = \frac{1}{5}\theta^3$ $\frac{1}{2}\pi < \theta < \frac{3}{2}\pi$.

Solve by iteration to 4 places

11 $\sin \theta + 0.5 = \theta$

13 $\cos \theta = 2\theta$

12 $\tan \theta = \theta$ $\theta \approx 7$

14* $4 \cos \theta = \theta$.

REVIEW EXERCISES

Solve on the interval $-\pi < \theta \le \pi$

1 $\sin 3\theta = -\frac{1}{2}\sqrt{3}$

2 $\tan 3\theta = -1$.

Find all solutions

3 $\sin 2\theta = -\sin \theta$

4 $\tan \theta = -\sin \theta$.

Solve to 4 places

5 $\sin^3 \theta = \theta^2 - 1$

6 $\tan \theta = 2\theta$ $0 < \theta < \frac{1}{2}\pi$.

9 TOPICS IN ANALYTIC GEOMETRY

1 POLAR COORDINATES

Until now, we have always used rectangular coordinates to locate points in the plane. Other coordinate systems exist as well. The most important of these is the system of polar coordinates.

In a rectangular coordinate system (Fig. 1a) two families of grid lines, $x = $ constant and $y = $ constant, fill the plane. Each point (a, b) is the intersection of two of these lines, $x = a$ and $y = b$.

In a polar coordinate system (Fig. 1b), the grid lines are (1) all circles centered at $(0, 0)$, and (2) all rays from $(0, 0)$. Each point (x, y) different from $(0, 0)$

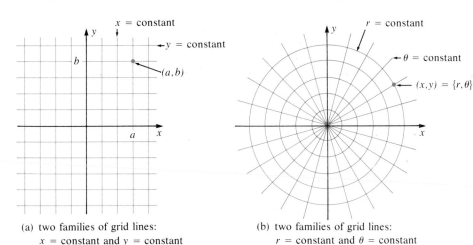

(a) two families of grid lines:
  $x = $ constant and $y = $ constant

(b) two families of grid lines:
  $r = $ constant and $\theta = $ constant

Fig. 1 Rectangular and polar coordinates

is the intersection of one circle and one ray. The circle is identified by a positive number r, its radius, and the ray is identified by a real number θ, its angle in radians from the positive x-axis. Thus (x, y) is assigned the polar coordinates $\{r, \theta\}$. Since θ is determined only up to a multiple of 2π, we agree that

$$\{r, \theta + 2\pi n\} = \{r, \theta\} \quad (n \text{ any integer}).$$

The point $(0, 0)$ does not determine an angle θ. Nonetheless, it is customary to say that any pair $\{0, \theta\}$ represents $(0, 0)$.

The idea of polar coordinates is quite natural. You identify a point by telling how far it is from the origin and in what direction. (This is the principle of the radar screen, where the origin is your current position.) In Figure 2, we plot a few points in polar coordinates.

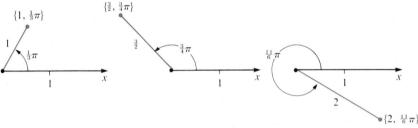

Fig. 2 Examples of polar coordinates

Given the polar coordinates $\{r, \theta\}$ of a point, what are its rectangular coordinates? The point is r units from the origin in the direction θ. Hence $x = r \cos \theta$, $y = r \sin \theta$. See Fig. 3a. Conversely, given the rectangular coordinates (x, y), what are the polar coordinates? Figure 3b shows that $r = \sqrt{x^2 + y^2}$, and that θ is determined by $\cos \theta = x/r$ and $\sin \theta = y/r$.

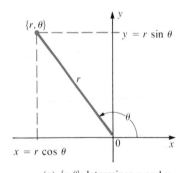

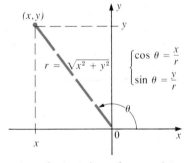

(a) $\{r, \theta\}$ determines x and y (b) (x, y) determines r and θ

Fig. 3

Polar to Rectangular	Rectangular to Polar
$x = r \cos \theta$	$r = \sqrt{x^2 + y^2}$
$y = r \sin \theta$	$\cos \theta = \dfrac{x}{\sqrt{x^2 + y^2}} = \dfrac{x}{r}$
	$\sin \theta = \dfrac{y}{\sqrt{x^2 + y^2}} = \dfrac{y}{r}$

EXAMPLE 1 (a)   Convert $(2, -2\sqrt{3})$ to polar coordinates.
(b)   Convert $\{3, \frac{1}{6}\pi\}$ to rectangular coordinates.

Solution (a) $r^2 = 4 + 12 = 16$, $r = 4$. Also $\cos\theta = \frac{2}{4} = \frac{1}{2}$ and $\sin\theta = \frac{1}{4}(-2\sqrt{3}) = -\frac{1}{2}\sqrt{3}$, so $\theta = \frac{5}{3}\pi$.

(b) $x = r\cos\theta = 3\cos\frac{1}{6}\pi = \frac{3}{2}\sqrt{3}$, and $y = r\sin\theta = 3\sin\frac{1}{6}\pi = \frac{3}{2}$.

Answer (a) $\{4, \frac{5}{3}\pi\}$ (b) $(\frac{3}{2}\sqrt{3}, \frac{3}{2})$

Remark   Some calculators have built-in keys for rectangular-polar conversions.

Negative r

In applications it is convenient to allow points $\{r, \theta\}$ with $r < 0$. For example, consider a ray and a point $\{r, \theta\}$ on the ray (Fig. 4a). Suppose the point moves towards $(0,0)$, through $(0,0)$, and keeps on going! Then r decreases, becomes 0, but then what? So that θ won't jump abruptly to $\theta + \pi$, we agree that θ remains constant, but r becomes negative (Fig. 4b). This amounts to agreeing that

$$\{-r, \theta\} = \{r, \theta + \pi\}.$$

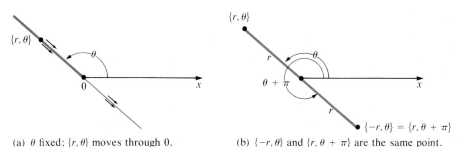

(a) θ fixed; $\{r, \theta\}$ moves through 0.   (b) $\{-r, \theta\}$ and $\{r, \theta + \pi\}$ are the same point.
Fig. 4   Negative r

Just for practice, let us plot a few specific points with r negative (Fig. 5).

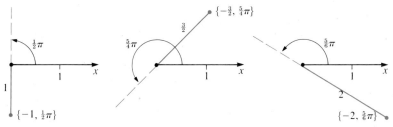

Fig. 5   Examples with $r < 0$

EXAMPLE 2 Convert to rectangular coordinates

(a) $\{-2, \frac{3}{4}\pi\}$ (b) $\{-5, -\frac{1}{3}\pi\}$.

Solution (a) $\{-2, \frac{3}{4}\pi\} = \{2, \frac{3}{4}\pi + \pi\} = \{2, \frac{7}{4}\pi\}$. Hence

$$x = 2\cos\frac{7}{4}\pi = \sqrt{2} \qquad y = 2\sin\frac{7}{4}\pi = -\sqrt{2}.$$

(b) $\{-5, -\frac{1}{3}\pi\} = \{5, -\frac{1}{3}\pi + \pi\} = \{5, \frac{2}{3}\pi\},$

$\qquad x = 5 \cos \frac{2}{3}\pi = -\frac{5}{2} \qquad y = 5 \sin \frac{2}{3}\pi = \frac{5}{2}\sqrt{3}$

Answer   (a) $(\sqrt{2}, -\sqrt{2})$ (b) $(-\frac{5}{2}, \frac{5}{2}\sqrt{3})$

Lines

Each curve $r = r_0$ is a circle with center $(0, 0)$, except for $r_0 = 0$. Each curve $\theta = \theta_0$ is a line through the origin (Fig. 6a).

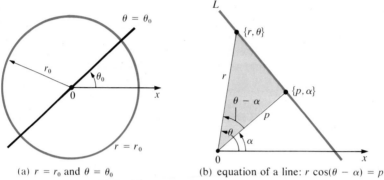

(a) $r = r_0$ and $\theta = \theta_0$ (b) equation of a line: $r \cos(\theta - \alpha) = p$

Fig. 6

What is the equation in polar coordinates for a line L not necessarily through the origin? Drop a perpendicular from $(0, 0)$ to L; let its foot have polar coordinates $\{p, \alpha\}$. See Fig. 6b. If $\{r, \theta\}$ is any point on L, then from the right triangle, $\cos(\theta - \alpha) = p/r$. Therefore

> The equation in polar coordinates of the general straight line is
>
> $\qquad r \cos(\theta - \alpha) = p,$
>
> where p and α are constants.

EXAMPLE 3   (a)   Express in rectangular coordinates $r \cos(\theta - \frac{1}{6}\pi) = 3$.

(b)   Express in polar coordinates $x + y = -5$.

Solution   (a)   Use the addition law for cosines:

$\qquad r \cos(\theta - \frac{1}{6}\pi) = 3 \qquad r(\cos\theta \cos\frac{1}{6}\pi + \sin\theta \sin\frac{1}{6}\pi) = 3.$

Replace $r \cos\theta$ by x and $r \sin\theta$ by y:

$\qquad x \cos\frac{1}{6}\pi + y \sin\frac{1}{6}\pi = 3 \qquad \frac{1}{2}\sqrt{3}\,x + \frac{1}{2}y = 3.$

(b)   Substitute $x = r \cos\theta$ and $y = r \sin\theta$:

$\qquad x + y = -5 \qquad r \cos\theta + r \sin\theta = -5.$

To express this relation in polar form, we must have the right-hand side positive and the left-hand side in the form $r \cos(\theta - \alpha)$. So we bring the minus sign to the left and express the result as a constant times a cosine (review p. 84):

$$r(-\cos \theta - \sin \theta) = 5$$
$$r(-\tfrac{1}{2}\sqrt{2} \cos \theta - \tfrac{1}{2}\sqrt{2} \sin \theta) = \tfrac{5}{2}\sqrt{2}$$
$$r(\cos \tfrac{5}{4}\pi \cos \theta + \sin \tfrac{5}{4}\pi \sin \theta) = \tfrac{5}{2}\sqrt{2}$$
$$r \cos(\theta - \tfrac{5}{4}\pi) = \tfrac{5}{2}\sqrt{2}.$$

Answer (a) $\tfrac{1}{2}x\sqrt{3} + \tfrac{1}{2}y = 3$ (b) $r \cos(\theta - \tfrac{5}{4}\pi) = \tfrac{5}{2}\sqrt{2}$.

Remark The polar form $r \cos(\theta - \alpha) = p$ is equivalent to the normal form

$$x \cos \alpha + y \sin \alpha = p.$$

For substitute $x = r \cos \theta$ and $y = r \sin \theta$ into the normal form and use the addition formula for cosine:

$$r \cos \theta \cos \alpha + r \sin \theta \sin \alpha = p \qquad r \cos(\theta - \alpha) = p.$$

Exercises

Plot the points

1 $\{3, \tfrac{1}{6}\pi\}$ 2 $\{2, \pi\}$ 3 $\{2, \tfrac{5}{4}\pi\}$

4 $\{1, \tfrac{9}{4}\pi\}$ 5 $\{-3, \tfrac{2}{3}\pi\}$ 6 $\{-2, \tfrac{5}{3}\pi\}$.

Sketch the region in the plane

7 $r > 1$ 8 $\tfrac{1}{2} \le r < \tfrac{3}{2}$

9 $\tfrac{1}{4}\pi \le \theta \le \tfrac{1}{2}\pi$ 10 $1 \le r \le 2$, $\tfrac{4}{3}\pi \le \theta \le \tfrac{3}{2}\pi$.

Express in rectangular coordinates

11 $\{1, \tfrac{1}{2}\pi\}$ 12 $\{1, -\tfrac{1}{2}\pi\}$ 13 $\{1, -\tfrac{1}{6}\pi\}$

14 $\{1, \tfrac{1}{3}\pi\}$ 15 $\{2, -\tfrac{3}{4}\pi\}$ 16 $\{2, \tfrac{5}{4}\pi\}$.

Express in polar coordinates

17 $(1, 1)$ 18 $(0, -1)$ 19 $(-1, 1)$

20 $(-\tfrac{1}{2}, \tfrac{1}{2}\sqrt{3})$ 21 $(\sqrt{3}, -1)$ 22 $(\sqrt{2}, -\sqrt{2})$.

Find the equation in polar coordinates

23 line through $(0, 0)$ and $\{3, \tfrac{1}{4}\pi\}$ 24 circle, center $(0, 0)$, radius 5

25 line through $\{1, 0\}$ and $\{1, \tfrac{1}{2}\pi\}$ 26 line $ax + by = c$.

2   POLAR GRAPHS

We now find the equation in polar coordinates for a circle tangent to the y-axis at the origin (Fig. 1a, next page). Let the radius of the circle be a, so its center is $(a, 0)$ (provided the circle lies to the right of the y-axis). As seen in Fig. 1b, the point $\{r, \theta\}$ on the circle and the two endpoints of the horizontal diameter determine a right triangle. (Recall that an angle inscribed in a semicircle is a right angle.) Its hypotenuse is $2a$ and the side adjacent to θ is r, hence

$$r = 2a \cos \theta.$$

This is the polar equation of the circle.

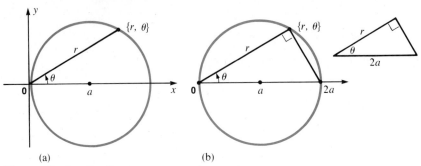

Fig. 1   Circle of radius a tangent to the y-axis at **0**

How does $\{r, \theta\}$ move on the circle $r = 2a \cos \theta$, as θ makes a complete revolution? If θ starts at 0, then r starts at $2a$. If θ increases to $\frac{1}{2}\pi$, then r decreases to 0. (Think of an arm rotating counterclockwise and shrinking.) Hence $\{r, \theta\}$ traces the upper half of the circle (Fig. 2a).

If θ increases from $\frac{1}{2}\pi$ to π, then r decreases from 0 to $-2a$. Since r is negative, the point $\{r, \theta\}$ is measured "backwards" and moves through the fourth quadrant, tracing the lower half of the circle (Fig. 2b).

Thus the full circle is described as θ runs from 0 to π. As θ runs from π to 2π, the same circle is traced again. For when θ is in the third quadrant, $r < 0$, so $\{r, \theta\}$ describes the semicircle in the first quadrant; when θ is in the fourth quadrant $r > 0$, so $\{r, \theta\}$ describes the semicircle in the fourth quadrant. Therefore, in one complete revolution of θ, the circle is traced *twice*.

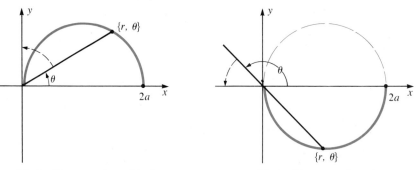

(a) As θ increases from 0 to $\frac{1}{2}\pi$,   (b) As θ increases from $\frac{1}{2}\pi$ to π,
  r decreases from $2a$ to 0. r decreases from 0 to $-2a$.

Fig. 2   The circle is traced by $\{r, \theta\}$ for $0 \leq \theta \leq \pi$

The graph of the equation

  $r = 2a \cos \theta$

is a circle of radius a and center $\{a, 0\}$. The circle is traced twice as θ makes a complete revolution.

Circles passing through the origin, but in other positions, can be handled similarly in polar coordinates.

Graphs of Functions

Graphing a function $r = f(\theta)$ in polar coordinates is tricky at first, because you must change your point of view. For $y = f(x)$, you think of x running along the horizontal axis, with the corresponding point (x, y) measured above or below. Basically your mental set is "left-right" and "up-down."

In polar coordinates, however, you must think of the angle θ swinging around (like a radar scope) and repeating after 2π. For each θ, you must measure forward from the origin a distance $f(\theta)$, or backward if $f(\theta) < 0$. Your mental set must be "round and round," "in and out," "forward or backward."

Because of the special nature of points $\{r, \theta\}$ where $r < 0$, pay close attention to the sign of $f(\theta)$ and be sure to plot points "backwards" if $f(\theta) < 0$.

Look for symmetries and periodicity. For example, if $f(\theta + 2\pi) = f(\theta)$, the polar graph $r = f(\theta)$ will repeat after 2π. There are many symmetries possible; we mention only two, $f(\theta)$ even and $f(\theta)$ odd. If $f(\theta)$ is even, that is, $f(-\theta) = f(\theta)$, then the point $\{r, -\theta\}$ is on the graph whenever $\{r, \theta\}$ is. The curve is symmetric in the x-axis (Fig. 3a). If $f(\theta)$ is odd, $f(-\theta) = -f(\theta)$, then the point $\{-r, -\theta\}$ is on the graph whenever $\{r, \theta\}$ is. The curve is symmetric in the vertical axis (Fig. 3b).

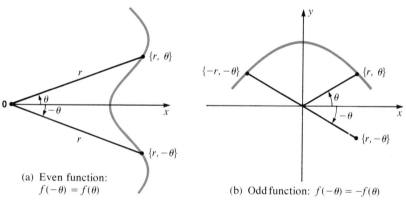

(a) Even function:
   $f(-\theta) = f(\theta)$

(b) Odd function: $f(-\theta) = -f(\theta)$

Fig. 3 Polar graphs of $r = f(\theta)$, where $f(\theta)$ is even or odd

EXAMPLE 1 Graph the **spiral of Archimedes,** $r = \theta$.

Solution As θ increases from 0 we see that r increases steadily. Hence the locus goes round and round, its distance from **0** greater and greater. The result is a spiral (Fig. 4a, next page). Since $f(\theta) = \theta$ is an odd function, we obtain the locus for $\theta < 0$ by reflection in the vertical axis (Fig. 4b).

EXAMPLE 2 Graph the **rose** $r = a \cos 2\theta$ where $a > 0$.

Solution Since $\cos 2(\theta + 2\pi) = \cos 2\theta$, the curve repeats every 2π, so we need plot it only for $0 \le \theta \le 2\pi$.

Since the sign of $\cos 2\theta$ fluctuates, it is advisable to make a preliminary sketch showing the proper sign (Fig. 5a, next page). As θ starts at 0 and increases to $\frac{1}{4}\pi$, we see that $\cos 2\theta$ starts at 1 and decreases to 0. Since $\cos 2\theta$ is an even function, this part of the graph is repeated below, forming a loop (Fig. 5b).

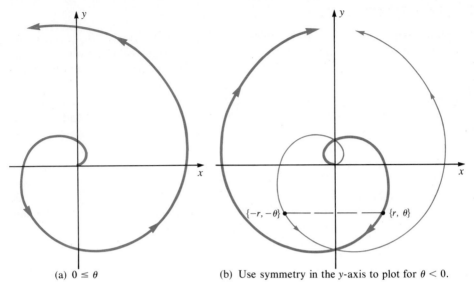

(a) $0 \le \theta$ (b) Use symmetry in the y-axis to plot for $\theta < 0$.

Fig. 4   The spiral $r = \theta$

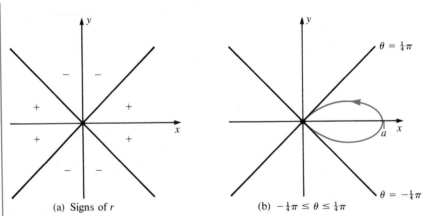

(a) Signs of r (b) $-\frac{1}{4}\pi \le \theta \le \frac{1}{4}\pi$

Fig. 5   Partial graph of $r = a \cos 2\theta$

As θ increases from $\frac{1}{4}\pi$ to $\frac{1}{2}\pi$ to $\frac{3}{4}\pi$, we see that $\cos 2\theta$ is negative and goes from 0 to -1 and back to 0. Thus we get another loop, but between $\frac{3}{4}\pi$ and $\frac{7}{4}\pi$. See Fig. 6a. For θ going from $\frac{3}{4}\pi$ to $\frac{5}{4}\pi$, we get a third loop plotted forward, and from $\frac{5}{4}\pi$ to $\frac{7}{4}\pi$ a fourth loop plotted backwards. The complete graph is shown in Fig. 6b.

For an accurate picture of the petals, plot some points. One thing we can say without plotting: the petals are rounded at their ends, not pointed. That stems from a property of the cosine: for small angles, $\cos 2\theta$ is very close to 1. Hence for θ small (near the tip of the petal to the right) the curve $r = a \cos 2\theta$ looks like the circle $r = a$.

Hindsight It is necessary to plot only one of the petals in Fig. 6b. Since $\cos(\theta + \pi) = -\cos\theta$, we have $\cos 2(\theta + \frac{1}{2}\pi) = -\cos 2\theta$. Thus $\cos 2\theta$ repeats it-

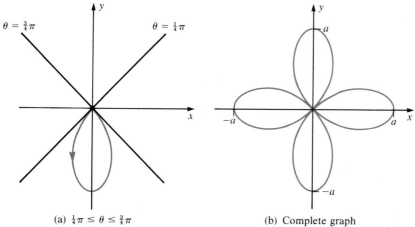

(a) $\frac{1}{4}\pi \le \theta \le \frac{3}{4}\pi$ (b) Complete graph

Fig. 6 Continuation of $r = a \cos 2\theta$

self negatively after $\frac{1}{2}\pi$. We conclude that the first loop we plotted (for $-\frac{1}{4}\pi \le \theta \le \frac{1}{4}\pi$) is repeated negatively as θ continues from $\frac{1}{4}\pi$ to $\frac{3}{4}\pi$. In other words, rotate the first loop backwards by $\frac{1}{2}\pi$; the result is another loop of the curve. Rotate again, and once again, and you have generated the whole curve.

Exercises

Find the equation in polar coordinates of the circle with

1   radius 2   center $(2, 0)$ **2**   radius 3   center $(-3, 0)$
3   radius 1   center $(0, 1)$ **4**   radius 1   center $(0, -1)$.

Graph

5 $r = 2\theta$ **6** $r = -\theta$ **7** $r = \theta^2$
8 $r = \theta^3$ **9** $r = \theta + \pi$ **10** $r = \pi/\theta$
11 $r = \sin 2\theta$ **12** $r = \cos 3\theta$ **13** $r = \cos 4\theta$
14 $r = \sin 4\theta$ **15** $r = \cos 5\theta$ **16** $r = \sin 5\theta$
17   the **cardioid** $r = 1 - \cos \theta$ **18**   the **lemniscate** $r^2 = \cos 2\theta$
19   the **limaçon** $r = 2 + \cos \theta$ **20**   the **limaçon** $r = 1 + 2 \cos \theta$
21   the **bifolium** $r = \sin \theta \cos^2\theta$ **22**   the **strophoid** $r = \cos 2\theta \sec \theta$
23   the **conchoid** $r = \csc \theta - 2$
24   the **cissoid** $r = \sin \theta \tan \theta$.

3   THE ELLIPSE

An **ellipse** is the locus of all points **x** such that the sum of the distances of **x** from two fixed points **p** and **q** is a constant greater than $\overline{pq}$, the distance from **p** to **q**. The points **p** and **q** are the **foci** (plural of **focus**) of the ellipse.

Our immediate goal is to derive the equation in rectangular coordinates of an ellipse whose foci lie on the x-axis. We shall do this in two steps; first we derive the equation in polar coordinates when one focus is at the origin, then we switch to rectangular coordinates.

Remark   The orbit of a planet around a fixed star is an ellipse with the star at one focus. Because in astronomy one measures angles rather than distances, it is natural to study the polar equation of an ellipse with one focus at the origin.

Polar Equation of an Ellipse

Let **p** and **q** be the foci of an ellipse, and set

$$2c = \overline{pq}$$

The ellipse, by definition, is the locus of all points **x** the sum of whose distances from **p** and **q** is a constant. We call this constant $2a$:

$$2a = \overline{xp} + \overline{xq}.$$

The ellipse is completely determined, except for *where* it is located, by these two constants a and c. We assume $0 < c < a$.

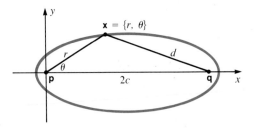

Fig. 1   Derivation of the polar equation of the ellipse

Now let us locate **p** at **0** and **q** on the positive x-axis (Fig. 1), so that **q** = $(2c, 0)$. Let $\{r, \theta\}$ be the polar coordinates of a typical point **x** of the ellipse. Then $\overline{xp} = r$, so the ellipse is defined by

$$r + \overline{xq} = 2a, \text{ that is, } \overline{xq} = 2a - r.$$

By the law of cosines,

$$\overline{xq}^2 = r^2 + (2c)^2 - 2r(2c) \cos \theta = r^2 + 4c^2 - 4rc \cos \theta.$$

Therefore the relation between r and θ is

$$(2a - r)^2 = r^2 + 4c^2 - 4rc \cos \theta,$$

that is,

$$4a^2 - 4ar + r^2 = r^2 + 4c^2 - 4rc \cos \theta.$$

Cancel r^2, divide by 4, and rearrange:

$$4a^2 - 4ar = 4c^2 - 4rc \cos \theta$$

$$a^2 - ar = c^2 - rc \cos \theta$$

$$r(a - c \cos \theta) = a^2 - c^2.$$

It is convenient to set $b^2 = a^2 - c^2, b > 0$.

Polar Equation   The polar equation of the ellipse with foci **p** = $(0, 0)$ and **q** = $(2c, 0)$ and length sum $2a$ is

$$r(a - c \cos \theta) = b^2$$

where $a^2 = b^2 + c^2$.

Rectangular Equation of the Ellipse

We now convert the polar equation $r(a - c \cos \theta) = b^2$ to rectangular coordinates. Since $r \cos \theta = x$, we can write the equation as

$$ar - cx = b^2, \quad \text{that is,} \quad ar = cx + b^2.$$

Square both sides and replace r^2 by $x^2 + y^2$:

$$a^2(x^2 + y^2) = (cx + b^2)^2 = c^2x^2 + 2cb^2x + b^4$$

$$(a^2 - c^2)x^2 - 2cb^2x + a^2y^2 = b^4$$

$$b^2x^2 - 2cb^2x + a^2y^2 = b^4$$

$$b^2(x^2 - 2cx) + a^2y^2 = b^4.$$

Complete the square:

$$b^2(x - c)^2 + a^2y^2 = b^4 + b^2c^2 = b^2(b^2 + c^2) = a^2b^2.$$

Finally divide by a^2b^2:

$$\frac{(x - c)^2}{a^2} + \frac{y^2}{b^2} = 1.$$

This is the rectangular equation for the ellipse with foci $(0, 0)$ and $(2c, 0)$ and length sum $2a$. If we translate it to the left by c units, we have the ellipse with foci $(-c, 0)$ and $(c, 0)$ and length sum still $2a$.

Rectangular Equation   The rectangular equation of the ellipse with foci $(-c, 0)$ and $(c, 0)$ and length sum $2a$ is

$$\frac{x^2}{a^2} + \frac{y^2}{b^2} = 1$$

where $a^2 = b^2 + c^2$.

Such an ellipse, whose foci are on the x-axis, symmetrically placed with respect to the origin, is said to be in **standard position** (Fig. 2).

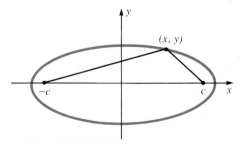

Fig. 2   Ellipse in standard position

Geometry of the Ellipse

We study the ellipse

$$\frac{x^2}{a^2} + \frac{y^2}{b^2} = 1 \qquad a^2 = b^2 + c^2$$

in standard position. Clearly for each point (x, y) on the ellipse

$$\frac{x^2}{a^2} \le \frac{x^2}{a^2} + \frac{y^2}{b^2} = 1, \quad \text{so} \quad x^2 \le a^2.$$

Therefore $-a \le x \le a$, and similarly $-b \le y \le b$; the ellipse is bounded by a rectangular box (Fig. 3a).

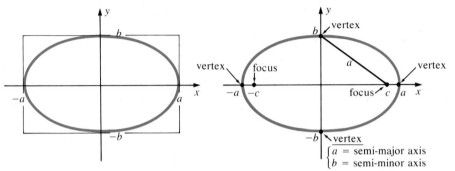

(a) The ellipse is bounded and symmetric.   (b) Center, axes, and vertices

Fig. 3   Geometry of the ellipse

If (x, y) is on the ellipse, then so are $(-x, y), (x, -y)$, and $(-x, -y)$, as follows readily from

$$\frac{(\pm x)^2}{a^2} + \frac{(\pm y)^2}{b^2} = \frac{x^2}{a^2} + \frac{y^2}{b^2} = 1.$$

Therefore the ellipse is symmetric with respect to both coordinate axes and with respect to the origin.

The points $(\pm a, 0)$ and $(0, \pm b)$ are on the ellipse, and are called its **vertices.**

The numbers a and b are known (historically) by the names **semi-major axis** and **semi-minor axis.** The midpoint of the foci is called the **center** of the ellipse. It is the origin for the ellipse in standard position.

Note from Fig. 3b that the distance from a focus to either of the points $(0, \pm b)$ is a; that is because $a^2 = b^2 + c^2$.

Remark 1   To construct an ellipse, tie a string of length $2a$ to two fixed pins $2c$ units apart $(a > c)$. Place your pencil against the string and move it so the string is taut. The locus generated is an ellipse. Why? If the pins are moved closer and closer together $(c \longrightarrow 0)$, the ellipse becomes more and more like a circle.

Remark 2   It is shown by calculus that on an elliptical billiard table, a ball shot in any direction from one focus will bounce once and pass through the other focus. (This accounts for the name focus, or focal point.) Whispering galleries are constructed in an elliptical shape because of this property.

Ellipses in Other Positions

By a translation of axes, we get

$$\frac{(x - h)^2}{a^2} + \frac{(y - k)^2}{b^2} = 1,$$

the equation of an ellipse centered at (h, k).

If $a > b$ then

$$\frac{x^2}{b^2} + \frac{y^2}{a^2} = 1$$

is an ellipse with major axis along the y-axis instead of the x-axis.

Remark If $a = b$, the equation $x^2/a^2 + y^2/b^2 = 1$ becomes

$$\frac{x^2}{a^2} + \frac{y^2}{a^2} = 1, \text{ that is, } x^2 + y^2 = a^2$$

a circle of radius a. Thus a circle can be considered as a limiting case of an ellipse, where $c = 0$. (The foci come together at one point, the center.)

EXAMPLE 1 Sketch the ellipse

$$\frac{(x - 2)^2}{9} + \frac{(y + 1)^2}{25} = 1$$

Locate the center, foci, and vertices.

Solution We have $a^2 = 25$ and $b^2 = 9$, so $a = 5$ and $b = 3$; and $c^2 = a^2 - b^2 = 16$, so $c = 4$. The center is $(2, -1)$ and the major axis is vertical. The foci are up and down $c = 4$ units from the center, so they are

$$(2, -1 + 4) = (2, 3) \quad \text{and} \quad (2, -1 - 4) = (2, -5).$$

The vertices on the major axis are up and down $a = 5$ units from the center, so they are

$$(2, -1 + 5) = (2, 4) \quad \text{and} \quad (2, -1 - 5) = (2, -6).$$

The vertices on the minor axis are left and right $b = 3$ units from the center, so they are

$$(2 + 3, -1) = (5, -1) \quad \text{and} \quad (2 - 3, -1) = (-1, -1).$$

Finally the curve is sketched in Fig. 4.

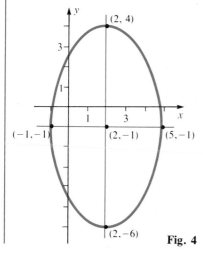

Fig. 4

Fig. 5 Foci: $(1, 2), (7, 2)$

EXAMPLE 2   Graph and find the equation of the ellipse with foci $(1, 2)$ and $(7, 2)$ and semi-minor axis 1.

Solution   The center is $(4, 2)$, the midpoint of the foci. The distance from the center to either focus is 3, so $c = 3$. Since the major axis is horizontal, the semi-minor axis is $b = 1$, and

$$a^2 = b^2 + c^2 = 1 + 9 = 10.$$

Consequently the equation is

$$\frac{(x - 4)^2}{10} + \frac{(y - 2)^2}{1} = 1$$

and the graph is sketched in Fig. 5, previous page.

Answer $\frac{1}{10}(x - 4)^2 + (y - 2)^2 = 1$

Eccentricity

Let us return to the polar form

$$r(a - c \cos \theta) = b^2$$

of the ellipse with foci $(0, 0)$ and $(2c, 0)$ and length sum $2a$. As usual, $b^2 = a^2 - c^2$. We can rewrite the equation in the form

$$r(1 - \frac{c}{a} \cos \theta) = \frac{b^2}{a}$$

We define the **eccentricity** of the ellipse to be the number $e = c/a$. We also set $p = b^2/ae$. Then the polar equation of the ellipse is

$$r(1 - e \cos \theta) = ep.$$

Since $0 < c < a$, the eccentricity satisfies $0 < e < 1$. The eccentricity determines the shape of the ellipse. If e is near zero, then c is small compared to a. That means that the foci are close together relative to the semi-major axis, hence the ellipse is circle-like. If e is near 1, the foci are relatively far apart and the ellipse is cigar-shaped. See Fig. 6. Once e is given, the scale factor p determines the size of the ellipse (as the radius does for a circle).

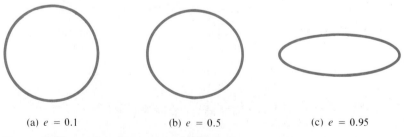

(a) $e = 0.1$ (b) $e = 0.5$ (c) $e = 0.95$

Fig. 6   Ellipses of various eccentricities

Parameterization of the Ellipse

Consider again the ellipse

$$\frac{x^2}{a^2} + \frac{y^2}{b^2} = 1 \qquad (a > b)$$

in standard position. Since

$$\left(\frac{x}{a}\right)^2 + \left(\frac{y}{b}\right)^2 = 1$$

there is an angle θ such that $x/a = \cos\theta$ and $y/b = \sin\theta$, so

$$(x, y) = (a\cos\theta, b\sin\theta).$$

As θ makes a complete revolution, the point $(a\cos\theta, b\sin\theta)$ traverses the ellipse once. See Fig. 7. Note that θ is *not* the polar angle of (x, y).

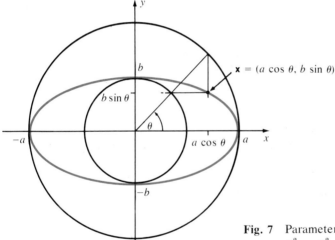

Fig. 7 Parameterization of the ellipse
$$\frac{x^2}{a^2} + \frac{y^2}{b^2} = 1$$

Exercises

Give the center, foci, major and minor semi-axes (a and b), and vertices, and sketch

1   $\frac{1}{25}x^2 + \frac{1}{9}y^2 = 1$ 　　　　　　　　　2 $x^2 + 4y^2 = 4$
3   $2(x + 1)^2 + (y - 2)^2 = 2$ 　　　　　　　4 $4x^2 + y^2 - 2y = 0$
5   $2x^2 + y^2 - 12x - 4y = -21$ 　　　　　　6 $x^2 + 2y^2 + 8y = 0$.

Write the equation of the ellipse

7   center at $(1, 4)$ vertices at $(10, 4)$ and $(1, 2)$
8   center at $(-2, -3)$ vertices at $(7, -3)$ and $(-2, -7)$
9   foci at $(2, 0)$ and $(8, 0)$ vertices at $(0, 0)$ and $(10, 0)$
10 foci at $(0, 3)$ and $(0, -3)$ semi-major axis $= 10$
11 foci at $(-1, 0)$ and $(3, 0)$ eccentricity $= \frac{1}{2}$
12 vertices at $(0, 2)$ and $(0, 6)$ eccentricity $= \frac{3}{4}$ major axis vertical.

13 Prove that the points on an ellipse farthest from its center are the two vertices on the major axis. [*Hint*   Use parameterization.]

14 Prove that the points on an ellipse nearest to its center are the two vertices on the minor axis. [*Hint*   Use parameterization.]

15 A 9-ft ladder slides along the floor as its top slides down a wall. If **x** is the point on the ladder $\frac{1}{3}$ of the way from its top, find the path of **x**.

16 A rod moves with one end on the *x*-axis and the other on the *y*-axis. If *P* is a point on the rod, prove that the locus of *P* is an ellipse.

17* Let **p** be one end of the major axis of an ellipse. Find the locus of the midpoint of **pq** as **q** traces the ellipse.

18* Let *E* be an ellipse and *L* a line. Prove that the midpoints of all chords of *E* parallel to *L* are collinear.

19 Let an ellipse have eccentricity *e*, and write $e = \cos \theta°$ with $0 < \theta° < 90°$. Then we speak of a **$\theta°$-ellipse.** Interpret geometrically. Draw ellipses of $15°, 30°, 45°$, and $60°$.

20 Fix *p* and *e* with $0 < p$ and $0 < e < 1$. Find the locus of all points *x* whose distance from 0 is *e* times its distance from the line $x = -p$. [*Hint*   Use polar coordinates.]

21 The orbit of the earth is approximately an ellipse with the sun at one focus and semi-major and semi-minor axes 9.3×10^7 and 9.1×10^7 miles respectively. Compute the eccentricity of the orbit.

22 (cont.) Find the distance from the sun to the other focus of the ellipse.

23 Show that

$$x = \frac{1 - t^2}{1 + t^2} \qquad y = \frac{2t}{1 + t^2}$$

parameterizes the unit circle by rational functions.

24* (cont.) Can you parameterize the ellipse $x^2/a^2 + y^2/b^2 = 1$ by rational functions? [*Hint*   Intersect the ellipse with the line $y = t(x + a)$. One intersection is $(-a, 0)$; express the other in terms of *t*.]

4 THE HYPERBOLA

A **hyperbola** is the locus of all points **x** such that the absolute value of the difference of the distances of **x** from two fixed points **p** and **q**, called the **foci,** is a constant, $2a$. Because **p**, **q**, and **x** form a triangle, $2a < \overline{pq}$. See Fig. 1.

Equation of the Hyperbola

Our derivation of the equation of a hyperbola parallels that of the ellipse, so we shall merely sketch the steps, and omit the details. First we work in polar coordinates with a hyperbola (Fig. 2) whose foci are

$$\mathbf{p} = (0, 0) \quad \text{and} \quad \mathbf{q} = (2c, 0).$$

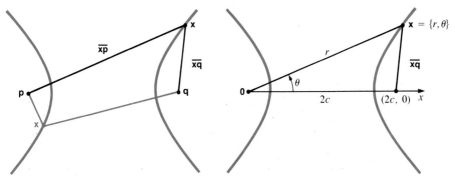

Fig. 1 The hyperbola:
$|\overline{\mathbf{xp}} - \overline{\mathbf{xq}}| = 2a$.
(Let the distance between the foci be
$2c = \overline{\mathbf{pq}}$.
Then $2a > 2c$ by the triangle
inequality, so $a > c$.)

Fig. 2 Set-up for deriving the
polar equation. Here the
foci are $\mathbf{p} = \mathbf{0}$ and
$\mathbf{q} = (2c, 0)$.

Then **x** is on the hyperbola provided

$$\overline{\mathbf{xq}} = r \pm 2a.$$

By the law of cosines

$$\overline{\mathbf{xq}}^2 = r^2 + 4c^2 - 4rc \cos \theta.$$

We eliminate $\overline{\mathbf{xq}}$ and simplify; the result is:

Polar Equation The polar equation of the hyperbola with foci $\mathbf{p} = (0,0)$
and $\mathbf{q} = (2c, 0)$ and absolute length difference $2a$ is

$$r(\pm a + c \cos \theta) = b^2$$

where $c^2 = a^2 + b^2$.

To derive the rectangular form, we first replace $r \cos \theta$ by x, transpose, square,
and replace r^2 by $x^2 + y^2$:

$$\pm ar + cx = b^2 \qquad \pm ar = b^2 - cx$$

$$a^2(x^2 + y^2) = (b^2 - cx)^2.$$

After we simplify and complete the square, the result is

$$\frac{(x - c)^2}{a^2} - \frac{y^2}{b^2} = 1.$$

This is the rectangular equation for the hyperbola with foci $(0,0)$ and $(2c, 0)$
and absolute length difference $2a$. If we translate this hyperbola to the left by c
units, we obtain the hyperbola with foci $(-c, 0)$ and $(c, 0)$ and absolute length
difference $2a$.

> **Rectangular Equation**   The rectangular equation of the hyperbola with foci $(-c, 0)$ and $(c, 0)$ and absolute length difference $2a$ is
>
> $$\frac{x^2}{a^2} - \frac{y^2}{b^2} = 1$$
>
> where $c^2 = a^2 + b^2$.

A hyperbola whose foci are on the x-axis, symmetrically placed with respect to the origin, is said to be in **standard position** (Fig. 3).

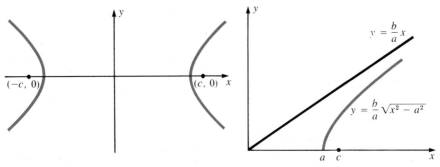

Fig. 3 Standard position of the hyperbola

Fig. 4 First quadrant portion of $\dfrac{x^2}{a^2} - \dfrac{y^2}{b^2} = 1$

Note   For the hyperbola in standard position, $a^2 = c^2 - b^2$, unlike for the ellipse in standard position, where $a^2 = b^2 + c^2$. Therefore both $a > b$ and $a \leq b$ are possible for the hyperbola, not just $a > b$ as for the ellipse.

Geometry of the Hyperbola

Let us sketch the hyperbola

$$\frac{x^2}{a^2} - \frac{y^2}{b^2} = 1.$$

We first note symmetry; if (x, y) satisfies this equation, then so do $(-x, y)$, $(x, -y)$, and $(-x, -y)$. Therefore the curve is symmetric in both axes and in the origin; we need plot it only in the first quadrant, then extend the curve to the other quadrants by symmetry.

We solve for y:

$$y = \frac{b}{a}\sqrt{x^2 - a^2}.$$

(The positive square root applies for the first quadrant.) The locus is defined only for $x \geq a$. When x starts at a and increases, y starts at 0 and increases. When x is very large, we suspect that y is slightly less than bx/a. To confirm this suspicion, we write

$$\frac{b}{a}x - y = \frac{b}{a}(x - \sqrt{x^2 - a^2}) = \frac{b}{a}(x - \sqrt{x^2 - a^2})\frac{x + \sqrt{x^2 - a^2}}{x + \sqrt{x^2 - a^2}}$$

$$= \left(\frac{b}{a}\right)\frac{x^2 - (x^2 - a^2)}{x + \sqrt{x^2 - a^2}} = \frac{ab}{x + \sqrt{x^2 - a^2}} < \frac{ab}{x}.$$

It follows that

$$\frac{b}{a}x - y \longrightarrow 0+ \quad \text{as} \quad x \longrightarrow \infty.$$

This means the curve approaches the line $y = bx/a$ (from below) as x increases; the line is an asymptote of the hyperbola. Now we can sketch the first quadrant portion (Fig. 4). We then extend the sketch to the other quadrants by symmetry (Fig. 5).

In dealing with hyperbolas, the following terminology is customary. A hyperbola consists of two **branches**: one where $\overline{xp} - \overline{xq} = 2a$, and the other where $\mathbf{xq} - \overline{xp} = 2a$. The point half-way between the foci is the **center** of a hyperbola. The line through the foci is the **principal axis,** and the line through the center perpendicular to the principal axis is the **conjugate axis.** The points where the hyperbola meets its principal axis are its **vertices.**

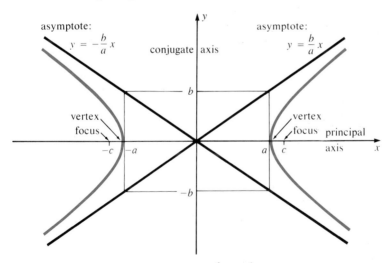

Fig. 5 Geometry of the hyperbola $\dfrac{x^2}{a^2} - \dfrac{y^2}{b^2} = 1$

The lines $y = \pm bx/a$ are the **asymptotes** of the hyperbola. A neat way to remember this is that the asymptotes of

$$\frac{x^2}{a^2} - \frac{y^2}{b^2} = 1 \quad \text{form the locus of} \quad \frac{x^2}{a^2} - \frac{y^2}{b^2} = 0.$$

This follows from the factorization

$$\frac{x^2}{a^2} - \frac{y^2}{b^2} = \left(\frac{x}{a} + \frac{y}{b}\right)\left(\frac{x}{a} - \frac{y}{b}\right).$$

The expression on the left is zero if and only if one of the factors on the right is zero, that is, if and only if $y = \pm bx/a$.

The asymptotes of the hyperbola

$$\frac{x^2}{a^2} - \frac{y^2}{b^2} = 1$$

are the lines $\quad y = \dfrac{b}{a}x \quad$ and $\quad y = -\dfrac{b}{a}x,$

or equivalently, the locus of the equation $\quad \dfrac{x^2}{a^2} - \dfrac{y^2}{b^2} = 0.$

Remark   The equations

$$\frac{x^2}{a^2} + \frac{y^2}{b^2} = 1 \quad \text{and} \quad \frac{x^2}{a^2} - \frac{y^2}{b^2} = 1$$

differ by a little minus sign, but that makes all the difference in the world. The first equation, where the sign is plus, requires $x^2 \le a^2$ and $y^2 \le b^2$; the locus is confined. The second imposes no such restriction; x^2/a^2 and y^2/b^2 can both be enormous, yet differ by 1.

By translation,

$$\frac{(x - h)^2}{a^2} - \frac{(y - k)^2}{b^2} = 1$$

is the equation of a hyperbola with center at (h, k) and horizontal principal axis.
By interchanging the roles of x and y, we see that the equation

$$-\frac{x^2}{a^2} + \frac{y^2}{b^2} = 1$$

defines a hyperbola with center at the origin, vertical principal axis, and foci at $(0, \pm c)$, where $c^2 = a^2 + b^2$.

EXAMPLE 1   Sketch the hyperbola

$$\frac{(x - 1)^2}{9} - \frac{(y + 2)^2}{16} = 1.$$

Locate the center, foci, vertices, and asymptotes.

Solution   Clearly the center is $(1, -2)$, the principal axis is horizontal, and $a = 3$ and $b = 4$, so $c^2 = a^2 + b^2 = 25$, or $c = 5$. Therefore the foci are

$$(1 \pm 5, -2) = (-4, -2), \quad (6, -2).$$

The asymptotes are the locus of

$$\frac{(x - 1)^2}{9} - \frac{(y + 2)^2}{16} = 0,$$

that is,

$$\frac{y + 2}{4} = \pm\frac{x - 1}{3} \quad \text{that is} \quad y + 2 = \pm\tfrac{4}{3}(x - 1).$$

They are the lines through $(1, -2)$ with slopes $\pm\tfrac{4}{3}$.

Since the principal axis is the horizontal line $y = -2$, the vertices are the points of the hyperbola for which $y = -2$, so

$$\frac{(x - 1)^2}{9} = 1, \quad \text{that is,} \quad x - 1 = \pm 3.$$

Hence the vertices are $(-2, -2)$ and $(4, -2)$. All of this information and a sketch of the curve are shown in Fig. 6.

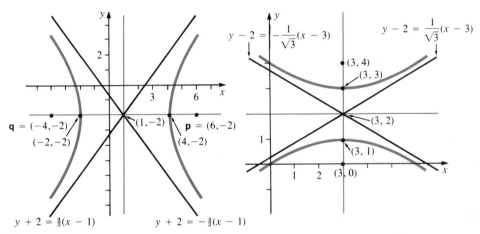

Fig. 6

Fig. 7

EXAMPLE 2   Find the equation of the hyperbola with foci $(3, 0)$ and $(3, 4)$ and a vertex at $(3, 3)$. Sketch the hyperbola and its asymptotes.

Solution   The center is the midpoint of the foci, so it is $(3, 2)$, and the equation of the hyperbola has the form

$$-\frac{(x - 3)^2}{a^2} + \frac{(y - 2)^2}{b^2} = 1.$$

The distance between the foci is 4, so $2c = 4$ and $c = 2$. Therefore

$$a^2 + b^2 = c^2 = 4.$$

But $(3, 3)$ is a vertex, hence on the hyperbola. This implies

$$-\frac{0}{a^2} + \frac{(3 - 2)^2}{b^2} = 1 \qquad b^2 = 1.$$

Hence $a^2 = 3$, so the equation of the hyperbola is

$$-\frac{(x - 3)^2}{3} + \frac{(y - 2)^2}{1} = 1.$$

To find its asymptotes, we set

$$-\frac{(x-3)^2}{3} + \frac{(y-2)^2}{1} = 0, \quad \text{that is,} \quad y - 2 = \pm\frac{1}{\sqrt{3}}(x-3).$$

Since $1/\sqrt{3} = \tan 30°$, we can make an accurate sketch (Fig. 7, previous page).

Answer $-\dfrac{(x-3)^2}{3} + (y-2)^2 = 1$

A hyperbola is called **rectangular** if its asymptotes are perpendicular. This happens when the slopes of the two asymptotes are negative reciprocals of each other.

$$\left(\frac{b}{a}\right)\left(-\frac{b}{a}\right) = -1 \qquad b^2 = a^2 \qquad b = a.$$

Thus the locus of

$$x^2 - y^2 = a^2$$

is a rectangular hyperbola.

Exercises

Find the center, foci, and asymptotes, and sketch

1   $\frac{1}{4}x^2 - \frac{1}{9}y^2 = 1$

2   $\frac{1}{9}x^2 - \frac{1}{4}y^2 = 1$

3   $-\frac{1}{9}x^2 + \frac{1}{4}y^2 = 1$

4   $-\frac{1}{4}x^2 + \frac{1}{9}y^2 = 1$

5   $(x+1)^2 - (y-1)^2 = 1$

6   $-(x-2)^2 + 4(y+1)^2 = 4$

7   $x^2 - y^2 = 1$

8   $-2x^2 + y^2 = 1$

9   $4x^2 - y^2 - 24x - 2y + 31 = 0$

10   $3x^2 - 3y^2 - 3x - 2y = \frac{31}{12}.$

Write the equation of the hyperbola with

11   foci $(0, \pm5)$ vertices $(0, \pm4)$

12   vertices $(\pm3, 0)$ foci $(\pm5, 0)$

13   asymptotes $y = \pm2x$ vertices $(\pm2, 0)$

14   foci $(1, 7), (1, -3)$ vertices $(1, 6), (1, -2)$

15   asymptotes $y = \pm(x-1)$ curve passes through $(3, 1)$

16   asymptotes $y = \pm2x$ curve passes through $(1, 1)$.

17   Show that the graph of $x^2 - y^2 + ax + by + c = 0$ is a rectangular hyperbola.

18   Show that the graph of $3x^2 - y^2 + ax + by + c = 0$ is a hyperbola whose asymptotes form a 60° angle.

19   Sketch the region $x^2/a^2 + y^2/b^2 < 1$ and the region $x^2/a^2 + y^2/b^2 > 1$.

20   Sketch the region $x^2/a^2 - y^2/b^2 < 1$ and the region $x^2/a^2 - y^2/b^2 > 1$.

21   Let C_1 and C_2 be circles external to each other. Prove that the locus of the center of a circle that touches C_1 and C_2 externally either is a branch of a hyperbola or a line.

22   A rifle at point **a** on level ground is shot at a target at point **b**. Find the locus of all observers who hear the shot and the impact of the shell simultaneously.

5 ROTATION OF AXES

In this section we shall study rotations of the coordinate system about the origin, and some applications.

Rotation is much easier in polar coordinates than in rectangular coordinates. If the polar axis is rotated through an angle α, a point $\{r, \theta\}$ acquires new polar coordinates $\{\bar{r}, \bar{\theta}\}$. The relation between the new and old coordinates of a point are seen directly from Fig. 1. Since **0** is unchanged, $\bar{r} = r$. From the figure, $\bar{\theta} = \theta - \alpha$.

Rotation (Polar Coordinates)   Suppose the polar axis is rotated by an angle α. Then the old polar coordinates $\{r, \theta\}$ and the new polar coordinates $\{\bar{r}, \bar{\theta}\}$ of a point are related by

$$\begin{cases} \bar{r} = r \\ \bar{\theta} = \theta - \alpha \end{cases} \quad \text{or equivalently} \quad \begin{cases} r = \bar{r} \\ \theta = \bar{\theta} + \alpha. \end{cases}$$

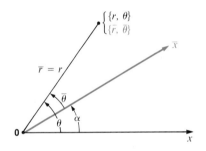

Fig. 1   Rotation in polar coordinates:
$\bar{r} = r \quad \bar{\theta} = \theta - \alpha$

Fig. 2   Snappy derivation of
the polar equation of a line

As an application, let us find the polar equation of the line L that is p units from the origin, perpendicular to the ray $\theta = \alpha$. See Fig. 2. Relative to the tilted axis, the line has equation $\bar{x} = p$ or $\bar{r} \cos \bar{\theta} = p$. Its r, θ-equation therefore is $r \cos (\theta - \alpha) = p$. This is a quick derivation of the result given on p. 464.

EXAMPLE 1   Find the equation in polar coordinates of the circle with center $(4, 3)$ and radius 5.

Solution   Since $3^2 + 4^2 = 5^2$, the circle passes through the origin (Fig. 3a, next page). Its diameter through $(0, 0)$ and its center $(4, 3)$ is inclined at angle α to the polar axis, where

$$\alpha = \text{arc tan } \tfrac{3}{4}.$$

Now rotate the polar axis through angle α. See Fig. 3b. In the new $\bar{r}, \bar{\theta}$-coordinates the circle has center $(5, 0)$ and radius 5, a situation we studied on p. 190. Its equation in the new coordinates is therefore

$$\bar{r} = 10 \cos \bar{\theta},$$

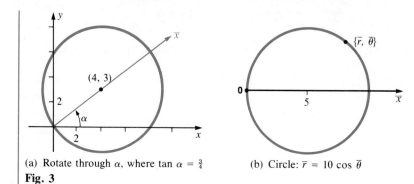

(a) Rotate through α, where $\tan \alpha = \frac{3}{4}$ (b) Circle: $\bar{r} = 10 \cos \bar{\theta}$

Fig. 3

hence in the old coordinates it is

$$r = 10 \cos(\theta - \alpha) = 10(\cos \alpha \cos \theta + \sin \alpha \sin \theta).$$

Clearly

$$\cos \alpha = \tfrac{4}{5} \quad \text{and} \quad \sin \alpha = \tfrac{3}{5},$$

so the equation is

$$r = 8 \cos \theta + 6 \sin \theta.$$

Check   When $\theta = \alpha$, then $r = 8 \cdot \frac{4}{5} + 6 \cdot \frac{3}{5} = \frac{32}{5} + \frac{18}{5} = \frac{50}{5} = 10$.

When $\theta = \alpha + \frac{1}{2}\pi$, then $r = 8 \cdot (-\frac{3}{5}) + 6 \cdot (\frac{4}{5}) = 0$.

Answer $r = 8 \cos \theta + 6 \sin \theta$

Rotation in Rectangular Coordinates

Now let's find how the rectangular coordinates of a point transform under a rotation about the origin. As before, let α denote the rotation angle (Fig. 4).

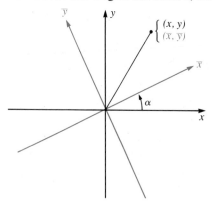

Fig. 4   Rotation in rectangular coordinates

A point has old coordinates (x, y) and new coordinates $(\bar{x}, \bar{y})$. We wish to express one set of coordinates in terms of the other and α. The simplest way to do this is to pass through polar coordinates, using

$$\begin{cases} x = r \cos \theta \\ y = r \sin \theta \end{cases} \quad \begin{cases} \bar{x} = \bar{r} \cos \bar{\theta} \\ \bar{y} = \bar{r} \sin \bar{\theta} \end{cases}$$

and the relations between the old and new polar coordinates:

$$\bar{r} = r \qquad \bar{\theta} = \theta - \alpha.$$

We have

$$\bar{x} = \bar{r} \cos \bar{\theta} = r \cos(\theta - \alpha)$$

$$= r \cos \theta \cos \alpha + r \sin \theta \sin \alpha = x \cos \alpha + y \sin \alpha$$

and

$$\bar{y} = \bar{r} \sin \bar{\theta} = r \sin(\theta - \alpha)$$

$$= r \sin \theta \cos \alpha - r \cos \theta \sin \alpha = y \cos \alpha - x \sin \alpha.$$

We have proved that

$$\bar{x} = x \cos \alpha + y \sin \alpha \qquad \bar{y} = -x \sin \alpha + y \cos \alpha.$$

These formulas express $\bar{x}$ and $\bar{y}$ in terms of x and y (and of course α). To express x and y in terms of $\bar{x}$ and $\bar{y}$ we can solve the formulas as a linear system. Alternatively we can argue that a rotation of $-\alpha$ takes the $\bar{x}, \bar{y}$-axes to the x, y-axes, hence

$$x = \bar{x} \cos(-\alpha) + \bar{y} \sin(-\alpha) = \bar{x} \cos \alpha - \bar{y} \sin \alpha$$

$$y = -\bar{x} \sin(-\alpha) + \bar{y} \cos(-\alpha) = \bar{x} \sin \alpha + \bar{y} \cos \alpha.$$

Rotation (Rectangular Coordinates) Suppose the plane is rotated through angle α and the x- and y-axes, under this rotation, go to the $\bar{x}$- and $\bar{y}$-axes. Then the x, y-coordinates and $\bar{x}, \bar{y}$-coordinates of any point are related by

$$\begin{cases} x = \bar{x} \cos \alpha - \bar{y} \sin \alpha \\ y = \bar{x} \sin \alpha + \bar{y} \cos \alpha \end{cases} \quad \begin{cases} \bar{x} = x \cos \alpha + y \sin \alpha \\ \bar{y} = -x \sin \alpha + y \cos \alpha. \end{cases}$$

Examples

(1) $\alpha = \frac{1}{4}\pi$ $\begin{cases} x = \frac{1}{2}\sqrt{2}\,\bar{x} - \frac{1}{2}\sqrt{2}\,\bar{y} \\ y = \frac{1}{2}\sqrt{2}\,\bar{x} + \frac{1}{2}\sqrt{2}\,\bar{y} \end{cases}$

(2) $\alpha = -\frac{1}{6}\pi$ $\begin{cases} x = \frac{1}{2}\sqrt{3}\,\bar{x} + \frac{1}{2}\bar{y} \\ y = -\frac{1}{2}\bar{x} + \frac{1}{2}\sqrt{3}\,\bar{y} \end{cases}$

Conics

We have learned how to graph quadratic equations of the form

$$ax^2 + cy^2 + dx + ey + f = 0 \qquad (a^2 + c^2 > 0).$$

By completing squares, we generally obtain one of the conic sections. Now we tackle the most general quadratic equation

$$ax^2 + bxy + cy^2 + dx + ey + f = 0.$$

It is the term bxy that makes life difficult. Where does it come from and how can we get rid of it? We can learn a good deal from an experiment.

EXAMPLE 2   Find the equation of the ellipse

$$\frac{x^2}{9} + \frac{y^2}{4} = 1$$

in the $\bar{x}, \bar{y}$-coordinate system that results from a $\frac{1}{4}\pi$ rotation of the x, y-coordinate system.

Solution   The rotation formulas are

$$x = \tfrac{1}{2}\sqrt{2}\,(\bar{x} - \bar{y}) \qquad y = \tfrac{1}{2}\sqrt{2}\,(\bar{x} + \bar{y}).$$

Substitute:

$$\frac{x^2}{9} + \frac{y^2}{4} = \frac{1}{2}\left[\frac{(\bar{x} - \bar{y})^2}{9} + \frac{(\bar{x} + \bar{y})^2}{4} \right] = \frac{1}{2}\left[\frac{13}{36}\bar{x}^2 + \frac{10}{36}\bar{x}\bar{y} + \frac{13}{36}\bar{y}^2 \right].$$

Answer $\frac{13}{72}\bar{x}^2 + \frac{10}{72}\bar{x}\bar{y} + \frac{13}{72}\bar{y}^2 = 1$ See Fig. 5.

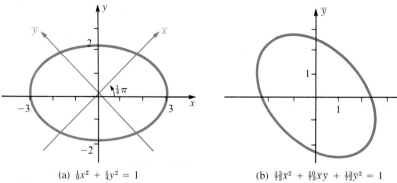

(a) $\frac{1}{9}x^2 + \frac{1}{4}y^2 = 1$ (b) $\frac{13}{72}\bar{x}^2 + \frac{10}{72}\bar{x}\bar{y} + \frac{13}{72}\bar{y}^2 = 1$

Fig. 5   An ellipse in rotated coordinates

The experiment suggests that the $\bar{x}\bar{y}$ term is due to the tilt of the coordinate axes relative to the axes of the ellipse. (The same should be true for hyperbolas.)

Example 5.2 suggests that an xy term occurs when the axes are placed obliquely. We suspect it can be eliminated by rotating the axes through a suitably chosen angle. Now the most general rotation of coordinates is

$$\begin{cases} x = \bar{x}\cos\alpha - \bar{y}\sin\alpha \\ y = \bar{x}\sin\alpha + y\cos\alpha. \end{cases}$$

This rotation changes a linear polynomial $dx + ey + f$ in x and y into a linear polynomial in $\bar{x}$ and $\bar{y}$. Of more interest to us is what happens to a quadratic polynomial $ax^2 + bxy + cy^2$. Substitute:

$$ax^2 + bxy + cy^2 = a(\bar{x}\cos\alpha - \bar{y}\sin\alpha)^2$$

$$+ b(\bar{x}\cos\alpha - \bar{y}\sin\alpha)(\bar{x}\sin\alpha + \bar{y}\cos\alpha) + c(\bar{x}\sin\alpha + \bar{y}\cos\alpha)^2.$$

Multiply out and collect terms in $\bar{x}^2$, $\bar{x}\bar{y}$, and $\bar{y}^2$; the result is a quadratic polynomial:

Under a rotation through an angle α, the quadratic polynomial

$$ax^2 + bxy + cy^2 + dx + ey + f$$

is changed into

$$\bar{a}\bar{x}^2 + \bar{b}\bar{x}\bar{y} + \bar{c}\bar{y}^2 + \bar{d}\bar{x} + \bar{e}\bar{y} + \bar{f}$$

where
$$\begin{cases} \bar{a} = a\cos^2\alpha + b\cos\alpha\sin\alpha + c\sin^2\alpha \\ \bar{b} = 2(c - a)\sin\alpha\cos\alpha + b(\cos^2\alpha - \sin^2\alpha) \\ \bar{c} = a\sin^2\alpha - b\sin\alpha\cos\alpha + c\cos^2\alpha. \end{cases}$$

For our purposes the most interesting of these formulas is the one for $\bar{b}$, which we can write in the form

$$\bar{b} = (c - a)\sin 2\alpha + b\cos 2\alpha.$$

From this formula we see that it is always possible to choose the rotation angle α so that $\bar{b} = 0$, that is, so that

$$(c - a)\sin 2\alpha + b\cos 2\alpha = 0.$$

For if $c = a$, then we simply choose $\alpha = \pm\frac{1}{4}\pi$; otherwise we choose α so that

$$\tan 2\alpha = \frac{b}{a - c}.$$

A quadratic locus

$$ax^2 + bxy + cy^2 + dx + ey + f = 0$$

is changed into a quadratic locus

$$\bar{a}\bar{x}^2 + \bar{c}\bar{y}^2 + \bar{d}\bar{x} + \bar{e}\bar{y} + \bar{f} = 0$$

without an $\bar{x}\bar{y}$ term by rotating the axes through angle α, where

$$\alpha = \pm\tfrac{1}{4}\pi \quad \text{if} \quad a = c,$$

$$\tan 2\alpha = \frac{b}{a - c} \quad \text{if} \quad a \neq c.$$

Because the tangent has period π, the angle 2α is determined up to a multiple of π; hence α is determined only up to a multiple of $\frac{1}{2}\pi$. Therefore we can always choose α in the first quadrant.

In numerical examples, we must compute $\bar{a}$ and $\bar{c}$, having a, b, c, and $\tan 2\alpha$. We write the formulas for $\bar{a}$ and $\bar{c}$ in the form

$$\begin{cases} \bar{a} = a\cos^2\alpha + \tfrac{1}{2}b\sin 2\alpha + c\sin^2\alpha \\ \bar{c} = a\sin^2\alpha - \tfrac{1}{2}b\sin 2\alpha + c\cos^2\alpha. \end{cases}$$

From $\tan 2\alpha$ we can find $\sin 2\alpha$ and $\cos 2\alpha$:

$$\sin 2\alpha = \frac{\pm\tan 2\alpha}{\sqrt{1 + \tan^2 2\alpha}} \qquad \cos 2\alpha = \frac{\pm 1}{\sqrt{1 + \tan^2 2\alpha}}.$$

From $\cos 2\alpha$ we can find $\cos^2\alpha$ and $\sin^2\alpha$:

$$\cos^2\alpha = \tfrac{1}{2}(1 + \cos 2\alpha) \qquad \sin^2\alpha = \tfrac{1}{2}(1 - \cos 2\alpha).$$

EXAMPLE 3   Describe the locus of $xy = 1$.

Solution   In this case $a = c = 0$, and $b = 1$. Therefore we choose $\alpha = \tfrac{1}{4}\pi$ to have $\bar{b} = 0$. The rotation is $x = \tfrac{1}{2}\sqrt{2}\,(\bar{x} - \bar{y})$ and $y = \tfrac{1}{2}\sqrt{2}\,(\bar{x} + \bar{y})$, so by direct computation.

$$xy = \tfrac{1}{2}(\bar{x} - \bar{y})(\bar{x} + \bar{y}) = \tfrac{1}{2}(\bar{x}^2 - \bar{y}^2).$$

The locus is $\tfrac{1}{2}(\bar{x}^2 - \bar{y}^2) = 1$.

Answer   Rectangular hyperbola   See Fig. 6.

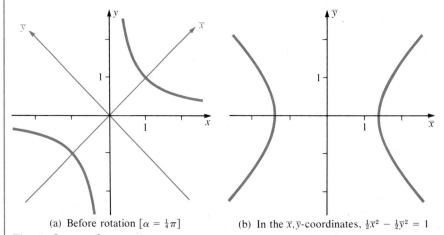

(a) Before rotation $[\alpha = \tfrac{1}{4}\pi]$ (b) In the $\bar{x}, \bar{y}$-coordinates, $\tfrac{1}{2}\bar{x}^2 - \tfrac{1}{2}\bar{y}^2 = 1$

Fig. 6   Locus of $xy = 1$

EXAMPLE 4   Describe the locus of $x^2 - 2xy + 3y^2 = 1$.

Solution   Rotate the axes through angle α where

$$\tan 2\alpha = \frac{b}{a - c} = \frac{-2}{-2} = 1 \qquad 2\alpha = \tfrac{1}{4}\pi \qquad \alpha = \tfrac{1}{8}\pi$$

In this case, $\sin 2\alpha = \cos 2\alpha = \tfrac{1}{2}\sqrt{2}$, so

$$\cos^2\alpha = \tfrac{1}{2}(1 + \tfrac{1}{2}\sqrt{2}) \qquad \sin^2\alpha = \tfrac{1}{2}(1 - \tfrac{1}{2}\sqrt{2}).$$

Substitute these values with $a = 1, b = -2, c = 3$ into the expressions preceding Example 3:

$$\begin{cases} \bar{a} = \frac{1}{2}(1 + \frac{1}{2}\sqrt{2}) - \frac{1}{2}\sqrt{2} + \frac{3}{2}(1 - \frac{1}{2}\sqrt{2}) = 2 - \sqrt{2}, \\ \bar{c} = \frac{1}{2}(1 - \frac{1}{2}\sqrt{2}) + \frac{1}{2}\sqrt{2} + \frac{3}{2}(1 + \frac{1}{2}\sqrt{2}) = 2 + \sqrt{2}. \end{cases}$$

Therefore, in the $\bar{x}, \bar{y}$-coordinate system, the locus is

$$(2 - \sqrt{2})\bar{x}^2 + (2 + \sqrt{2})\bar{y}^2 = 1.$$

Because $2 - \sqrt{2}$ and $2 + \sqrt{2}$ are both positive, this is an ellipse (Fig. 7). We may write

$$\frac{\bar{x}^2}{a^2} + \frac{\bar{y}^2}{b^2} = 1,$$

where $a^2 = 1/(2 - \sqrt{2})$ and $b^2 = 1/(2 + \sqrt{2})$. Clearly $a > b$.

Answer   Ellipse with $a = \dfrac{1}{\sqrt{2 - \sqrt{2}}}$ and $b = \dfrac{1}{\sqrt{2 + \sqrt{2}}}$.

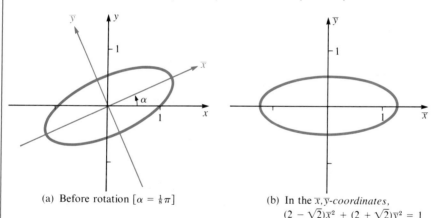

(a) Before rotation $[\alpha = \frac{1}{8}\pi]$ (b) In the $\bar{x}, \bar{y}$-*coordinates*,
$(2 - \sqrt{2})\bar{x}^2 + (2 + \sqrt{2})\bar{y}^2 = 1$

Fig. 7   Locus of $x^2 - 2xy + 3y^2 = 1$

Exercises

Find the equation in polar coordinates of the circle with

1   center $(1, 1)$ radius $\sqrt{2}$ **2**   center $(-1, 1)$ radius $\sqrt{2}$
3   center $(1, \sqrt{3})$ radius 2 **4**   center $(\sqrt{3}, -1)$ radius 2
5   center $(5, 12)$ radius 13 **6**   center $(-12, -5)$ radius 13.

Make a suitable rotation and write the $\bar{x}, \bar{y}$-equation (without an $\bar{x}\bar{y}$ term)

7 $x^2 - xy = 1$ **8** $xy - y^2 = 1$
9 $xy + y^2 = 1$ **10** $2xy + y^2 = 1$.

Determine the type of the conic, the directions of its axes, and its equation in the rotated axes

11 $x^2 + xy + y^2 = 1$ **12** $x^2 - xy + y^2 = 1$
13 $x^2 + xy - y^2 = 1$ **14** $x^2 - xy - y^2 = 1$

15 $x^2 + xy + 2y^2 = 1$ **16** $x^2 - xy + 2y^2 = 1$

17 $2x^2 - 6xy + y^2 = 1$ **18** $x^2 + 3xy - y^2 = 1$.

19 Let $x = \bar{x} \cos \alpha - \bar{y} \sin \alpha$ and $y = \bar{x} \sin \alpha + \bar{y} \cos \alpha$. Compute $x^2 + y^2$. Explain your answer.

20 Let (x_1, y_1) and (x_2, y_2) be two points in the x, y-coordinate system. Let $(\bar{x}_1, \bar{y}_1)$ and $(\bar{x}_2, \bar{y}_2)$ be their coordinates in the $\bar{x}, \bar{y}$-coordinate system obtained by a rotation. Compute $x_1 x_2 + y_1 y_2$ in terms of $\bar{x}_1, \bar{x}_2, \bar{y}_1, \bar{y}_2$, and α, the angle of rotation. Explain your answer.

Suppose a rotation converts $ax^2 + bxy + cy^2$ into $\bar{a}\bar{x}^2 + \bar{b}\bar{x}\bar{y} + \bar{c}\bar{y}^2$. Prove

21 $a + c = \bar{a} + \bar{c}$ **22*** $4ac - b^2 = 4\bar{a}\bar{c} - \bar{b}^2$.

REVIEW EXERCISES

Find the polar coordinates of

1 $(1, -1)$ **2** $(1, -\sqrt{3})$.

Find the rectangular coordinates of

3 $\{2, \frac{5}{3}\pi\}$ **4** $\{-2, \frac{3}{3}\pi\}$.

5 Find the polar equation for the circle with center $(3, 4)$ and radius 5.

Graph

6 $r = 2 \sin 3\theta$ **7** $r = 1 + \sin \theta$

8 $y + 2 = \frac{1}{4}(x - 3)^2$ **9** $\frac{1}{36}x^2 + \frac{1}{4}y^2 = 1$

10 $x^2 + 9y^2 - 2x + 36y = -28$.

Graph and find the asymptotes

11 $-\frac{1}{16}x^2 + \frac{1}{4}y^2 = 1$ **12** $(x + 1)^2 - 9(y - 3)^2 = 9$.

13 Graph and find all intersections of the ellipse $\frac{1}{4}x^2 + y^2 = 1$ and the hyperbola $x^2 - y^2 = 1$.

14 Graph the conic $x^2 + 2xy + 2y^2 = 1$.

10 VECTORS

1 VECTOR ALGEBRA

In this section we introduce and discuss addition of vectors and multiplication of vectors by numbers. In the next section, we introduce the inner product of vectors, which has many geometrical and physical applications.

The idea of vectors grew out of the study of forces in physics. A force applied at a point was represented by an arrow; it pointed in the direction of the force and its length represented the magnitude of the force. The concept of a quantity having direction and magnitude turned out to be a powerful tool not only in physics and engineering, but also in non-physical fields such as statistics, numerical analysis, and mathematical economics.

We choose a point **0** in the plane and call it the **origin.** A **vector** is a directed line segment from **0** to a point of the plane (Fig. 1a). The vector that goes from **0** to **0** (a degenerate line segment) we call the vector **0** .

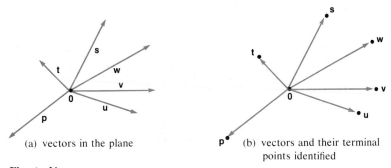

(a) vectors in the plane

(b) vectors and their terminal points identified

Fig. 1 Vectors

The vector from **0** to a point **z** is completely determined by its terminal point. For this reason, we shall often identify a vector and its terminal point as one and the same thing (Fig. 1b).

We have defined vectors without reference to coordinate systems. This is as it should be, for in real life a force is a force and a velocity is a velocity, regardless of how we choose to set up coordinate axes. Nevertheless, to compute with vectors we shall need coordinates. Given a cartesian coordinate system in the plane, we assign to each vector **v** the coordinates of its terminal point, and write

$$\mathbf{v} = (x, y).$$

The numbers x and y are called the **x-component** and the **y-component** of **v**.

Addition

We now define the sum **v** + **w** of two vectors **v** and **w**. Draw a directed line segment parallel to **w**, starting from the terminal point of **v**. See Fig. 2a. Then the vector from **0** to the end point of this parallel segment is the vector **v** + **w**. See Fig. 2b.

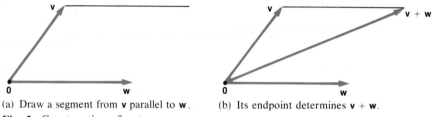

(a) Draw a segment from **v** parallel to **w**.   (b) Its endpoint determines **v** + **w**.

Fig. 2 Construction of **v** + **w**

By closing the parallelogram, we see that the roles of **v** and **w** can be interchanged in the definition (Fig. 3). The result is

$$\mathbf{v} + \mathbf{w} = \mathbf{w} + \mathbf{v}.$$

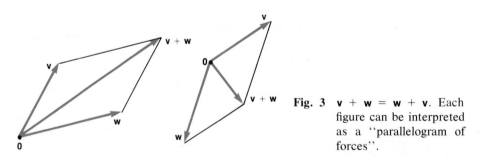

Fig. 3 **v** + **w** = **w** + **v**. Each figure can be interpreted as a "parallelogram of forces".

The simplest interpretation of addition of vectors is in terms of force vectors. If two forces **v** and **w** are applied at the same point, the resultant force is **v** + **w**; that is the law of "parallelogram of forces".

We need a practical way to compute **v** + **w** when the coordinates of **v** and **w** are given. Suppose

$$\mathbf{v} = (a, b) \quad \text{and} \quad \mathbf{w} = (c, d).$$

The answer is

$$\mathbf{v} + \mathbf{w} = (a + c, b + d).$$

In words, vectors are added *componentwise;* the x-component of $\mathbf{v} + \mathbf{w}$ is the sum of the x-components of $\mathbf{v}$ and $\mathbf{w}$—ditto the y-component.

To see why, we go back to the geometric definition of $\mathbf{v} + \mathbf{w}$ and project everything onto the x-axis (Fig. 4). The segment from $\mathbf{v}$ to $\mathbf{v} + \mathbf{w}$ is parallel to $\mathbf{w}$, so its projection on the x-axis has the same length and direction as the projection of $\mathbf{w}$, that is, c. Hence the x-component of $\mathbf{v} + \mathbf{w}$ is obtained by starting at a and adding c to it; that is, the x-component of $\mathbf{v} + \mathbf{w}$ is $a + c$. Similarly the y-component is $b + d$.

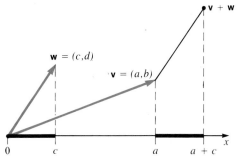

Fig. 4 The x-component of $\mathbf{v} + \mathbf{w}$ is $a + c$. **Fig. 5** Construction of $\mathbf{v} - \mathbf{w}$

Sum of Vectors If $\mathbf{v} = (a, b)$ and $\mathbf{w} = (c, d)$, then

$$\mathbf{v} + \mathbf{w} = (a + c, b + d).$$

The **difference $\mathbf{v} - \mathbf{w}$** of two vectors $\mathbf{v}$ and $\mathbf{w}$ is that vector that satisfies

$$(\mathbf{v} - \mathbf{w}) + \mathbf{w} = \mathbf{v}.$$

Thus $\mathbf{v} - \mathbf{w}$ is the other side of a parallelogram, one side of which is $\mathbf{w}$ and the diagonal of which is $\mathbf{v}$. See Fig. 5. In components:

If $\mathbf{v} = (a, b)$ and $\mathbf{w} = (c, d)$, then $\mathbf{v} - \mathbf{w} = (a - c, b - d)$.

Examples

$$(2, 7) + (1, 3) = (3, 10) \qquad (2, 7) - (1, 3) = (1, 4)$$

$$(2, 7) + (1, -3) = (3, 4) \qquad (2, 7) - (1, -3) = (1, 10)$$

$$(2, 7) + (-5, 0) = (-3, 7) \qquad (2, 7) - (-5, 0) = (7, 7).$$

Multiplication by a Scalar

In vector algebra, the word **scalar** means real number. We combine a scalar a and a vector $\mathbf{v}$ to form a new vector $a\mathbf{v}$ called the **scalar multiple** of a and $\mathbf{v}$.

Multiplication by a Scalar   If a is a scalar and $\mathbf{v}$ is a vector, then the scalar multiple $a\mathbf{v}$ is the vector defined by

$$a\mathbf{v} = \begin{cases} \mathbf{0} & \text{if } a = 0. \\[4pt] \text{a vector in the same direction as } \mathbf{v} \text{ with length } a \\ \text{times the length of } \mathbf{v} & \text{if } a > 0. \\[4pt] \text{a vector in the opposite direction from } \mathbf{v} \text{ with length} \\ |a| \text{ times the length of } \mathbf{v} & \text{if } a < 0. \end{cases}$$

In components, if $\mathbf{v} = (x, y)$, then $a\mathbf{v} = (ax, ay)$.

The formula $a(x, y) = (ax, ay)$ can be verified by an argument using similar triangles (Fig. 6). We omit the details since the figure is pretty convincing.

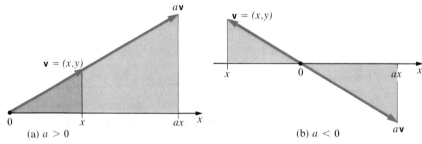

(a) $a > 0$ (b) $a < 0$

Fig. 6   Proof by similar triangles that $a(x, y) = (ax, ay)$.

One can interpret scalar multiplication physically in several ways. For example, a velocity vector $\mathbf{v}$ has direction and magnitude (speed). If the direction remains the same, but the speed triples, then the new velocity vector is $3\mathbf{v}$.

Examples

$$1(x, y) = (x, y) \qquad\qquad -3(5, -7) = (-15, 21)$$
$$3(0, 0) = (0, 0) = \mathbf{0} \qquad\qquad 0(5, -1) = (0, 0) = \mathbf{0}$$
$$-2(1, 0) = (-2, 0) \qquad\qquad 2(3, -2) = (6, -4).$$

We denote by $-\mathbf{v}$ the vector $(-1)\mathbf{v}$, opposite in direction but equal in magnitude to $\mathbf{v}$. Thus $-(3, -4) = (-3, 4)$. It is easily seen that

$$\mathbf{v} - \mathbf{w} = \mathbf{v} + (-\mathbf{w}).$$

For instance $(5, -7) - (3, -4) = (5, -7) + (-3, 4) = (2, -3)$.

The basic rules of vector algebra follow easily from the coordinate formulas for scalar multiplication and addition:

Rules of Vector Algebra

$$1 \cdot \mathbf{v} = \mathbf{v} \qquad\qquad \mathbf{v} + (-\mathbf{v}) = \mathbf{0}$$

$$\mathbf{v} + \mathbf{w} = \mathbf{w} + \mathbf{v} \qquad\qquad (\text{commutative law})$$

$$\mathbf{u} + (\mathbf{v} + \mathbf{w}) = (\mathbf{u} + \mathbf{v}) + \mathbf{w} \qquad (\text{associative law})$$

$$\begin{cases} a(\mathbf{v} + \mathbf{w}) = a\mathbf{v} + a\mathbf{w} \\ (a + b)\mathbf{v} = a\mathbf{v} + b\mathbf{v} \end{cases} \qquad (\text{distributive laws})$$

$$a(b\mathbf{v}) = (ab)\mathbf{v}.$$

Exercises

Compute

1 $3(0, 1)$ **2** $-5(2, 0)$ **3** $2(1, -1)$

4 $7(3, -2)$ **5** $4(\frac{1}{2}, 3)$ **6** $-\frac{1}{3}(4, -2)$

7 $(3, -1) + (0, 1)$ **8** $(4, 2) + (3, 1)$ **9** $(-1, 1) + (1, -1)$

10 $(2, -2) + (3, 3)$ **11** $(4, -1) + (-3, 2)$ **12** $(7, 11) + (11, 7)$

13 $3(1, 1) + (7, 1)$ **14** $(2, -2) + 8(1, 2)$ **15** $-2(1, 3) + 4(5, 4)$

                 16 $3(2, 0) + 2(4, 5)$.

Plot the vectors and compute their sum graphically by the "parallelogram of forces." Check your answer:

17 $(4, 1) + (3, 4)$ **18** $(-3, 2) + (4, 4)$ **19** $(1, 5) + (-5, -1)$

                 20 $(2, -3) + (-1, -3)$.

21   Find a vector parallel to and having the same length as the directed segment from $(-2, 4)$ to $(-1, -6)$.

22   Find a point $\mathbf{x}$ such that the directed segment from $(2, 3)$ to $\mathbf{x}$ is parallel to the directed segment from $(1, 1)$ to $(-4, 0)$ and of the same length.

Find a and b; check your answer graphically:

23 $(-4, 5) = a(2, 0) + b(0, 1)$ **24** $(4, 2) = a(1, 0) + b(1, 1)$.

2   LENGTH AND INNER PRODUCT

We start with some basic properties of the length of a vector. We denote the length of $\mathbf{v}$ by $\|\mathbf{v}\|$.

If $\mathbf{v} = (x, y)$, then

(1) $\|\mathbf{v}\|^2 = \|(x, y)\|^2 = x^2 + y^2$

(2) $\|\mathbf{0}\| = 0$, $\|\mathbf{v}\| > 0$ if $\mathbf{v} \neq \mathbf{0}$

(3) $\|a\mathbf{v}\| = |a| \cdot \|\mathbf{v}\|$

(4) $\|\mathbf{v} + \mathbf{w}\| \leq \|\mathbf{v}\| + \|\mathbf{w}\|$ (triangle inequality).

The first property comes directly from the distance formula, and the second follows from the first. The third is also proved by direct calculation:

$$\|a(x,y)\|^2 = \|(ax, ay)\|^2 = (ax)^2 + (ay)^2$$
$$= a^2(x^2 + y^2) = a^2\|(x,y)\|^2,$$

therefore

$$\|a\mathbf{v}\|^2 = a^2\|\mathbf{v}\|^2 \qquad \|a\mathbf{v}\| = |a| \cdot \|\mathbf{v}\|.$$

A useful fact: if $\mathbf{v}$ is any non-zero vector, then $\mathbf{v}/\|\mathbf{v}\|$ is a unit vector (length 1) in the same direction. Just apply (3) with $a = 1/\|\mathbf{v}\|$. [We used this device previously; we stretched (or shrank) a vector (a, b) to the unit circle by dividing its coordinates by $\sqrt{a^2 + b^2}$. This amounted to making the vector $\mathbf{v} = (a, b)$ into a unit vector by dividing it by $\|\mathbf{v}\| = \sqrt{a^2 + b^2}$.]

The triangle inequality (4) expresses a basic geometric fact: the length of each side of a triangle cannot exceed the sum of the lengths of the other two sides (Fig. 1).

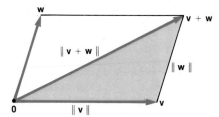

Fig. 1 Geometric proof that $\|\mathbf{v} + \mathbf{w}\| \leq \|\mathbf{v}\| + \|\mathbf{w}\|$

Inner Product

We now define another basic operation on vectors, the **inner product** (also called **dot product**). The inner product combines two vectors to produce a scalar. Suppose $\mathbf{v}$ and $\mathbf{w}$ are non-zero vectors, and $\cos \theta$ is the cosine of the angle between them. We define

$$\mathbf{v} \cdot \mathbf{w} = \|\mathbf{v}\| \cdot \|\mathbf{w}\| \cdot \cos \theta.$$

There are four possibilities for "the angle between $\mathbf{v}$ and $\mathbf{w}$." If one is θ, then the others are $-\theta$, $2\pi - \theta$, and $\theta - 2\pi$. But

$$\cos \theta = \cos(-\theta) = \cos(2\pi - \theta) = \cos(\theta - 2\pi),$$

so there is only one possibility for "the cosine of the angle between $\mathbf{v}$ and $\mathbf{w}$." If either $\mathbf{v} = \mathbf{0}$ or $\mathbf{w} = \mathbf{0}$, the angle between $\mathbf{v}$ and $\mathbf{w}$ is not defined. In this case we define

$$\mathbf{0} \cdot \mathbf{w} = \mathbf{v} \cdot \mathbf{0} = 0.$$

Notice that the inner product of two vectors is a real number, not a vector.

Three special cases that arise all the time follow directly from the definition of inner product:

> (1) $\mathbf{v} \cdot \mathbf{v} = \|\mathbf{v}\|^2$.
>
> (2) Vectors $\mathbf{v}$ and $\mathbf{w}$ are perpendicular if and only if $\mathbf{v} \cdot \mathbf{w} = 0$.
>
> (3) If $\mathbf{v}$ and $\mathbf{w}$ are unit vectors ($\|\mathbf{v}\| = \|\mathbf{w}\| = 1$), then $\mathbf{v} \cdot \mathbf{w} = \cos \theta$.

Some examples of dot products are shown in Fig. 2.

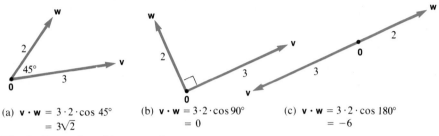

(a) $\mathbf{v} \cdot \mathbf{w} = 3 \cdot 2 \cdot \cos 45°$
      $= 3\sqrt{2}$

(b) $\mathbf{v} \cdot \mathbf{w} = 3 \cdot 2 \cdot \cos 90°$
      $= 0$

(c) $\mathbf{v} \cdot \mathbf{w} = 3 \cdot 2 \cdot \cos 180°$
      $= -6$

Fig. 2 Examples of inner products

There is another important geometric interpretation of the inner product. Suppose $\mathbf{v}$ and $\mathbf{w}$ are non-zero vectors and θ is the angle between them (we can always take θ so that $0 \le \theta \le \pi$). See Fig. 3. The quantity $\|\mathbf{v}\| \cdot \cos \theta$ is the length of the projection of $\mathbf{v}$ on $\mathbf{w}$ if $0 \le \theta \le \frac{1}{2}\pi$, and the negative of the length if $\frac{1}{2}\pi \le \theta \le \pi$. We call $\|\mathbf{v}\| \cdot \cos \theta$ the **signed projection** of $\mathbf{v}$ on $\mathbf{w}$. Its sign is $+$ when $\mathbf{v}$ projects directly onto $\mathbf{w}$, and $-$ when $\mathbf{v}$ projects onto $\mathbf{w}$ extended backwards. Since $\mathbf{v} \cdot \mathbf{w} = (\|\mathbf{v}\| \cos \theta) \cdot \|\mathbf{w}\| = (\|\mathbf{w}\| \cos \theta) \cdot \|\mathbf{v}\|$,

> $\mathbf{v} \cdot \mathbf{w} = (\text{signed projection of } \mathbf{v} \text{ on } \mathbf{w}) \cdot \|\mathbf{w}\|$
>
> $= (\text{signed projection of } \mathbf{w} \text{ on } \mathbf{v}) \cdot \|\mathbf{v}\|$.

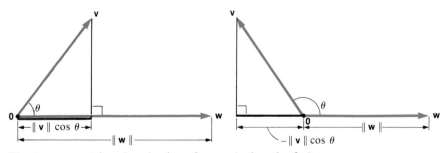

Fig. 3 $\mathbf{v} \cdot \mathbf{w} = (\text{signed projection of } \mathbf{v} \text{ on } \mathbf{w}) \cdot (\text{length of } \mathbf{w})$

Computations with Inner Products

If we are given the lengths of two vectors and the angle between them, we can compute their inner product directly from the definition.

Example $\|\mathbf{v}\| = 8.000$ $\|\mathbf{w}\| = 11.00$ $\theta = 26.00°$

$\mathbf{v} \cdot \mathbf{w} = 8 \cdot 11 \cdot \cos 26° \approx 79.09$.

But what if we are given the components of **v** and **w**? Then how do we find **v·w**?

Inner Product   If $\mathbf{v} = (a, b)$ and $\mathbf{w} = (c, d)$, then

$\mathbf{v \cdot w} = ac + bd$.

Proof   Consider the triangle in Fig. 4. By the law of cosines,

$$\|\mathbf{v} - \mathbf{w}\|^2 = \|\mathbf{v}\|^2 + \|\mathbf{w}\|^2 - 2\|\mathbf{v}\| \cdot \|\mathbf{w}\| \cdot \cos \theta = \|\mathbf{v}\|^2 + \|\mathbf{w}\|^2 - 2\mathbf{v \cdot w}.$$

Therefore

$$\begin{aligned}
2\mathbf{v \cdot w} &= \|\mathbf{v}\|^2 + \|\mathbf{w}\|^2 - \|\mathbf{v} - \mathbf{w}\|^2 \\
&= (a^2 + b^2) + (c^2 + d^2) - [(a - c)^2 + (b - d)^2] \\
&= (a^2 + b^2 + c^2 + d^2) - (a^2 - 2ac + c^2) - (b^2 - 2bd + d^2) \\
&= 2(ac + bd).
\end{aligned}$$

Consequently $\mathbf{v \cdot w} = ac + bd$.

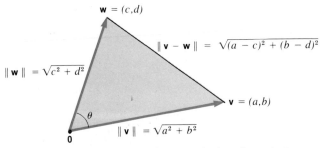

Fig. 4 Use of the law of cosines to obtain a formula for **v·w**

Now let $\mathbf{v} = (a, b)$ and $\mathbf{w} = (c, d)$. Then

$$\mathbf{v \cdot w} = \|\mathbf{v}\| \cdot \|\mathbf{w}\| \cdot \cos \theta = ac + bd$$

and

$$\|\mathbf{v}\| = \sqrt{a^2 + b^2} \qquad \|\mathbf{w}\| = \sqrt{c^2 + d^2}.$$

By putting these facts together we derive a formula for $\cos \theta$ in terms of the components of **v** and **w**:

If $\mathbf{v} = (a, b)$ and $\mathbf{w} = (c, d)$ are non-zero vectors, and θ is the angle between them, then

$$\cos \theta = \frac{\mathbf{v \cdot w}}{\|\mathbf{v}\| \cdot \|\mathbf{w}\|} = \frac{ac + bd}{\sqrt{a^2 + b^2} \sqrt{c^2 + d^2}}.$$

EXAMPLE 1   Find the acute angle between $\mathbf{v} = (4, 3)$ and $\mathbf{w} = (1, 5)$.

Solution $\|\mathbf{v}\| = \sqrt{4^2 + 3^2} = 5$, $\|\mathbf{w}\| = \sqrt{1^2 + 5^2} = \sqrt{26}$, and

$$\mathbf{v \cdot w} = 4 \cdot 1 + 3 \cdot 5 = 19.$$

Hence

$$\cos \theta = \frac{19}{5\sqrt{26}} \qquad \theta = \text{arc cos } \frac{19}{5\sqrt{26}}.$$

Answer   arc cos $\dfrac{19}{5\sqrt{26}} \approx 41.82°$

The main properties of the inner product look just like the familiar rules of algebra.

Rules for Inner Products

(1) $\mathbf{v} \cdot \mathbf{w} = \mathbf{w} \cdot \mathbf{v}$

(2) $(a\mathbf{v}) \cdot \mathbf{w} = \mathbf{v} \cdot (a\mathbf{w}) = a(\mathbf{v} \cdot \mathbf{w})$

(3) $(\mathbf{u} + \mathbf{v}) \cdot \mathbf{w} = \mathbf{u} \cdot \mathbf{w} + \mathbf{v} \cdot \mathbf{w}$

(4) $\mathbf{u} \cdot (\mathbf{v} + \mathbf{w}) = \mathbf{u} \cdot \mathbf{v} + \mathbf{u} \cdot \mathbf{w}.$

These rules are easily proved by using components; the proofs are left as exercises.

Perpendicular Vectors

If $\mathbf{v} = (a, b)$ is a non-zero vector, then $(-b, a)$ is perpendicular to $\mathbf{v}$ and has the same length as $\mathbf{v}$:

$$(a, b) \cdot (-b, a) = -ab + ba = 0$$

$$\|(-b, a)\| = \sqrt{(-b)^2 + a^2} = \sqrt{a^2 + b^2} = \|(a, b)\|.$$

Therefore $(-b, a)$ is one of the two vectors shown in Fig. 5a. A direct check of the four quadrants shows that $(-b, a)$ is always the one of these two that is $\frac{1}{2}\pi$ counterclockwise from $\mathbf{v}$. See Fig. 5b.

The result of rotating a vector $\mathbf{v} = (a, b)$ counterclockwise $\frac{1}{2}\pi$ radians is $(-b, a)$.

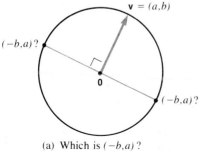

(a) Which is $(-b,a)$?   (b) Answer: the choice $\frac{1}{2}\pi$ *forward* of $\mathbf{v}$

Fig. 5   Rotation

Applications

Physical quantities such as force, velocity, momentum, and magnetic field strength are vectors. Hence it is natural that vector algebra is useful in problems involving such quantities.

> *EXAMPLE 2*   A weight W hangs from a pulley that slides on a fixed rope (Fig. 6a). Find the tension T in the fixed rope for (a) $\theta = 10.0°$ and (b) $\theta = 5.0°$.
>
> *Solution*   Take the origin **0** at the pulley (Fig. 6b). Three forces act on the pulley: a downward force **w** with $\|\mathbf{w}\| = W$ and two forces along the fixed rope, $\mathbf{t}_1$ left and $\mathbf{t}_2$ right, with $\|\mathbf{t}_1\| = \|\mathbf{t}_2\| = T$, the tension. Since the pulley is in equilibrium, the three forces sum to zero:

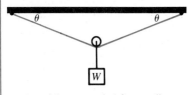
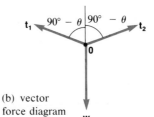

(a) weight suspended from pulley   (b) vector force diagram   **w**

Fig. 6

$$\mathbf{w} + \mathbf{t}_1 + \mathbf{t}_2 = \mathbf{0}, \quad \text{that is,} \quad \mathbf{w} = -(\mathbf{t}_1 + \mathbf{t}_2).$$

The vertical components of $\mathbf{t}_1$ and $\mathbf{t}_2$ are both equal to

$$\|\mathbf{t}_1\| \cos(90° - \theta) = T \sin \theta.$$

Therefore, the condition for equilibrium is

$$2T \sin \theta = W, \quad \text{that is,} \quad T = \frac{W}{2 \sin \theta}.$$

(a) $\theta = 10.0°$ $T = \dfrac{W}{2 \sin 10.0°} \approx 2.88\ W$

(b) $\theta = 5.0°$ $T = \dfrac{W}{2 \sin 5.0°} \approx 5.74\ W$.

Answer   (a) $T \approx 2.88\ W$ (b) $T \approx 5.74\ W$

The next example explains how a sailboat can sail "into the wind." The model is simplified by assuming (1) that the sail is straight (actually it is a more efficient airfoil) and (2) that the keel prevents any sideways motion.

> *EXAMPLE 3*   On a sailboat, the sail is set at $15°$ to the keel (Fig. 7a). The wind blows into the sail from forward, at an angle $45°$ to the keel, and exerts a total force of 78 kg against the sail. Find the effective wind force driving the boat forward.

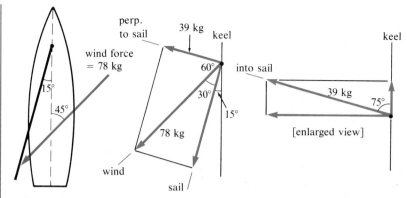

(a) Boat, sail, wind   (b) Find the wind component   (c) The part of the wind into the sail
                           perpendicular to the sail   that actually drives the boat forward

Fig. 7   Simplified model of sailboat

Solution   The problem must be solved in two steps. First the wind force is re-
solved into two components, one parallel to the sail and the other perpendicular
to the sail. The wind parallel to the sail slips past, giving no push to the boat.
Therefore, only the component of the wind perpendicular to the sail pushes the
boat.

   The second step is to resolve this (perpendicular) component further into two
components, one parallel to the keel and the other perpendicular to the keel. The
component perpendicular to the keel does nothing because the boat can move
only in the direction of the keel. Therefore, the parallel component is the effective force pushing the boat forward.

   By Figure 7b, the wind strikes the sail at 30°, so the component of the wind
perpendicular to the sail has magnitude

   78 sin 30° = 39 kg .

By Figure 7c, this 39 kg force acts at angle 75° to the keel; hence its component
in the direction of the keel has magnitude

   39 cos 75° ≈ 10 kg .

Answer   10 kg

Exercises

Compute

1 $\|(12, 5)\|$

2 $\|(15, -8)\|$

3 $\|(-5, 6)\|$

4 $\|(3, 11)\|$

5 $(2, 0) \cdot (3, 1)$

6 $(-1, -1) \cdot (1, 1)$

7 $(4, 1) \cdot (-1, 4)$

8 $(2, 2) \cdot (-1, 2)$

9 $(0, 3) \cdot (1, 1)$

10 $(2, 3) \cdot (-3, -1)$

11 $(3.5, 4.1) \cdot (7.2, 1.5)$

12 $(-5.8, 6.1) \cdot (3.0, -7.8)$.

Compute in two ways

13 $(2, 3) \cdot [(1, 4) + (2, 2)]$

14 $(-1, 2) \cdot [(1, 3) + (1, 4)]$

15 $[(-1, 7) - (2, 3)] \cdot (4, 5)$

16 $[(-2, -3) - (-1, 4)] \cdot (-2, 3)$

17 $[3(4, 5)] \cdot (-1, 2)$

18 $[-2(3, -1)] \cdot (5, -6)$.

19 (1) on p. 223   **20** (2) on p. 223

21 (3) on p. 223   **22** (4) on p. 223

23 $\|v + w\|^2 = \|v\|^2 + 2v \cdot w + \|w\|^2$ **24** $\|v + w\|^2 + \|v - w\|^2 = 2\|v\|^2 + 2\|w\|^2$

25 $\|v + w\|^2 - \|v - w\|^2 = 4v \cdot w$ **26*** $|v \cdot w| \le \|v\| \cdot \|w\|$.

Compute the smallest positive angle between the vectors

27 $(3, 4)$ $(-4, 3)$ **28** $(1, -2)$ $(-2, 1)$

29 $(1, 3)$ $(3, 1)$ **30** $(1, -3)$ $(3, 6)$

31 $(1, 1)$ $(-1, 7)$ **32** $(3, 4)$ $(-5, -2)$.

33 Prove that the diagonals of a rhombus are perpendicular.

34 Prove that the midpoints of the sides of any quadrilateral are the vertices of a parallelogram.

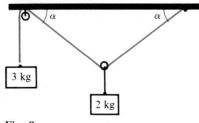 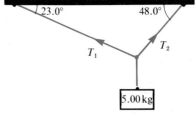

Fig. 8   **Fig. 9**

35 Estimate $\alpha°$ to 2 places for the system of pulleys and weights in Fig. 8.

36 Find the tensions T_1 and T_2 in the ropes in Fig. 9. There are no pulleys, only knots.

37 Suppose the sail is set at 45° to the keel and the wind comes in directly from the side, that is, perpendicular to the keel, with total force W. Find the effective force pushing the sailboat forward.

38 Work Ex. 37 with the wind directly from behind.

39 A boat in open water is headed $26\frac{1}{2}°$ east of north at a water speed of 23 knots (nautical miles per hour). After 3 hours it is observed that the boat has actually moved 58 nautical miles 19° east of north. Find the current speed and direction.

40 A plane is headed due west at an air speed of 187 mph, hoping to reach a point 150 miles west. After $\frac{1}{2}$ hour the pilot finds he has actually gone 112 miles in a direction 9° south of west. Find the wind vector.

3 NORMAL FORM

Using inner products, we obtain a useful vector equation for a straight line. Let L be any line (Fig. 1). Let n be either of the two unit vectors perpendicular to L. Let pn be the intersection of L and the line along n. Clearly $p > 0$ if n points towards L. Also $p < 0$ if n points away from L, and $p = 0$ if L passes through 0.

Now let x be any point on L. Then the signed projection of x on n is pn because L and n are perpendicular. Hence

$$x \cdot n = pn \cdot n = p$$

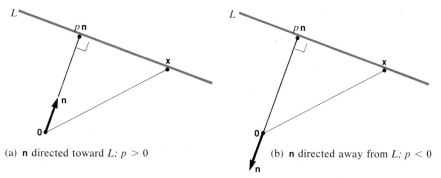

(a) **n** directed toward L; $p > 0$ (b) **n** directed away from L; $p < 0$

Fig. 1   Normal form of a line

since $\mathbf{n} \cdot \mathbf{n} = 1$. This simple relation, $\mathbf{x} \cdot \mathbf{n} = p$, is the equation for L that we want. To find its scalar form, set $\mathbf{n} = (a, b)$ and $\mathbf{x} = (x, y)$. Then

$$\mathbf{x} \cdot \mathbf{n} = (x, y) \cdot (a, b) = ax + by = p.$$

Normal Form   Each line in the plane has an equation

$$\mathbf{x} \cdot \mathbf{n} = p,$$

where $\mathbf{n}$ is a unit vector and p is a scalar. This equation is called the **normal form** of the line. The scalar version of this normal form is

$$ax + by = p \qquad a^2 + b^2 = 1.$$

·Remark   Actually each line has two normal forms because there are two choices, $\mathbf{n}$ and $-\mathbf{n}$, for the unit perpendicular vector, hence two equivalent equations

$$\mathbf{x} \cdot \mathbf{n} = p \qquad \text{and} \qquad \mathbf{x} \cdot (-\mathbf{n}) = -p.$$

We try to ignore this ambiguity and we refer (incorrectly) to either as "the" normal form of the line.

EXAMPLE 1   Find the normal form of the line

(a) $3x + 4y = 10$ (b) $x - y = 1$.

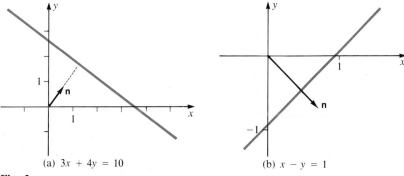

(a) $3x + 4y = 10$ (b) $x - y = 1$

Fig. 2

Solution   (a)   See Fig. 2a. Write $3x + 4y = 10$ in the vector form

$$(3, 4) \cdot (x, y) = 10.$$

This is almost normal form, except that $(3, 4)$ is not a unit vector. But

$$\|(3, 4)\| = \sqrt{3^2 + 4^2} = \sqrt{25} = 5$$

so $\frac{1}{5}(3, 4)$ is a unit vector. Thus

$$\tfrac{1}{5}(3, 4) \cdot (x, y) = \tfrac{1}{5} \cdot 10 \qquad (\tfrac{3}{5}, \tfrac{4}{5}) \cdot (x, y) = 2.$$

(b)   See Fig. 2b. We write $x - y = 1$ in the form

$$(1, -1) \cdot (x, y) = 1$$

and divide both sides by

$$\|(1, -1)\| = \sqrt{1^2 + (-1)^2} = \sqrt{2}:$$

$$\frac{1}{\sqrt{2}}(1, -1) \cdot (x, y) = \frac{1}{\sqrt{2}} \qquad (\tfrac{1}{2}\sqrt{2}, -\tfrac{1}{2}\sqrt{2}) \cdot (x, y) = \tfrac{1}{2}\sqrt{2}.$$

Answer   (a) $(\tfrac{3}{5}, \tfrac{4}{5}) \cdot \mathbf{x} = 2$ (b) $(\tfrac{1}{2}\sqrt{2}, -\tfrac{1}{2}\sqrt{2}) \cdot \mathbf{x} = \tfrac{1}{2}\sqrt{2}$

Remark 1   Sometimes it is convenient to write the normal form of a line as

$$x \cos \alpha + y \sin \alpha = p.$$

The point is that the unit vector $\mathbf{n}$ in $\mathbf{x} \cdot \mathbf{n} = p$ can be written as $\mathbf{n} = (\cos \alpha, \sin \alpha)$, where α is determined up to $\alpha + 2\pi n$. See Fig. 3.

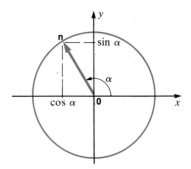

Fig. 3

Remark 2   The normal form is easy for horizontal lines and for vertical lines. The horizontal line $y = p$ is in scalar normal form. In vector form,

$$(0, 1) \cdot (x, y) = p \qquad \|(0, 1)\| = 1.$$

Similarly the vertical line $x = p$ is in scalar normal form, and its vector normal form is

$$(1, 0) \cdot (x, y) = p \qquad \|(1, 0)\| = 1.$$

Distance from a Point to a Line

As a first application of normal form, we derive a formula for the distance from a point $\mathbf{x}_0$ to a line L given in normal form $\mathbf{x} \cdot \mathbf{n} = p$.

Let $\mathbf{x}$ be the point of L nearest to $\mathbf{x}_0$. Then $\mathbf{x} - \mathbf{x}_0$ is perpendicular to L, hence parallel to $\mathbf{n}$. See Fig. 4. This means that

$$\mathbf{x} - \mathbf{x}_0 = c\mathbf{n} \qquad c \text{ a scalar}.$$

The distance we want is

$$\|\mathbf{x} - \mathbf{x}_0\| = \|c\mathbf{n}\| = |c|.$$

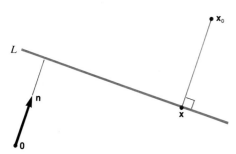

Fig. 4   **x** is the point of L closest to $\mathbf{x}_0$, so the vector $\mathbf{x} - \mathbf{x}_0$ is parallel to **n**.

But **x** is in L, so $\mathbf{x} \cdot \mathbf{n} = p$ and we have

$$(\mathbf{x} - \mathbf{x}_0) \cdot \mathbf{n} = c\mathbf{n} \cdot \mathbf{n} = c$$

$$\mathbf{x} \cdot \mathbf{n} - \mathbf{x}_0 \cdot \mathbf{n} = c$$

$$p - \mathbf{x}_0 \cdot \mathbf{n} = c.$$

Therefore the distance is

$$|c| = |p - \mathbf{x}_0 \cdot \mathbf{n}|.$$

Suppose $\mathbf{x} \cdot \mathbf{n} = p$ is the normal form of a line L, and $\mathbf{x}_0$ is a point of the plane. Then the distance from $\mathbf{x}_0$ to L is

$$|p - \mathbf{x}_0 \cdot \mathbf{n}|.$$

If $ax + by = p$ is the normal form in coordinates, then the distance is

$$|p - ax_0 - by_0|.$$

EXAMPLE 2   Find the distance from $(4, -1)$ to the line

(a) $\frac{3}{5}x - \frac{4}{5}y = 6$ (b) $y = 3x - 2$.

Solution   (a)   Since $\mathbf{n} = (\frac{3}{5}, -\frac{4}{5})$ is a unit vector, the equation is in normal form, with $p = 6$. Therefore the distance from $\mathbf{x}_0 = (4, -1)$ to the line is

$$|p - \mathbf{x}_0 \cdot \mathbf{n}| = |6 - (4, -1) \cdot (\tfrac{3}{5}, -\tfrac{4}{5})|$$

$$= |6 - (\tfrac{12}{5} + \tfrac{4}{5})| = |6 - \tfrac{16}{5}| = \tfrac{14}{5}.$$

(b)   We must first put $y = 3x - 2$ into normal form. We write it as

$$-3x + y = -2 \qquad (-3, 1) \cdot (x, y) = -2$$

and divide by $\|(-3, 1)\| = \sqrt{10}$:

$$\frac{1}{\sqrt{10}}(-3, 1) \cdot (x, y) = \frac{-2}{\sqrt{10}} \qquad \tfrac{1}{10}\sqrt{10}(-3, 1) \cdot (x, y) = -\tfrac{1}{5}\sqrt{10}.$$

Hence $\mathbf{n} = \tfrac{1}{10}\sqrt{10}(-3, 1)$ and $p = -\tfrac{1}{5}\sqrt{10}$. The distance from $\mathbf{x}_0 = (4, -1)$ to the line is

$$|p - \mathbf{x}_0 \cdot \mathbf{n}| = |-\tfrac{1}{5}\sqrt{10} - (4, -1) \cdot [\tfrac{1}{10}\sqrt{10}(-3, 1)]|$$

$$= |-\tfrac{1}{5}\sqrt{10} - \tfrac{1}{10}\sqrt{10}(-12 - 1)|$$

$$= \sqrt{10}|-\tfrac{1}{5} + \tfrac{13}{10}| = \tfrac{11}{10}\sqrt{10}$$

Answer   (a) $\tfrac{14}{5}$ (b) $\tfrac{11}{10}\sqrt{10}$

Angle Between Lines

Let M and N be intersecting lines (Fig. 5a). They determine two angles, θ and $\pi - \theta$. But $\cos\theta = \cos(\pi - \theta)$, so the "cosine of their angle of intersection" is a single number—no ambiguity. Let

$$M: \mathbf{x}\cdot\mathbf{m} = p \qquad N: \mathbf{x}\cdot\mathbf{n} = q$$

be the normal forms of the lines. Then the unit vector $\mathbf{m}$ is perpendicular to M and the unit vector $\mathbf{n}$ is perpendicular to N. It follows (Fig. 5b) that the angle between $\mathbf{m}$ and $\mathbf{n}$ is one of the two angles between M and N. Therefore $\mathbf{m}\cdot\mathbf{n} = \cos\theta$.

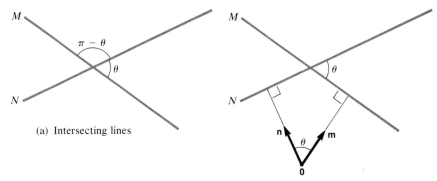

(a) Intersecting lines

(b) The angle between their normals equals one of the angles between the lines.

Fig. 5 Angle between lines

If $\mathbf{x}\cdot\mathbf{m} = p$ and $\mathbf{x}\cdot\mathbf{n} = q$ are two lines in normal form, then

$$\mathbf{m}\cdot\mathbf{n} = \cos\theta,$$

where θ is one of the angles of intersection of the lines.

EXAMPLE 3   Find the acute angle θ between the lines $3x - 4y = 2$ and $x + y = 1$.

Solution   Write the lines in the form

$$(3, -4)\cdot(x, y) = 2 \qquad \text{and} \qquad (1, 1)\cdot(x, y) = 1.$$

Now

$$\|(3, -4)\| = 5 \quad \text{and} \quad \|(1, 1)\| = \sqrt{2}$$

so

$$\mathbf{m} = \tfrac{1}{5}(3, -4) \quad \text{and} \quad \mathbf{n} = \frac{1}{\sqrt{2}}(1, 1) = \tfrac{1}{2}\sqrt{2}(1, 1)$$

are corresponding normals to M and to N. Hence $\cos\theta = \mathbf{m}\cdot\mathbf{n}$ or $\cos\theta = -\mathbf{m}\cdot\mathbf{n}$, whichever is positive, since θ is acute. Clearly

$$\cos\theta = -\mathbf{m}\cdot\mathbf{n} = -\tfrac{1}{10}\sqrt{2}(3-4)\cdot(1,1) = \tfrac{1}{10}\sqrt{2}$$

$$\theta = \text{arc cos }\tfrac{1}{10}\sqrt{2}.$$

Answer $\theta = \text{arc cos }\tfrac{1}{10}\sqrt{2} \approx 81.87°$

Remark 1 If two lines M and N are parallel (or coincide), then $\mathbf{m} = \pm\mathbf{n}$ so $\mathbf{m}\cdot\mathbf{n} = \pm1$. But $\cos\theta = \pm1$ means $\theta = 0$ or $\theta = \pi$. The result $\mathbf{m}\cdot\mathbf{n} = \cos\theta$ is thus correct provided we interpret $\theta = 0$ or $\theta = \pi$ to mean parallel (or coincident) lines.

Remark 2 Two lines M and N are perpendicular if $\mathbf{m}\cdot\mathbf{n} = 0$, that is, $\theta = \tfrac{1}{2}\pi$. Suppose the lines are given in the function forms

$$M: y = mx + b \qquad N: y = nx + c,$$

where m and n are their slopes. We may rewrite these as

$$(-m, 1)\cdot(x, y) = b \qquad \text{and} \qquad (-n, 1)\cdot(x, y) = c.$$

Clearly the vectors $(-m, 1)$ and $(-n, 1)$ are parallel to $\mathbf{m}$ and $\mathbf{n}$ respectively, so $\mathbf{m}\cdot\mathbf{n} = 0$ if and only if

$$(-m, 1)\cdot(-n, 1) = 0 \qquad mn + 1 = 0 \qquad mn = -1.$$

Thus we have derived the result that two non-vertical lines are perpendicular if and only if their slopes are negative reciprocals of each other.

Exercises

Find the normal form

1 $y = x$ 2 $y = -x$ 3 $y = \tfrac{3}{4}x$

4 $y = -\tfrac{12}{5}x$ 5 $3x + 4y = 2$ 6 $4x - 3y = 3$

7 $12x + 5y = 1$ 8 $-8x + 15y = 17$ 9 $7x - 24y = 25$

10 $24x + 7y = 1$ 11 $2x - 3y = 4$ 12 $3x + 2y = 2$

13 $y = 2x + 1$ 14 $y = -3x - 2$ 15 $y = ax + b$

16 $\dfrac{x}{a} + \dfrac{y}{b} = 1.$

Find the distance from $\mathbf{x}_0$ to L

	$\mathbf{x}_0$	L		$\mathbf{x}_0$	L
17	$(1, 0)$	$y = -2$	18	$(0, -1)$	$x = 3$
19	$(1, 1)$	$\tfrac{4}{5}x - \tfrac{3}{5}y = -2$	20	$(-2, 3)$	$\tfrac{3}{5}x + \tfrac{4}{5}y = 1$
21	$(0, -2)$	$4x + 3y = 7$	22	$(1, -1)$	$3x - 4y = 1$
23	$(10, 5)$	$12x + 5y = 26$	24	$(1, 3)$	$-5x + 12y = 13$
25	$(1, 0)$	$x + y = 2$	26	$(2, -1)$	$x = 4y - 3$

Find the acute angle between the lines M and N

27 $y = x$ $x = 0$ 28 $y = -x$ $y = 0$

29 $y = \tfrac{1}{2}x$ $y = 2x$ 30 $y = 3x$ $y = -x$

31 $3x + 4y = 1$ $5x + 12y = 1$ **32** $-3x + 4y = 1$ $12x - 5y = 1$

33 $x + y = 1$ $x + 2y = 2$ **34** $2x - y = 3$ $x + 3y = 2.$

35 Find the normal form of the line through $(1, 0)$ and perpendicular to $x - 2y = 6$.

36 Find the normal form of the line through $(-2, -3)$ and perpendicular to the line through $(1, 0)$ and $(5, 4)$.

37 Find the normal form of the perpendicular bisector of the line segment between $(0, 1)$ and $(5, -2)$.

38 Find the normal form of the perpendicular bisector of the line segment between $(1, 3)$ and $(-4, 4)$.

REVIEW EXERCISES

Compute

1 $-4(3, 1) + 7(2, 2)$ **2** $(8, -5) - (5, -8)$

3 $0(1, -1) - 3(-1, 1)$ **4** $2(3, 4) + 5(6, 7).$

5 Let $\mathbf{v} = (\cos 20°, \sin 20°)$ and $\mathbf{w} = (\cos 80°, \sin 80°)$. Find $\mathbf{v} \cdot \mathbf{w}$ and $\|\mathbf{v}\| + 2\|\mathbf{w}\|$.

6 Let $\mathbf{v} = (4, -3)$ and $\mathbf{w} = (-5, 12)$. Find $\mathbf{v} \cdot \mathbf{w}$ and $\|\mathbf{v}\|/\|\mathbf{w}\|$.

7 Compute the smallest positive angle between $\mathbf{v} = (3, -1)$ and $\mathbf{w} = (3 + \sqrt{3}, 3\sqrt{3} - 1)$.

8 If there were no current, the drifting float would be pushed due NE at 3 mph by the wind. But there is a 7 mph current moving due NW. Find the speed and direction of the float.

Find the normal form

9 $40x - 9y = 123$ **10** $x = -5y + 5.$

11 Find the distance from $(1, -2)$ to $\frac{3}{5}x - \frac{4}{5}y = 10$.

12 Find the acute angle between the lines $x - y = 2$ and $\sqrt{3}\, x + y = 1$.

11 COMPLEX NUMBERS

1 COMPLEX ARITHMETIC

As we know, not every polynomial with real coefficients has a real zero.* Even the simple quadratic polynomial $x^2 + 1$ has no real zero. For if r is any real number, then $r^2 \geq 0$ so $r^2 + 1 > 0$, not $r^2 + 1 = 0$.

The most general quadratic equation

$$ax^2 + bx + c = 0 \qquad (a \neq 0)$$

with real coefficients has no real solution if its discriminant is negative, that is, if $D = b^2 - 4ac < 0$. Imagine that there exists a number system containing the real number system, in which the special quadratic equation $x^2 + 1 = 0$ has a root, $\sqrt{-1}$. Then the general quadratic equation can be solved even when $D < 0$. Its roots are

$$\frac{-b \pm \sqrt{D}}{2a} = \frac{-b \pm \sqrt{-D}\sqrt{-1}}{2a}$$

provided the rules of ordinary arithmetic are valid in the extended number system.

We shall now construct such a new number system. It will contain the real number system as a part; it will contain a square root of -1; and it will obey the usual rules of arithmetic.

We start with the set of real numbers and a symbol i, which will play the role of $\sqrt{-1}$. The new number system, called the **complex number system,** consists of all formal expressions

$$a + bi,$$

---

* A **zero** of a function $f(x)$ is a root of the equation $f(x) = 0$, that is, a real number r in the domain of $f(x)$ such that $f(r) = 0$.

where a and b are real numbers. We must now say how to operate with these formal symbols. First of all, since i is a new sort of object, it is natural to say that two complex numbers $a + bi$ and $c + di$ are equal if and only if $a = c$ and $b = d$.

Equality of Complex Numbers   Two complex numbers $a + bi$ and $c + di$ are equal if and only if $a = c$ and $b = d$.

If ordinary rules of arithmetic are to hold, we must have

$$(a + bi) + (c + di) = (a + c) + (b + d)i.$$

This we take as the *definition* of addition in the complex number system.

Example $(3 - 7i) + (5 + 2i) = (3 + 5) + [(-7) + 2]i = 8 - 5i$

Since we want to have $i^2 = -1$, we agree to replace i^2 by -1. Thus it seems natural to compute the product of $a + bi$ and $c + di$ as follows:

$$(a + bi)(c + di) = ac + a(di) + (bi)c + (bi)(di)$$

$$= ac + (ad)i + (bc)i + (bd)i^2$$

$$= [ac + (bd)i^2] + [ad + bc]i$$

$$= (ac - bd) + (ad + bc)i.$$

This we take as the *definition* of multiplication in the complex number system.

Example $(-4 + 3i)(2 + 5i)$

$$= [(-4)(2) - (3)(5)] + [(-4)(5) + (3)(2)]i = -23 - 14i$$

Addition and Multiplication of Complex Numbers   The **sum** of two complex numbers is defined by

$$(a + bi) + (c + di) = (a + c) + (b + d)i.$$

The **product** of two complex numbers is defined by

$$(a + bi)(c + di) = (ac - bd) + (ad + bc)i.$$

These definitions attempt to breathe life into the set of formal symbols $a + bi$. We shall show that they succeed; the complex number system obeys all the usual rules of arithmetic valid for the real numbers. The crucial definition is the product. It includes the relation $i^2 = -1$ that we wanted:

$$i^2 = (0 + 1i)(0 + 1i) = (0 \cdot 0 - 1 \cdot 1) + (0 \cdot 1 + 1 \cdot 0)i = -1.$$

We shall *identify* the complex number $a + 0i$ with the real number a. This is perfectly reasonable since complex numbers $a + 0i$ and $b + 0i$ add and multiply just as do the real numbers a and b:

$$(a + 0i) + (b + 0i) = (a + b) + (0 + 0)i = (a + b) + 0i,$$

$$(a + 0i)(b + 0i) = (ab - 0 \cdot 0) + (a \cdot 0 + 0 \cdot b)i = ab + 0i.$$

Thus, the complex number system contains a subsystem that we can identify with the real number system. In other words, the complex number system is an *extension* of the real number system, and its arithmetic is *consistent* with that of the real number system.

Notation Sometimes it is convenient to write $a + ib$ instead of $a + bi$. This is particularly useful with expressions such as $\cos \theta + i \sin \theta$ and $-1 + i\sqrt{3}$. In electrical engineering i is used for current and the symbol j is used for the complex number we are calling i.

Rules of Complex Arithmetic

We have referred to the "usual rules of arithmetic" being valid in the complex number system. Let us be precise now. There are two types of "rules": identities and existence of inverses.

Rules of Complex Arithmetic: Identities
Let α, β, γ be complex numbers. Then

Commutative Laws

$$\alpha + \beta = \beta + \alpha \quad \alpha\beta = \beta\alpha$$

Associative Laws

$$\alpha + (\beta + \gamma) = (\alpha + \beta) + \gamma \quad \alpha(\beta\gamma) = (\alpha\beta)\gamma$$

Distributive Law

$$\alpha(\beta + \gamma) = \alpha\beta + \alpha\gamma$$

Identity Laws

$$\alpha + 0 = \alpha \quad \alpha \cdot 1 = \alpha$$

Each of these rules is a consequence of the rules of arithmetic for *real* numbers. For instance, let

$$\alpha = a + bi \quad \text{and} \quad \beta = c + di.$$

Then

$$\alpha\beta = (a + bi)(c + di) = (ac - bd) + (ad + bc)i$$
$$\beta\alpha = (c + di)(a + bi) = (ca - db) + (cb + da)i.$$

By the commutative laws for real numbers,

$$ac - bd = ca - db \quad \text{and} \quad ad + bc = cb + da,$$

hence $\alpha\beta = \beta\alpha$. This proves the commutative law for complex multiplication. Proofs of the remaining laws are left as exercises.

Example $\alpha = 1 - i \quad \beta = 2 + 3i \quad \gamma = -2 + i$

$$\alpha\beta = (1 - i)(2 + 3i) = 5 + i$$

$$\beta\gamma = (2 + 3i)(-2 + i) = -7 - 4i$$

$$(\alpha\beta)\gamma = (5 + i)(-2 + i) = -11 + 3i$$

$$\alpha(\beta\gamma) = (1 - i)(-7 - 4i) = -11 + 3i.$$

Hence $(\alpha\beta)\gamma = \alpha(\beta\gamma)$, as predicted by the associative law for multiplication.

Rules of Complex Arithmetic: Inverses

Additive Inverse If $\alpha = a + bi$, set $-\alpha = (-a) + (-b)i$. Then

$$\alpha + (-\alpha) = 0.$$

Multiplicative Inverse If $\alpha = a + bi \neq 0$, set

$$\alpha^{-1} = \left(\frac{a}{a^2 + b^2}\right) + \left(\frac{-b}{a^2 + b^2}\right)i = \frac{a - bi}{a^2 + b^2}.$$

Then $\alpha\alpha^{-1} = 1$.

The first is rather obvious; for instance

$$(-3 + 4i) + (3 - 4i) = 0 + 0i = 0.$$

The second says that each non-zero complex number has a reciprocal. It looks complicated, but it follows easily from the previous rules:

$$\alpha\alpha^{-1} = (a + bi)\left(\frac{a - bi}{a^2 + b^2}\right) = \frac{(a + bi)(a - bi)}{a^2 + b^2} = \frac{a^2 + b^2}{a^2 + b^2} = 1.$$

Just as for real numbers, division by a complex number β is defined as multiplication by β^{-1}.

$$\frac{\alpha}{\beta} = \alpha\beta^{-1} \qquad (\beta \neq 0).$$

Just as for real numbers $(\beta^{-1})^2 = (\beta^2)^{-1}$ since

$$(\beta^2)(\beta^{-1})^2 = (\beta\beta)(\beta^{-1}\beta^{-1}) = (\beta\beta^{-1})(\beta\beta^{-1}) = 1.$$

Thus we can define β^{-2}, and similarly β^{-3}, β^{-4}, $\cdots$ so that the usual rules of exponents hold for integer powers (positive and negative) of complex numbers.

Examples $\alpha = 1 - i$ $\beta = 2 + 3i$

(1) $\beta^{-1} = (2 + 3i)^{-1} = \dfrac{2}{4 + 9} + \dfrac{-3}{4 + 9}i = \frac{2}{13} - \frac{3}{13}i$.

Check $\beta\beta^{-1} = (2 + 3i)(\frac{2}{13} - \frac{3}{13}i) = \frac{13}{13} + \frac{0}{13}i = 1$.

(2) $\dfrac{\alpha}{\beta} = \alpha\beta^{-1} = (1 - i)(2 + 3i)^{-1}$

$\qquad = (1 - i)(\frac{2}{13} - \frac{3}{13}i) = -\frac{1}{13} - \frac{5}{13}i$.

Check $\beta(\alpha/\beta) = (2 + 3i)(-\frac{1}{13} - \frac{5}{13}i) = \frac{13}{13} - \frac{13}{13}i = \alpha$.

(3) $\beta^{-2} = (\beta^{-1})^2 = (\frac{2}{13} - \frac{3}{13}i)(\frac{2}{13} - \frac{3}{13}i) = -\frac{5}{169} - \frac{12}{169}i$.

Complex Conjugates and Absolute Values

Complex arithmetic is richer than real arithmetic. One feature is an operation, conjugation, that has no real counterpart.

Complex Conjugate   Let $\alpha = a + bi$. Its **complex conjugate** is defined to be

$$\bar{\alpha} = a - bi.$$

The operation of complex conjugation satisfies the rules

$$\bar{\bar{\alpha}} = \alpha \qquad \overline{\alpha + \beta} = \bar{\alpha} + \bar{\beta} \qquad \overline{\alpha\beta} = \bar{\alpha}\bar{\beta}.$$

Examples

$$\overline{4} = 4 \qquad \overline{i} = -i \qquad \overline{-3i} = 3i \qquad \overline{4 - 3i} = 4 + 3i$$

The first two rules are left to prove as exercises. The third rule, $\overline{\alpha\beta} = \bar{\alpha}\bar{\beta}$, is by no means obvious; given the complicated definition of the complex product, it should come as a surprise. To prove it, set $\alpha = a + bi$ and $\beta = c + di$. Then

$$\overline{\alpha\beta} = \overline{(ac - bd) + (ad + bc)i} = (ac - bd) - (ad + bc)i$$

and

$$\bar{\alpha}\bar{\beta} = (a - bi)(c - di) = (ac - bd) - (ad + bc)i$$

So $\overline{\alpha\beta} = \bar{\alpha}\bar{\beta}$.

The rules extend to any number of summands or factors; for instance,

$$\overline{\alpha + \beta + \gamma} = \bar{\alpha} + \bar{\beta} + \bar{\gamma} \qquad \overline{\alpha\beta\gamma} = \bar{\alpha}\bar{\beta}\bar{\gamma} \qquad \text{etc.}$$

Each real number a has an absolute value, a nonnegative real $|a|$ that measures the magnitude of a. (Note that $|a| = \sqrt{a^2}$.) So does each complex number.

Absolute Value   The **absolute value** (or **modulus**) of $\alpha = a + bi$ is

$$|\alpha| = \sqrt{a^2 + b^2}.$$

Examples

$$|3| = 3 \qquad |-5i| = 5 \qquad |-12 + 5i| = \sqrt{(-12)^2 + 5^2} = \sqrt{169} = 13$$

Rules for Absolute Values

Let α and β be complex numbers. Then

$$|\alpha| \geq 0; \qquad |\alpha| = 0 \quad \text{if and only if} \quad \alpha = 0.$$

$$|\alpha\beta| = |\alpha| \cdot |\beta| \qquad \left|\frac{\alpha}{\beta}\right| = \frac{|\alpha|}{|\beta|} \qquad |\bar{\alpha}| = |\alpha| \qquad |\alpha|^2 = \alpha\bar{\alpha}.$$

Triangle Inequality $|\alpha + \beta| \leq |\alpha| + |\beta|.$

Except for the triangle inequality, which will be discussed in the next section, the proofs of these properties are left as exercises.

Conjugates and absolute values help simplify division of complex numbers. Since $\alpha\bar{\alpha} = |\alpha|^2$, we can divide both sides by $|\alpha|^2$:

$$\alpha \cdot \frac{\bar{\alpha}}{|\alpha|^2} = 1.$$

Therefore

$$\alpha^{-1} = \frac{\bar{\alpha}}{|\alpha|^2} = \frac{a - bi}{a^2 + b^2} \qquad (\alpha \neq 0)$$

(This agrees with the formula for the multiplicative inverse given earlier.) It follows that

$$\frac{\alpha}{\beta} = \alpha\beta^{-1} = \frac{\alpha\bar{\beta}}{|\beta|^2} = \frac{\alpha\bar{\beta}}{\beta\bar{\beta}}.$$

Thus, to evaluate α/β, multiply both numerator and denominator by $\bar{\beta}$. For example,

$$\frac{2 + 3i}{1 + 2i} = \frac{2 + 3i}{1 + 2i} \frac{1 - 2i}{1 - 2i} = \frac{8 - i}{5} = \tfrac{8}{5} - \tfrac{1}{5}i$$

Remark The introduction of complex numbers was one of the greatest advances ever made in mathematics. It took hundreds of years to develop the idea once the need was felt, so don't expect to learn it perfectly in a few minutes. You should read and reread this section until you are confident you understand it.

Exercises

Express in the form $a + bi$

1 $(3 + 2i) + (6 - i)$ **2** $(1 - i) + (4 + 3i)$ **3** $2(1 + 4i) - 3(2 + i)$
4 $(-2 + 3i) + 5(1 - i)$ **5** $i(2 - 3i)$ **6** $i(8i + 5)$
7 $(1 + i)(3 - 4i)$ **8** $(1 - i)(2 + 7i)$ **9** $(5 + 4i)(3 + 2i)$
10 $(1 - 6i)(3 + i)$ **11** $(1 + i)^2$ **12** $(2 - i)(2 + i)$

13 $\dfrac{1}{3 + 4i}$ **14** $\dfrac{1}{6 - i}$ **15** $\dfrac{3 + i}{2 - i}$

16 $\dfrac{2 + 7i}{1 - 3i}$ **17** $\dfrac{1}{(2 + 3i)(5 - 4i)}$ **18** $\dfrac{1}{(1 - i)(8 + 3i)}$

19 $\dfrac{1 - 3i}{(2 + i)(2 + 5i)}$ **20** $\dfrac{(1 + i)(1 + 2i)}{(3 - 2i)(4 - 3i)}$ **21** $\dfrac{i}{2 + i} + \dfrac{3 + i}{4 + i}$

22 $\dfrac{1}{4 - 3i} + \dfrac{5 + 3i}{2 - i}$.

Compute $\overline{\alpha\beta}$ and $\bar{\alpha}\bar{\beta}$

23 $\alpha = 1 - i$ $\beta = 3i$ **24** $\alpha = 1 + i$ $\beta = 1 - i$
25 $\alpha = 2 - i$ $\beta = 3 + 2i$ **26** $\alpha = -1 - 2i$ $\beta = 3 - i$
27 $\alpha = (1 + i)^2$ $\beta = 1 + 2i$ **28** $\alpha = (1 - i)^2$ $\beta = 1 - 2i$.

Compute

29 $|3 + 4i|$ **30** $|3 - 4i|$ **31** $|(1 + i)(2 - i)|$
32 $|(2 - 3i)(6 + i)|$ **33** $|i(2 + i)| + |1 + 4i|$ **34** $|6(3 - 2i)| + |2 - i|$

35 $\left| \dfrac{1}{2 - i} \right|$ **36** $\left| \dfrac{2 + 5i}{3 + i} \right|$ **37** $\left| \dfrac{1 + i}{1 - i} \right|$ **38** $\left| \dfrac{3 + 4i}{4 - 3i} \right|$.

39 Simplify $i + i^2 + i^3 + \cdots + i^{11}$.

40 Compute $(1 + i)^2$ and use the result to compute $(1 + i)^{12}$.

Compute

41 $\left(\dfrac{-1 + i\sqrt{3}}{2} \right)^3$ **42** $\left(\dfrac{1 + i}{\sqrt{2}} \right)^8$.

Prove

43 $\alpha + \beta = \beta + \alpha$ **44** $\alpha + (\beta + \gamma) = (\alpha + \beta) + \gamma$

45 $\alpha(\beta + \gamma) = \alpha\beta + \alpha\gamma$ **46** $\alpha(\beta\gamma) = (\alpha\beta)\gamma$

47 $\bar{\bar{\alpha}} = \alpha$ **48** $\overline{\alpha + \beta} = \bar{\alpha} + \bar{\beta}$

49 $|\alpha| \geq 0$ **50** $|\alpha| = 0$ implies $\alpha = 0$

51 $|\bar{\alpha}| = |\alpha|$ **52** $|\alpha|^2 = \alpha\bar{\alpha}$

53 $|\alpha\beta| = |\alpha||\beta|$ **54** $|\alpha/\beta| = |\alpha|/|\beta|$.

2 THE COMPLEX PLANE

Each complex number $\alpha = a + bi$ is completely determined by the ordered pair (a, b) of real numbers. We call a the **real part** of α and b the **imaginary part** of α, and we write

$$a = \text{Re}(\alpha) \quad \text{and} \quad b = \text{Im}(\alpha).$$

If $\alpha = a + bi$, then $\bar{\alpha} = a - bi$. Therefore

$$\alpha + \bar{\alpha} = (a + bi) + (a - bi) = 2a$$

$$\alpha - \bar{\alpha} = (a + bi) - (a - bi) = 2bi.$$

From these equations follow the useful relations

$$\boxed{\text{Re}(\alpha) = \frac{\alpha + \bar{\alpha}}{2} \quad \text{Im}(\alpha) = \frac{\alpha - \bar{\alpha}}{2i}.}$$

Example $\alpha = -2 + 5i$

$$\bar{\alpha} = -2 - 5i$$

$$\text{Re}(\alpha) = -2 \quad \text{Im}(\alpha) = 5$$

$$\frac{\alpha + \bar{\alpha}}{2} = \frac{(-2 + 5i) + (-2 - 5i)}{2} = -2 = \text{Re}(\alpha)$$

$$\frac{\alpha - \bar{\alpha}}{2i} = \frac{(-2 + 5i) - (-2 - 5i)}{2i} = \frac{10i}{2i} = 5 = \text{Im}(\alpha)$$

Geometric Interpretation of Complex Numbers

The correspondence

$$\alpha = a + bi \longleftrightarrow (a, b)$$

between complex numbers and ordered pairs of real numbers strongly suggests a geometric representation for complex numbers. Indeed, we identify the *complex number* $\alpha = a + bi$ with the *point* (a, b) in the plane (Fig. 1).

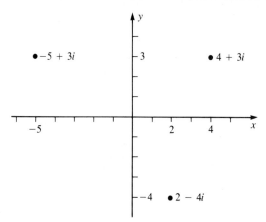

Fig. 1   Representation of complex numbers as points in the plane

This gives an entirely new interpretation to the cartesian (coordinate) plane: as the set of all complex numbers. When looked at this way, the plane is called the **complex plane** (also **Gaussian plane** after K.F. Gauss).

The definition of addition of complex numbers,

$$(a + bi) + (c + di) = (a + c) + (b + d)i,$$

has a useful geometric interpretation. To add $\beta = c + di$ to α, we must increase the x-coordinate of α by c and the y-coordinate of α by d. (See Fig. 2a.) Therefore we take the little rectangle built on β and move it lock, stock, and barrel, so that 0 is moved to α and β is moved to $\alpha + \beta$. Thus *the segment from α to $\alpha + \beta$ is the parallel displacement of the segment from 0 to β.*

This means that the three segments from 0 to β, from 0 to α, and from α to $\alpha + \beta$ make up three sides of a parallelogram. We complete the parallelogram

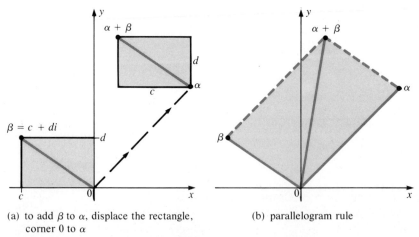

(a) to add β to α, displace the rectangle, corner 0 to α

(b) parallelogram rule

Fig. 2   Geometric meaning of addition of complex numbers

(Fig. 2b), and we conclude that complex numbers can be added *graphically* by (what is called) the **parallelogram rule.**

Remark The parallelogram rule of addition is the same as the rule for adding forces, or vectors in general. See p. 216.

Absolute values and complex conjugates also have geometric interpretations. The absolute value $|\alpha|$ is the distance of $\alpha = a + bi$ from 0, because $|\alpha|^2 = a^2 + b^2$. The complex conjugate $\bar{\alpha} = a - bi$ is the reflection (mirror image) of α in the x-axis (Fig. 3a).

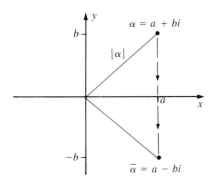

(a) geometric meaning of absolute value (b) proof of the triangle inequality
    and complex conjugate $|\alpha + \beta| \leq |\alpha| + |\beta|$

Fig. 3

The triangle inequality

$$|\alpha + \beta| \leq |\alpha| + |\beta|$$

follows directly from the parallelogram rule for addition. We simply observe (Fig. 3b) that a triangle exists whose side lengths are $|\alpha|$, $|\beta|$, and $|\alpha + \beta|$. But in a triangle, each side is at most the sum of the other two sides (even if the triangle collapses), so in particular $|\alpha + \beta| \leq |\alpha| + |\beta|$.

Thus $\alpha + \beta$, $|\alpha|$, and $\bar{\alpha}$ have simple and useful geometric interpretations. The product $\alpha\beta$ also has a geometric interpretation, but that is more easily seen in polar coordinates.

Polar Form

Let $\alpha = a + bi$ be a non-zero complex number. The point (a, b) has polar coordinates $\{r, \theta\}$, where $r > 0$,

$$a = r \cos \theta \quad \text{and} \quad b = r \sin \theta.$$

Therefore $\alpha = a + bi = (r \cos \theta) + (r \sin \theta)i$, that is,

$$\alpha = r(\cos \theta + i \sin \theta).$$

The representation of α in the form $r(\cos \theta + i \sin \theta)$ is called the **polar form** of the complex number α. See Fig. 4a, next page. (This contrasts with the **rectangular form** $\alpha = a + bi$.)

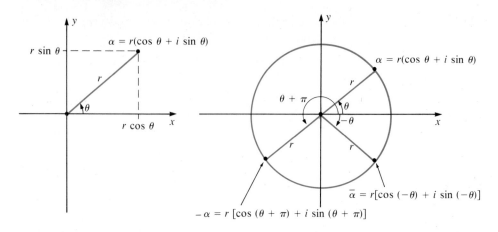

(a) polar form   (b) polar forms of α, $\bar{\alpha}$, and $-\alpha$

Fig. 4

When a complex number α is expressed in the polar form $\alpha = r(\cos \theta + i \sin \theta)$, the positive number r is the absolute value of α. Just check it:

$$|\alpha|^2 = \alpha\bar{\alpha} = [r(\cos \theta + i \sin \theta)][r(\cos \theta - i \sin \theta)]$$

$$= r^2(\cos^2 \theta + \sin^2 \theta) = r^2$$

$$|\alpha| = r.$$

The angle θ in $\alpha = r(\cos \theta + i \sin \theta)$ is called the **argument** of α, and written

$$\theta = \arg \alpha.$$

Warning The angle θ is only determined up to an integral multiple of 2π. Therefore, to be accurate, we should write $\arg \alpha = \theta + 2\pi n$ and thus let the symbol $\arg \alpha$ denote the whole family of possible angles θ such that $\alpha = r(\cos \theta + i \sin \theta)$. No one does this; the inaccurate notation $\theta = \arg \alpha$ is commonplace. But formulas involving $\arg \alpha$ must always be read with a grain of salt; you can always tack on $2\pi n$.

> **Polar Form** Each non-zero complex number can be written in the polar form
>
> $$\alpha = r(\cos \theta + i \sin \theta)$$
>
> where $r = |\alpha|$ is the modulus (absolute value) of α and the angle $\theta = \arg \alpha$ is the argument of α.

How do we convert back and forth from rectangular to polar form? One way is easy. If

$$\alpha = r(\cos \theta + i \sin \theta)$$

is given in polar form, then

$$\alpha = (r \cos \theta) + i(r \sin \theta)$$

is rectangular form.

Example $\alpha = 5(\cos 30° + i \sin 30°) = (5 \cos 30°) + (5 \sin 30°)i$

$\qquad = \frac{5}{2}\sqrt{3} + \frac{5}{2}i$

If

$\qquad \alpha = a + bi \neq 0$

is given in rectangular form, then from $\alpha = r(\cos \theta + i \sin \theta)$ we have

$\qquad r \cos \theta = a \qquad r \sin \theta = b$

$\qquad \cos \theta = \dfrac{a}{r} \qquad \sin \theta = \dfrac{b}{r}$

We know $r = |\alpha| = \sqrt{a^2 + b^2}$, so we can solve for θ. It's the old story all over again of rectangular to polar coordinates.

EXAMPLE 1 Convert into polar form

(a) $1 + i$ (b) $1 + i\sqrt{3}$.

Solution (a) $|1 + i| = \sqrt{1^2 + 1^2} = \sqrt{2}$.

Therefore

$\qquad \cos \theta = \dfrac{a}{r} = \dfrac{1}{\sqrt{2}} \qquad \sin \theta = \dfrac{b}{r} = \dfrac{1}{\sqrt{2}}.$

One choice for θ is $\theta = \frac{1}{4}\pi$, so $1 + i = \sqrt{2}(\cos \frac{1}{4}\pi + i \sin \frac{1}{4}\pi)$.

(b) $|1 + i\sqrt{3}| = \sqrt{1 + 3} = 2$

$\qquad \cos \theta = \dfrac{a}{r} = \frac{1}{2} \qquad \sin \theta = \dfrac{b}{r} = \frac{1}{2}\sqrt{3}.$

One choice for θ is $\theta = \frac{1}{3}\pi$, so $1 + i\sqrt{3} = 2(\cos \frac{1}{3}\pi + i \sin \frac{1}{3}\pi)$.

Answer (a) $\sqrt{2}(\cos \frac{1}{4}\pi + i \sin \frac{1}{4}\pi)$ (b) $2(\cos \frac{1}{3}\pi + i \sin \frac{1}{3}\pi)$

We next derive the polar forms for $\bar{\alpha}$ and $-\alpha$. Suppose we start with $\alpha = r(\cos \theta + i \sin \theta)$. Then

$\qquad |\bar{\alpha}| = |-\alpha| = |\alpha| = r$

so $\bar{\alpha}$ and $-\alpha$ have the same modulus as α. By Fig. 4b,

$\qquad \arg \bar{\alpha} = -\theta \quad \text{and} \quad \arg(-\alpha) = \theta + \pi.$

If $\alpha = r(\cos \theta + i \sin \theta)$, then

$\bar{\alpha} = r[\cos(-\theta) + i \sin(-\theta)] \qquad -\alpha = r[\cos(\theta + \pi) + i \sin(\theta + \pi)].$

Multiplication

Suppose

$\qquad \alpha_1 = r_1(\cos \theta_1 + i \sin \theta_1) \quad \text{and} \quad \alpha_2 = r_2(\cos \theta_2 + i \sin \theta_2)$

are non-zero complex numbers in polar form. Then

$\alpha_1\alpha_2$

$$= (r_1r_2)(\cos\theta_1 + i\sin\theta_1)(\cos\theta_2 + i\sin\theta_2)$$
$$= (r_1r_2)[(\cos\theta_1\cos\theta_2 - \sin\theta_1\sin\theta_2) + i(\cos\theta_1\sin\theta_2 + \sin\theta_1\cos\theta_2)].$$

But by the addition laws for sine and cosine,

$$\cos\theta_1\cos\theta_2 - \sin\theta_1\sin\theta_2 = \cos(\theta_1 + \theta_2)$$

and

$$\cos\theta_1\sin\theta_2 + \sin\theta_1\cos\theta_2 = \sin(\theta_1 + \theta_2).$$

Therefore

$$\alpha_1\alpha_2 = (r_1r_2)[\cos(\theta_1 + \theta_2) + i\sin(\theta_1 + \theta_2)].$$

Polar Form of Complex Multiplication

Given non-zero complex numbers

$$\alpha_1 = r_1(\cos\theta_1 + i\sin\theta_1) \quad \text{and} \quad \alpha_2 = r_2(\cos\theta_2 + i\sin\theta_2)$$

in polar form, the polar form of their product is

$$\alpha_1\alpha_2 = r_1r_2[\cos(\theta_1 + \theta_2) + i\sin(\theta_1 + \theta_2)].$$

Therefore

$$|\alpha_1\alpha_2| = |\alpha_1| \cdot |\alpha_2| \qquad \arg(\alpha_1\alpha_2) = \arg\alpha_1 + \arg\alpha_2.$$

In words, to multiply two complex numbers, multiply their moduli and add their arguments.

Examples (1) $\alpha = 1 + i \qquad \beta = 1 + i\sqrt{3}$

By Example 1,

$$\alpha = \sqrt{2}(\cos\tfrac{1}{4}\pi + i\sin\tfrac{1}{4}\pi) \qquad \beta = 2(\cos\tfrac{1}{3}\pi + i\sin\tfrac{1}{3}\pi).$$

Therefore $|\alpha\beta| = 2\sqrt{2} \qquad \arg(\alpha\beta) = \tfrac{1}{4}\pi + \tfrac{1}{3}\pi = \tfrac{7}{12}\pi,$

$$\alpha\beta = 2\sqrt{2}(\cos\tfrac{7}{12}\pi + i\sin\tfrac{7}{12}\pi).$$

(2) $\alpha = i \qquad \beta = r(\cos\theta + i\sin\theta)$

Then

$$|i| = 1 \qquad \arg i = \tfrac{1}{2}\pi \qquad i = \cos\tfrac{1}{2}\pi + i\sin\tfrac{1}{2}\pi$$

Therefore

$$i\beta = r[\cos(\theta + \tfrac{1}{2}\pi) + i\sin(\theta + \tfrac{1}{2}\pi)].$$

Thus multiplication by i simply rotates a complex number $\frac{1}{2}\pi$ radians counter-clockwise about the origin (Fig. 5). Likewise, multiplication by $-i$ rotates it clockwise through a right angle.

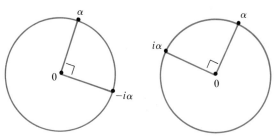

Fig. 5 Multiplication by i and by $-i$

Remark The factor 0 can be included in our recipes provided we agree that

$$|0| = 0 \qquad \arg 0 = \text{any } \theta \qquad 0 = 0(\cos \theta + i \sin \theta)$$

Exercises

Express in polar form

1 $-3i$ **2** -5 **3** $1 - i$ **4** $3 + 3i$
5 $-\sqrt{3} - i$ **6** $-2 + 2i\sqrt{3}$ **7** $3\sqrt{3} + 3i$ **8** $4 - 4i$.

Express the factors and the product in polar form

9 $(1 + i)(1 - i)$ **10** $(1 - i)(\sqrt{3} + i)$ **11** $(1 + i\sqrt{3})^2$
12 $(2 + 2i)(4\sqrt{3} - 4i)$.

Find $\operatorname{Re} \alpha$, $\operatorname{Im} \alpha$, $|\alpha|$, and $\arg \alpha$

13 $\alpha = \sqrt{3} + 3i$ **14** $\alpha = 1 - i$

Compute both $|\alpha| \cdot |\beta|$ and $|\alpha\beta|$ and compare

15 $\alpha = 1 - i$ $\beta = 3i$
16 $\alpha = 1 + i$ $\beta = -1 + i$
17 $\alpha = 2 - i$ $\beta = 3 + 2i$
18 $\alpha = -1 + 2i$ $\beta = 3 - i$
19 $\alpha = (1 + i)^2$ $\beta = 1 + 2i$
20 $\alpha = (1 - i)^2$ $\beta = 1 - 2i$.
21* Use complex numbers to prove the identity

$$(x^2 + y^2)(u^2 + v^2) = (xu - yv)^2 + (xv + yu)^2.$$

22 (cont.) From $13 = 2^2 + 3^2$ and $37 = 1^2 + 6^2$, express 481 as a sum of two perfect squares.

23 Justify geometrically the formula $\operatorname{Re}(\alpha) = \frac{1}{2}(\alpha + \bar{a})$.

24 Calculate $(1 + i)(1 + i\sqrt{3})$ in two different ways to prove

$$\cos \tfrac{7}{12}\pi = \tfrac{1}{4}(\sqrt{2} - \sqrt{6}) \qquad \sin \tfrac{7}{12}\pi = \tfrac{1}{4}(\sqrt{2} + \sqrt{6})$$

Compare Example 1(a), page 83.

3 QUOTIENTS AND SQUARE ROOTS

From the rule for multiplication of complex numbers in polar form follows a companion rule for division. First we note the form of the reciprocal of a complex number:

$$\frac{1}{r(\cos\theta + i\sin\theta)} = \frac{1}{r(\cos\theta + i\sin\theta)}\frac{\cos\theta - i\sin\theta}{\cos\theta - i\sin\theta}$$

$$= \frac{\cos\theta - i\sin\theta}{r(\cos^2\theta + \sin^2\theta)} = \frac{1}{r}(\cos\theta - i\sin\theta)$$

$$= \frac{1}{r}[\cos(-\theta) + i\sin(-\theta)].$$

Reciprocal in Polar Form   Let $\alpha = r(\cos\theta + i\sin\theta)$ Then

$$\frac{1}{\alpha} = \frac{1}{r}[\cos(-\theta) + i\sin(-\theta)].$$

Therefore $\left|\dfrac{1}{\alpha}\right| = \dfrac{1}{|\alpha|}$ and $\arg\dfrac{1}{\alpha} = -\arg\alpha$.

We can now derive the polar form of a quotient α_1/α_2. As before, we start with the polar forms of α_1 and α_2:

$$\alpha_1 = r_1(\cos\theta_1 + i\sin\theta_1) \qquad \alpha_2 = r_2(\cos\theta_2 + i\sin\theta_2).$$

Then

$$\frac{\alpha_1}{\alpha_2} = \alpha_1\alpha_2^{-1} \qquad \left|\frac{\alpha_1}{\alpha_2}\right| = |\alpha_1||\alpha_2^{-1}| = r_1\cdot\frac{1}{r_2} = \frac{r_1}{r_2}$$

$$\arg\frac{\alpha_1}{\alpha_2} = \arg(\alpha_1\alpha_2^{-1}) = \arg\alpha_1 + \arg\alpha_2^{-1}$$

$$= \arg\alpha_1 - \arg\alpha_2 = \theta_1 - \theta_2.$$

We fit the pieces together:

Quotient in Polar Form   Let

$$\alpha_1 = r_1(\cos\theta_1 + i\sin\theta_2) \quad\text{and}\quad \alpha_2 = r_2(\cos\theta_2 + i\sin\theta_2) \neq 0.$$

Then $\dfrac{\alpha_1}{\alpha_2} = \dfrac{r_1}{r_2}[\cos(\theta_1 - \theta_2) + i\sin(\theta_1 - \theta_2)].$

Therefore $\left|\dfrac{\alpha_1}{\alpha_2}\right| = \dfrac{|\alpha_1|}{|\alpha_2|}$ $\arg\dfrac{\alpha_1}{\alpha_2} = \arg\alpha_1 - \arg\alpha_2$.

In words, to divide two non-zero complex numbers, divide their moduli and subtract their arguments.

Example $\alpha = 1 + i$ $\beta = 1 + i\sqrt{3}$

By Example 1,

$$\alpha = \sqrt{2}\,(\cos \tfrac{1}{4}\pi + i \sin \tfrac{1}{4}\pi) \qquad \beta = 2(\cos \tfrac{1}{3}\pi + i \sin \tfrac{1}{3}\pi).$$

Therefore

$$\left|\frac{\alpha}{\beta}\right| = \frac{\sqrt{2}}{2} = \tfrac{1}{2}\sqrt{2} \qquad \arg \frac{\alpha}{\beta} = \tfrac{1}{4}\pi - \tfrac{1}{3}\pi = -\tfrac{1}{12}\pi,$$

$$\frac{\alpha}{\beta} = \tfrac{1}{2}\sqrt{2}\,[\cos(-\tfrac{1}{12}\pi) + i \sin(-\tfrac{1}{12}\pi)]$$

Squares and Square Roots

If $\alpha = r(\cos \theta + i \sin \theta) \neq 0$, then the multiplication rule implies

$$\alpha^2 = r^2(\cos 2\theta + i \sin 2\theta).$$

Therefore, to square a complex number, you square its modulus and double its argument.

It stands to reason that to find a complex square root of α, you go backwards, *i.e.*, you take the square root of the modulus and half of the argument. But there are *two* geometric angles whose doubles are θ (plus a multiple of 2π), namely $\tfrac{1}{2}\theta$ and $\tfrac{1}{2}(\theta + 2\pi) = \tfrac{1}{2}\theta + \pi$.

Each non-zero complex number has two complex square roots:

$$\sqrt{r(\cos \theta + i \sin \theta)} = \begin{cases} \sqrt{r}\,(\cos \tfrac{1}{2}\theta + i \sin \tfrac{1}{2}\theta) \\ \sqrt{r}\,[\cos(\tfrac{1}{2}\theta + \pi) + i \sin(\tfrac{1}{2}\theta + \pi)]. \end{cases}$$

Each square root is the negative of the other.

EXAMPLE 1   Find the square roots of $9i$.

Solution   In polar form,

$$9i = 9(\cos \tfrac{1}{2}\pi + i \sin \tfrac{1}{2}\pi).$$

According to the preceding formulas, one square root is

$$3(\cos \tfrac{1}{4}\pi + i \sin \tfrac{1}{4}\pi) = 3(\tfrac{1}{2}\sqrt{2} + \tfrac{1}{2}i\sqrt{2})$$

and the other square root is the negative of this number.

Check $[\tfrac{3}{2}\sqrt{2}(1 + i)]^2 = (\tfrac{3}{2}\sqrt{2})^2(1 + i)^2$

$$= \tfrac{9}{2}(1 + 2i + i^2) = \tfrac{9}{2}(1 + 2i - 1)$$

$$= 9i.$$

Answer $\pm\tfrac{3}{2}\sqrt{2}(1 + i)$.

Warning   If r is a positive *real* number, the symbol $\sqrt{r}$ denotes the unique positive square root of r. But if α is a non-zero *complex* number, the symbol $\sqrt{\alpha}$ denotes *two* complex numbers, each a square root of α. We really should always write $\pm\sqrt{\alpha}$ to remember this ambiguity.

Be very careful of "rules" like $\sqrt{\alpha_1}\sqrt{\alpha_2} = \sqrt{\alpha_1\alpha_2}$. This is true only up to sign. For instance,

$$-1 = i \cdot i = \sqrt{-1}\sqrt{-1} = \sqrt{(-1)(-1)} = \sqrt{1} = 1$$

is nonsense.

Exercises

Express numerator, denominator, and quotient in polar form

1 $\dfrac{\sqrt{3} - i}{1 + i}$

2 $\dfrac{-\sqrt{3} + i}{3 + 3i}$

3 $\dfrac{1 - i}{2 + 2i\sqrt{3}}$

4 $\dfrac{3 - 3i}{\sqrt{3} - i}$.

Find Re α, Im α, $|\alpha|$, and arg α

5 $\alpha = \dfrac{\sqrt{3} - i}{1 + i}$

6 $\alpha = \dfrac{4 - 4i}{2\sqrt{3} - 2i}$.

Compute $|\alpha|/|\beta|$ and $|\alpha/\beta|$

7 $\alpha = 1 + i \quad \beta = 2i$

8 $\alpha = -i \quad \beta = 2 - i$

9 $\alpha = 4 \quad\quad \beta = 3 - i$

10 $\alpha = 1 + 2i \quad \beta = 3i$

11 $\alpha = 3 - i \quad \beta = 2 + 5i$

12 $\alpha = 2 - 5i \quad \beta = -3 - i$.

Solve

13 $z^2 = -i$

14 $z^2 = -9$

15 $z^2 = 1 + i$

16 $z^2 = 2 - 2i$

17 $z^2 = \sqrt{3} - i$

18 $z^2 = 1 + i\sqrt{3}$

19 $z^2 = (1 + i)^3$

20 $z^2 = (1 - i)^3$.

21 Justify geometrically the formula Im $(\alpha) = (\alpha - \bar{\alpha})/2i$.

22 Calculate $(1 + i\sqrt{3})/(1 + i)$ in two different ways to prove

$$\cos\tfrac{1}{12}\pi = \tfrac{1}{4}(\sqrt{6} + \sqrt{2}) \qquad \sin\tfrac{1}{12}\pi = \tfrac{1}{4}(\sqrt{6} - \sqrt{2}).$$

4 ZEROS OF POLYNOMIALS

Let $ax^2 + bx + c$ be any quadratic polynomial with real coefficients and $a \neq 0$. The quadratic formula

$$r = \frac{-b \pm \sqrt{b^2 - 4ac}}{2a}$$

gives the zeros of the polynomial. If $b^2 - 4ac \geq 0$, the zeros are real; if $b^2 - 4ac < 0$, they are complex.

Example $2x^2 - x + 1$

The discriminant is

$$b^2 - 4ac = 1 - 8 = -7,$$

so the zeros are

$$\frac{1 + i\sqrt{7}}{4} \quad \text{and} \quad \frac{1 - i\sqrt{7}}{4}.$$

To check, set $r = \frac{1}{4}(1 \pm i\sqrt{7})$. Then

$$4r - 1 = \pm i\sqrt{7} \quad (4r - 1)^2 = -7 \quad 16r^2 - 8r + 1 = -7$$

$$16r^2 - 8r + 8 = 0 \quad 2r^2 - r + 1 = 0.$$

At this point we can assert at least this much: we have succeeded in enlarging the real number system to the complex number system, and in this new system each *quadratic* polynomial with *real* coefficients has zeros.

The emphasis on *quadratic* and *real* suggests two very natural questions.

(1) Does each quadratic polynomial with *complex* coefficients have complex zeros?
(2) Does each real polynomial (real coefficients)

$$a_n x^n + a_{n-1} x^{n-1} + \cdots + a_1 x + a_0$$

of any degree have complex zeros?

If the answer to either question were no, we would be faced with enlarging the complex number system (perhaps over and over again) to handle more and more complicated polynomials. Fortunately, the answer to both questions is yes. The complex number system is big enough; it is the "right" system for polynomials.

The answer to question (1) is easy because of the quadratic formula. Suppose α, β, and γ are complex. Then the equation

$$\alpha z^2 + \beta z + \gamma = 0 \quad (\alpha \neq 0)$$

has solutions

$$z = \frac{-\beta \pm \sqrt{\beta^2 - 4\alpha\gamma}}{2\alpha}.$$

This formula is meaningful since, as we have seen, complex numbers have complex square roots.

Example $z^2 + \sqrt{3}z + i = 0$

$$\alpha = 1 \quad \beta = \sqrt{3} \quad \gamma = i$$

$$\beta^2 - 4\alpha\gamma = 3 - 4i = (2 - i)^2$$

$$z = \frac{-\sqrt{3} \pm (2 - i)}{2} = \begin{cases} (1 - \tfrac{1}{2}\sqrt{3}) - \tfrac{1}{2}i \\ (-1 - \tfrac{1}{2}\sqrt{3}) + \tfrac{1}{2}i. \end{cases}$$

The proof that question (2) has a positive answer is very deep. It was given by K.F. Gauss and constitutes one of the most remarkable achievements in mathematics. Here is a precise statement of the result, in a generality that goes beyond questions (1) and (2).

Fundamental Theorem of Algebra   Let

$$f(z) = \alpha_n z^n + \alpha_{n-1} z^{n-1} + \cdots + \alpha_1 z + \alpha_0 \qquad (\alpha_n \neq 0)$$

be any polynomial with complex coefficients and degree $n \geq 1$. Then there is a complex number β such that $f(\beta) = 0$.

There are many proofs, all beyond the scope of this course. Note that the theorem only guarantees the *existence* of a zero; it does not say a word about *how to find* one. However, there are numerical methods for approximating zeros to any degree of accuracy, some well suited for computers.

The Remainder and Factor Theorems

We shall formulate for complex polynomials two basic results, which you may have studied previously for real polynomials.

Suppose a polynomial $f(z)$ is divided by a linear polynomial of the form $g(z) = z - \beta$. Then

$$f(z) = (z - \beta)q(z) + r(z),$$

where $r(z)$ is zero or a polynomial of degree zero, that is, a constant. Thus

$$f(z) = (z - \beta)q(z) + \gamma.$$

What is the constant γ? Just set $z = \beta$ in the formula;

$$f(\beta) = 0 + \gamma,$$

hence $\gamma = f(\beta)$. Thus the value of a polynomial $f(z)$ for $z = \beta$ is the remainder when $f(z)$ is divided by $z - \beta$. This fact is called the remainder theorem.

Remainder Theorem   If $f(z)$ is a polynomial and β is a complex number, then $f(\beta)$ is the remainder when $f(z)$ is divided by $z - \beta$. Thus

$$f(z) = (z - \beta)q(z) + f(\beta),$$

where $q(z)$ is a polynomial.

We deduce from the remainder theorem an important test for whether β is a root of the equation $f(z) = 0$. From the formula

$$f(z) = (z - \beta)q.(z) + f(\beta)$$

we see that $f(\beta) = 0$ if and only if $f(z) = (z - \beta)q(z)$, that is, if and only if $z - \beta$ is a factor of $f(z)$.

> **Factor Theorem**   Let $f(z)$ be a polynomial of degree $n \geq 1$ with complex coefficients. Suppose β is a complex zero of $f(z)$. Then $z - \beta$ divides $f(z)$, that is, $f(z) = (z - \beta)g(z)$, where $g(z)$ is a polynomial of degree $n - 1$.

Complete Factorization

Let $f(z)$ be a complex polynomial of positive degree. By the fundamental theorem, $f(z)$ has a complex zero β_1, and by the factor theorem

$$f(z) = (z - \beta_1)g(z),$$

where $\deg g = \deg f - 1$. If $g(z)$ has positive degree, then by the same reasoning

$$g(z) = (z - \beta_2)h(z),$$

where $\deg h = \deg g - 1 = \deg f - 2$. Clearly

$$f(z) = (z - \beta_1)(z - \beta_2)h(z).$$

The process can be repeated:

$$f(z) = (z - \beta_1)(z - \beta_2)(z - \beta_3)k(z) \qquad \deg k = \deg f - 3$$

etc., until we reach a constant factor, n steps in all.

> **Complete Factorization**   Let $f(z) = \alpha z^n + \cdots$ be a complex polynomial of degree $n \geq 1$. Then there are complex numbers $\beta_1, \cdots, \beta_n$ such that
> $$f(z) = \alpha(z - \beta_1) \cdots (z - \beta_n).$$

Multiply out the last expression; αz^n really is the leading term of $f(z)$. Since $\deg f \geq 1$ is given, it is tacitly assumed that $\alpha \neq 0$.

Note   The numbers $\beta_1, \cdots, \beta_n$ are the **zeros** of $f(z)$, and there can be repetitions among them. We could write the complete factorization of $f(z)$ in the form

$$f(z) = \alpha(z - \beta_1)^{m_1} \cdots (z - \beta_k)^{m_k},$$

where $\beta_1, \cdots, \beta_k$ are distinct from each other. The exponents $m_1, \cdots, m_k$ are positive integers with $m_1 + m_2 + \cdots + m_k = \deg f(z)$. We call m_j the **multiplicity** of the zero β_j. If $m_j = 1$, we call β_j a **simple zero** of $f(z)$.

Complex Zeros of Real Polynomials

Suppose $f(z) = az^2 + bz + c$ has *real* coefficients, $a \neq 0$. By the quadratic formula, its zeros are

$$\beta = -\frac{b}{2a} + \frac{1}{2a}\sqrt{D} \quad \text{and} \quad \gamma = -\frac{b}{2a} - \frac{1}{2a}\sqrt{D},$$

where $D = b^2 - 4ac$. The complete factorization is

$$f(z) = a(z - \beta)(z - \gamma).$$

(It can be checked directly in this case.) If $D \geq 0$, then β and γ are both real. But if $D < 0$, then $D = -d^2$, where d is real, $\sqrt{D} = \sqrt{-d^2} = di$, and

$$\beta = -\frac{b}{2a} + \frac{d}{2a}i \qquad \gamma = -\frac{b}{2a} - \frac{d}{2a}i.$$

By inspection, $\gamma = \overline{\beta}$ so

$$f(z) = a(z - \beta)(z - \overline{\beta}).$$

Thus the zeros of a real *quadratic* polynomial either are both real, or are a pair of non-real complex conjugates. What is remarkable is that this result can be generalized to real polynomials of any degree.

Let $f(z) = a_n z^n + \cdots + a_0$ be a polynomial with *real coefficients*. If β is a complex zero of $f(z)$ *that is not real*, then $\overline{\beta}$ also is a zero of $f(z)$.

In simple words, non-real zeros of real polynomials occur in conjugate pairs. How can we prove this? In the quadratic case we used a formula for the zeros to make a proof. Of course, for polynomials of higher degree, there are no formulas for zeros in general. So we need an entirely different idea.

The idea is to use the rules

$$\overline{\alpha + \beta} = \overline{\alpha} + \overline{\beta} \qquad \overline{\alpha\beta} = \overline{\alpha}\,\overline{\beta} \qquad \overline{\alpha^n} = \overline{\alpha}^n\,.$$

Now suppose $f(\beta) = 0$. Then

$$a_n\beta^n + a_{n-1}\beta^{n-1} + \cdots + a_0 = 0.$$

Take conjugates on both sides:

$$\overline{a_n\beta^n + a_{n-1}\beta^{n-1} + \cdots + a_0} = \overline{0}$$

$$\overline{a_n\beta^n} + \overline{a_{n-1}\beta^{n-1}} + \cdots + \overline{a_0} = 0$$

But a_j is real, so $\overline{a_j} = a_j$. Therefore $\overline{a_j\beta^j} = \overline{a_j}\,\overline{\beta}^j = a_j\overline{\beta}^j$, hence

$$a_n\overline{\beta}^n + a_{n-1}\overline{\beta}^{n-1} + \cdots + a_0 = \overline{0} = 0.$$

This says $f(\overline{\beta}) = 0$. Done!

An immediate consequence is the factorization

$$f(z) = (z - \beta)(z - \overline{\beta})h(z),$$

where $\deg h(z) = n - 2$. Now observe that

$$(z - \beta)(z - \overline{\beta}) = z^2 - (\beta + \overline{\beta})z + \beta\overline{\beta} = z^2 - 2[\operatorname{Re}(\beta)]z + |\beta|^2.$$

Therefore $g(z) = (z - \beta)(z - \overline{\beta})$ is a *real* quadratic. Since $f(z) = g(z)h(z)$, the polynomial $h(z) = f(z)/g(z)$ is the quotient of two real polynomials, hence real itself, since the long division of real polynomials can only lead to real polynomials. The same reduction can now be repeated on $h(z)$, etc.

Now let us fit all the pieces together. Start with any non-constant real polynomial $f(x)$. It has some (maybe no) real zeros and some (maybe no) conjugate pairs of non-real zeros. Each real zero yields a real linear factor $x - r$; each pair of non-real conjugate zeros yields a real quadratic factor.

Complete Factorization of Real Polynomials

Let $f(x)$ be a real polynomial of degree $n \geq 1$. Then

$$f(x) = a_n(x - r_1) \cdots (x - r_k)g_1(x) \cdots g_s(x),$$

where $r_1, \cdots, r_k$ are real and

$$g_j(x) = x^2 + b_jx + c_j$$

with b_j and c_j real and $b_j^2 - 4c_j < 0$. Here $k \geq 0$, $s \geq 0$, and $k + 2s = n$.

The quadratic factors $g_1(x), \cdots, g_s(x)$ cannot be split into real linear factors; we call such factors **irreducible.** Now we can restate the result above as follows:

Each real polynomial is the product of real irreducible linear and quadratic factors.

EXAMPLE 1 Express $f(x) = x^5 - 3x^4 + 2x^3 - 6x^2 + x - 3$ as the product of real irreducible factors.

Solution By trial and error we find $f(3) = 0$. Hence $x - 3$ is a factor, and by division

$$f(x) = (x - 3)(x^4 + 2x^2 + 1).$$

But $x^4 + 2x^2 + 1 = (x^2 + 1)^2$, and $x^2 + 1$ is irreducible.

Answer $(x - 3)(x^2 + 1)^2$

Remark The irreducible factor $x^2 + 1$ corresponds to the complex zeros $\pm i$ since $x^2 + 1 = (x - i)(x + i)$. The factorization of $f(x)$ into complex linear factors is

$$f(x) = (x - 3)(x - i)^2(x + i)^2.$$

Exercises

Solve

1 $z^2 - 2z + 5 = 0$	**2** $z^2 + 4z + 13 = 0$	**3** $z^2 + z + 6 = 0$
4 $2z^2 - 3z + 10 = 0$	**5** $z^4 + 5z^2 + 4 = 0$	**6** $z^4 + 4z^2 + 29 = 0$
7 $z^3 - 1 = 0$	**8** $z^3 + 8 = 0$	**9** $z^4 - 1 = 0$
	10 $z^3 + z - 2 = 0.$	

11 Factor $x^3 + 1$ into real irreducible factors.

12 Factor $x^4 - 1$ into real irreducible factors.

13 Show from the factorization

$$z^5 + 1 = (z + 1)(z^4 - z^3 + z^2 - z + 1)$$

that -1 is a simple zero of $z^5 + 1$.

14 Show from the factorization

$$z^n - 1 = (z - 1)(z^{n-1} + z^{n-2} + \cdots + z + 1)$$

that $+1$ is a simple zero of $z^n - 1$.

15 Use the methods of this section to prove that a real polynomial of odd degree has a real zero.

16 If a polynomial $f(z)$ has zeros $\pm i$, prove that it is divisible by $z^2 + 1$.

Write down the most general real polynomial satisfying the given conditions

17 degree 4 and zeros ± 2, $1 \pm i$

18 degree 5 and a real zero of multiplicity 4

19 degree 4 and no real zeros

20 degree 6 and zeros $\pm i$, $\pm 2i$, and 0 with multiplicity 2.

The next four exercises deal with the set **T** of all *real* numbers of the form $\alpha = a + b\sqrt{2}$, where a and b are *rational* numbers.

21* Show that the sum, difference, and product of numbers in **T** are in **T**.

22 For $\alpha = a + b\sqrt{2}$ in **T**, set $\alpha' = a - b\sqrt{2}$. Show that $(\alpha \pm \beta)' = \alpha' \pm \beta'$ and $(\alpha\beta)' = \alpha'\beta'$.

23* If $\alpha = a + b\sqrt{2}$ is in **T** and $\alpha \neq 0$, show that $\alpha\alpha'$ is a non-zero rational number and

$$\alpha^{-1} = \frac{a}{\alpha\alpha'} - \frac{b}{\alpha\alpha'}\sqrt{2} = \frac{\alpha'}{\alpha\alpha'} \quad \text{is in } \mathbf{T}.$$

24* Suppose

$$f(x) = a_n x^n + a_{n-1}x^{n-1} + \cdots + a_0$$

has rational coefficients and $f(\alpha) = 0$, where α belongs to **T**. Show that $f(\alpha') = 0$. [*Hint* Proceed by analogy with the corresponding statement for complex zeros of real polynomials.]

5 DE MOIVRE'S THEOREM

De Moivre's Theorem provides an important formula for the integer powers of a complex number.

De Moivre's Theorem   Let

$$\alpha = r(\cos \theta + i \sin \theta) \neq 0$$

be a complex number in polar form, and let n be an integer. Then

$$\alpha^n = r^n(\cos n\theta + i \sin n\theta).$$

In other words,

$$|\alpha^n| = |\alpha|^n \qquad \arg(\alpha^n) = n \arg \alpha.$$

Proof If $n = 0$, then $\alpha^n = 1$ by definition and $r^0(\cos 0 + i \sin 0) = 1$, so the formula is true. Clearly it is true for $n = 1$. We now prove it for $n \geq 1$ by mathematical induction. Suppose it is true for some $n \geq 1$. Then

$$\alpha^{n+1} = \alpha \cdot \alpha^n = [r(\cos\theta + i\sin\theta)][r^n(\cos n\theta + i\sin n\theta)]$$
$$= r^{n+1}[\cos(n+1)\theta + i\sin(n+1)\theta].$$

Therefore the formula also holds for $n+1$. By the principle of mathematical induction, it holds for all $n \geq 1$.

When $n < 0$, the formula follows from the positive case and the expression of the reciprocal in polar form. Set $n = -m$ with $m > 0$. Then

$$\alpha^n = [r(\cos\theta + i\sin\theta)]^n$$
$$= \{[r(\cos\theta + i\sin\theta)]^{-1}\}^m = \{r^{-1}[\cos(-\theta) + i\sin(-\theta)]\}^m$$
$$= (r^{-1})^m[\cos m(-\theta) + i\sin m(-\theta)]$$
$$= r^{-m}[\cos(-m\theta) + i\sin(-m\theta)] = r^n(\cos n\theta + i\sin n\theta).$$

EXAMPLE 1   Find $(1+i)^{15}$ in rectangular form.

Solution   In polar form,

$$1 + i = \sqrt{2}(\cos\tfrac{1}{4}\pi + i\sin\tfrac{1}{4}\pi).$$

By De Moivre's Theorem,

$$(1+i)^{15} = (\sqrt{2})^{15}(\cos\tfrac{15}{4}\pi + i\sin\tfrac{15}{4}\pi).$$

But

$$(\sqrt{2})^{15} = (\sqrt{2})^{2\times 7}(\sqrt{2}) = 2^7 \cdot \sqrt{2} = 128\sqrt{2}$$

and $\tfrac{15}{4}\pi = \tfrac{16}{4}\pi - \tfrac{1}{4}\pi = 4\pi - \tfrac{1}{4}\pi$.   Therefore

$$(1+i)^{15} = 128\sqrt{2}[\cos(-\tfrac{1}{4}\pi) + i\sin(-\tfrac{1}{4}\pi)]$$
$$= 128\sqrt{2}(\tfrac{1}{2}\sqrt{2} - \tfrac{1}{2}i\sqrt{2}) = 128 - 128i.$$

Answer $128 - 128i$

Roots of Unity

We now discuss the zeros of the polynomial

$$f(z) = z^n - 1.$$

They are called **n-th roots of unity** because they are the roots of

$$z^n - 1 = 0, \quad \text{that is,} \quad z^n = 1.$$

Clearly the only *real* n-th roots of unity are $+1$ and -1 if n is even, only $+1$ if n is odd. So most n-th roots of unity are non-real; how do we find them?

Suppose $\zeta^n = 1$. Then $|\zeta|^n = |\zeta^n| = 1$, so $|\zeta| = 1$ and the polar form of ζ is

$$\zeta = \cos\theta + i\sin\theta.$$

By De Moivre's Theorem

$$\zeta^n = \cos n\theta + i\sin n\theta.$$

Therefore, to solve the equation $\zeta^n = 1$, we must find all possible angles θ such that $n\theta$ equals the argument of 1. But clearly $\arg 1 = 0$. Not so fast! Arg is determined up to multiples of 2π; any angle $2\pi k$, where k is an integer, is an argument of 1. That is,

$$\cos 2\pi k + i \sin 2\pi k = 1,$$

and these angles $2\pi k$ are the only arguments of 1. Hence $\zeta^n = 1$ if and only if

$$n\theta = 2\pi k \qquad (k \text{ an integer}).$$

Therefore θ must take one of the values

$$\theta = \frac{2\pi k}{n} \qquad k = 0, \pm 1, \pm 2, \pm 3, \cdots .$$

While these look like a lot of values of θ, they represent only n distinct angles. After n consecutive integers, the same values appear increased by 2π:

$$\frac{2\pi(k + n)}{n} = \frac{2\pi k}{n} + 2\pi .$$

Thus we can restrict θ to the values

$$\theta = \frac{2\pi k}{n} \qquad k = 0, 1, 2, \cdots , n - 1.$$

Other choices of k will produce a value of $2\pi k/n$ that differs from one of the above numbers by a multiple of 2π.

It follows that the n-th roots of unity are the n complex numbers

$$\zeta_k = \cos \frac{2\pi k}{n} + i \sin \frac{2\pi k}{n} \qquad k = 0, 1, 2, \cdots , n - 1.$$

Geometrically, they are n equally spaced points on the unit circle; hence they form the vertices of a regular n-gon (Fig. 1).

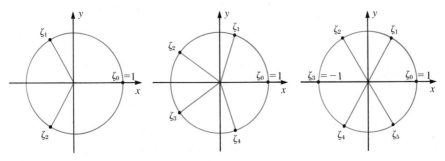

(a) cube roots of unity   (b) fifth roots of unity   (c) sixth roots of unity

Fig. 1 Roots of unity

Somewhat more is true. If we set

$$\zeta = \zeta_1 = \cos \frac{2\pi}{n} + i \sin \frac{2\pi}{n}$$

then by De Moivre's Theorem

$$\zeta^k = \left(\cos \frac{2\pi}{n} + i \sin \frac{2\pi}{n} \right)^k = \cos \frac{2\pi k}{n} + i \sin \frac{2\pi k}{n} = \zeta_k.$$

Therefore the n-th roots of unity are the powers of ζ:

$$1, \zeta, \zeta^2, \zeta^3, \cdots, \zeta^{n-1}.$$

These are the zeros of $z^n - 1$, so

$$z^n - 1 = (z - 1)(z - \zeta)(z - \zeta^2) \cdots (z - \zeta^{n-1}).$$

---

n-th Roots of Unity   The zeros of $z^n - 1$ are the complex numbers

$$\zeta_k = \cos \frac{2\pi k}{n} + i \sin \frac{2\pi k}{n} \qquad k = 0, 1, 2, \cdots, n - 1.$$

If $\zeta = \zeta_1 = \cos(2\pi/n) + i \sin(2\pi/n)$, then $\zeta_k = \zeta^k$, and

$$z^n - 1 = (z - 1)(z - \zeta)(z - \zeta^2) \cdots (z - \zeta^{n-1}).$$

EXAMPLE 2   Compute the cube roots of unity and express each in the form $a + bi$.

Solution   Use the formula with $n = 3$ and $k = 0, 1, 2$:

$$k = 0 \qquad z_0 = \cos 0 + i \sin 0 = 1$$

$$k = 1 \qquad z_1 = \cos \frac{2\pi}{3} + i \sin \frac{2\pi}{3} = -\tfrac{1}{2} + \tfrac{1}{2}i\sqrt{3}$$

$$k = 2 \qquad z_2 = \cos \frac{4\pi}{3} + i \sin \frac{4\pi}{3} = -\tfrac{1}{2} - \tfrac{1}{2}i\sqrt{3}.$$

Answer $1 \qquad -\tfrac{1}{2} + \tfrac{1}{2}i\sqrt{3} \qquad -\tfrac{1}{2} - \tfrac{1}{2}i\sqrt{3}$

Remark   Note that there is a conjugate pair of non-real roots. We could have predicted that, since the polynomial $z^3 - 1$ has real coefficients. Also note that it is not obvious that the cube of $-\tfrac{1}{2} \pm \tfrac{1}{2}i\sqrt{3}$ is 1; you should check it.

General n-th Roots

The solution of the equation

$$z^n = \alpha \qquad (\alpha \neq 0)$$

is similar to that of the special case $z^n = 1$. Write α in polar form:

$$\alpha = r_0(\cos \theta_0 + i \sin \theta_0) \qquad (r_0 > 0),$$

and write the unknown z as

$$z = r(\cos \theta + i \sin \theta) \qquad (r > 0).$$

Then $z^n = r^n(\cos n\theta + i \sin n\theta)$ by De Moivre's Theorem, so we must have

$$r^n(\cos n\theta + i \sin n\theta) = r_0(\cos \theta_0 + i \sin \theta_0).$$

This equation requires

$$r^n = r_0 \qquad n\theta = \theta_0 + 2\pi k \qquad k \text{ an integer.}$$

Therefore we choose

$$r = r_0^{1/n} \qquad \theta = \frac{1}{n}\theta_0 + \frac{2\pi k}{n} \qquad k = 0, 1, \cdots, n-1.$$

(As with roots of unity, there is no use taking other values of k; they yield no further angles.)

The n roots of the equation

$$z^n = r_0(\cos\theta_0 + i\sin\theta_0) \qquad (r_0 > 0)$$

are

$$\beta_k = r_0^{1/n}\left[\cos\left(\frac{1}{n}\theta_0 + \frac{2\pi k}{n}\right) + i\sin\left(\frac{1}{n}\theta_0 + \frac{2\pi k}{n}\right)\right]$$

$$k = 0, 1, \cdots, n-1.$$

They form the vertices of a regular n-gon centered at 0.

EXAMPLE 3   Solve the equation $z^3 = 2i$.

Solution   Write $z = r(\cos\theta + i\sin\theta)$ and $2i = 2(\cos\frac{1}{2}\pi + i\sin\frac{1}{2}\pi)$. Then $z^3 = 2i$ means

$$r^3(\cos 3\theta + i\sin 3\theta) = 2(\cos\tfrac{1}{2}\pi + i\sin\tfrac{1}{2}\pi).$$

Therefore

$$r^3 = 2 \quad \text{and} \quad 3\theta = \tfrac{1}{2}\pi + 2\pi k \qquad k = 0, 1, 2.$$

Consequently

$$r = \sqrt[3]{2} \quad \text{and} \quad \theta = \tfrac{1}{6}\pi + \tfrac{2}{3}\pi k \qquad k = 0, 1, 2.$$

The three values of θ are $\tfrac{1}{6}\pi, \tfrac{1}{6}\pi + \tfrac{2}{3}\pi = \tfrac{5}{6}\pi$, and $\tfrac{1}{6}\pi + \tfrac{4}{3}\pi = \tfrac{3}{2}\pi$. The corresponding roots of $z^3 = 2i$ are

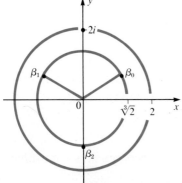

Fig. 2   Roots of $z^3 = 2i$

$$\begin{cases} \beta_0 = \sqrt[3]{2}(\tfrac{1}{2}\sqrt{3} + \tfrac{1}{2}i) \\ \beta_1 = \sqrt[3]{2}(-\tfrac{1}{2}\sqrt{3} + \tfrac{1}{2}i) \\ \beta_2 = -i\sqrt[3]{2} \end{cases}$$

$$\beta_0 = \sqrt[3]{2}(\cos \tfrac{1}{6}\pi + i \sin \tfrac{1}{6}\pi) = \sqrt[3]{2}(\tfrac{1}{2}\sqrt{3} + \tfrac{1}{2}i)$$

$$\beta_1 = \sqrt[3]{2}(\cos \tfrac{5}{6}\pi + i \sin \tfrac{5}{6}\pi) = \sqrt[3]{2}(-\tfrac{1}{2}\sqrt{3} + \tfrac{1}{2}i)$$

$$\beta_2 = \sqrt[3]{2}(\cos \tfrac{3}{2}\pi + i \sin \tfrac{3}{2}\pi) = -i\sqrt[3]{2}.$$

See Fig. 2.

Answer $\sqrt[3]{2}(\pm\tfrac{1}{2}\sqrt{3} + \tfrac{1}{2}i)$ $-i\sqrt[3]{2}$

Exercises

Compute by De Moivre's theorem; give answer in rectangular form if possible

1 $(1 - i)^{11}$ **2** $(1 - i)^{27}$ **3** $(\sqrt{3} + i)^6$

4 $(1 + i\sqrt{3})^{-17}$ **5** $[\tfrac{1}{2}(-1 + i\sqrt{3})]^{13}$ **6** $(\cos \tfrac{1}{7}\pi + i \sin \tfrac{1}{7}\pi)^{50}$

7 $(\cos 15° + i \sin 15°)^{25}$ **8** $(\cos 12° + i \sin 12°)^{65}$ **9** $(\sqrt{3} - i)^{-10}$

10 $(\sqrt{3} - i)^{-28}$ **11** $(\cos \tfrac{1}{9}\pi - i \sin \tfrac{1}{9}\pi)^{-11}$ **12** $(\sqrt{3} + i)^{-20}.$

Solve completely

13 $z^4 = 1$ **14** $z^5 = 1$ **15** $z^6 = 1$

16 $z^8 = 1$ **17** $z^3 = 8$ **18** $z^4 = 16$

19 $z^4 = -25$ **20** $z^4 = i$ **21** $z^3 = i$

22 $z^4 = -i$ **23** $z^3 = -8i$ **24** $z^6 = -27$

25 $z^3 = 1 + i$ **26** $z^4 = -1 + i$ **27** $z^3 = \tfrac{1}{2} + \tfrac{1}{2}i\sqrt{3}$

28 $z^3 = -\tfrac{1}{2}\sqrt{3} + \tfrac{1}{2}i$ **29** $z^4 = 4\sqrt{3} - 4i$ **30** $z^5 = 3 - 3i.$

31 Let α be an n-th root of unity and $\alpha \neq 1$. Show that

$$\alpha^{n-1} + \alpha^{n-2} + \cdots + \alpha + 1 = 0.$$

32 (cont.) Solve $z^4 + z^3 + z^2 + z + 1 = 0$.

33 Let $\alpha = \cos \tfrac{2}{5}\pi + i \sin \tfrac{2}{5}\pi$. Set $\beta = \alpha + \alpha^{-1}$. Prove that $\beta^2 + \beta - 1 = 0$.

34 (cont.) Prove that $\beta = \tfrac{1}{2}(-1 + \sqrt{5})$.

35 (cont.) Prove that $\alpha^2 - \beta\alpha + 1 = 0$.

36 (cont.) Prove that $\alpha = \tfrac{1}{2}(\beta + i\sqrt{4 - \beta^2})$. Conclude that

$$\cos 72° = \frac{-1 + \sqrt{5}}{4} \qquad \sin 72° = \frac{\sqrt{10 + 2\sqrt{5}}}{4}.$$

Find all solutions (z, w) of the system

37* $\begin{cases} z^2 = w \\ w^2 = z \end{cases}$ **38*** $\begin{cases} z^2 = w \\ w^3 = z \end{cases}$

39 Verify that $\alpha = 2 - i$ is a root of the equation $z^4 = -7 - 24i$. Show that all roots are $\alpha, i\alpha, -\alpha,$ and $-i\alpha$, that is, α multiplied by each of the 4-th roots of unity.

40 Prove $1 + \cos \dfrac{2\pi}{n} + \cos \dfrac{4\pi}{n} + \cdots + \cos \dfrac{2(n - 1)\pi}{n} = 0,$

$$\sin \frac{2\pi}{n} + \sin \frac{4\pi}{n} + \cdots + \sin \frac{2(n - 1)\pi}{n} = 0.$$

[*Hint* Use Ex. 31.]

A number α is a **primitive** n-th root of unity if $\alpha^n = 1$, but $\alpha^m \neq 1$ for $0 < m < n$. Find the primitive n-th roots of unity for

41 $n = 4$ **42** $n = 5$ **43** $n = 6$

44 $n = 7$ **45** $n = 8$ **46** $n = 9$

47 $n = 10$ **48** $n = 12$

49* Show that each primitive 6-th root of unity satisfies $z^2 - z + 1 = 0$.

50* Show that each primitive 8-th root of unity satisfies $z^4 + 1 = 0$.

51* Show that each primitive 9-th root of unity satisfies $z^6 + z^3 + 1 = 0$.

52* Show that each primitive 12-th root of unity satisfies $z^4 - z^2 + 1 = 0$.

REVIEW EXERCISES

Express in the form $a + bi$

1 $\dfrac{2 - 5i}{1 + 3i}$ **2** $\dfrac{(4 - i)(2 + i)}{3 + 4i}$ **3** $\overline{(1 + i)(3 - 2i)}$

4 $(1 + i)^2(3 + 2i)$ **5** i^{73}.

Express in polar form

6 $\dfrac{2 + 2i}{\sqrt{3} + i}$ **7** $-5i$ **8** $\cos \tfrac{1}{7}\pi - i \sin \tfrac{1}{7}\pi$.

9 Show that $z^3 + 1$ has three simple zeros by finding them.

10 Solve $z^4 + z^2 + 1 = 0$.

11 Compute $(\sqrt{3} + i)^{14}$.

12 Find the square roots of $1 - i\sqrt{3}$.

13 Plot in the complex plane the number $\alpha = 1 + i$ and all solutions of the equation $z^3 = \alpha$.

14 Find all 6-th roots of -1.

15 The point α moves counterclockwise around the circle $|z| = 2$ in the complex plane. Describe the corresponding motion of α^{-1}.

16 Let d_1, d_2, d_3, d_4 be the distances from a point z in the complex plane to the points $2, -2, 2i, -2i$. Show that $d_1 \cdot d_2 \cdot d_3 \cdot d_4 = |z^4 - 16|$.

17 If α, β, γ are the vertices of an equilateral triangle in the complex plane, prove that $\gamma - \alpha = \omega(\beta - \alpha)$, where ω is a 6-th root of unity.

18 Factor $z^4 + 4$ into

    (a) linear factors   (b) irreducible real quadratic factors.

12 EXPONENTIAL AND LOGARITHM FUNCTIONS

1 REVIEW OF EXPONENTS

In previous chapters we studied the trigonometric functions and their inverse functions. In this chapter we study another important class of functions, the exponential functions, and their inverses, the logarithm functions. Like the trigonometric functions, they cannot be expressed solely in terms of addition, subtraction, multiplication, division, and taking roots (radicals). This sets them apart from simpler functions, such as polynomials and rational functions.

We begin by reviewing the algebra of exponents. First we recall the definition of "a to the n-th power", where n is a positive integer (the **exponent**).

$$a^1 = a \quad a^2 = a \cdot a \quad a^3 = a \cdot a \cdot a \quad \cdots \quad a^n = \underbrace{a \cdot a \cdot a \cdot a \cdot a \cdots a}_{n \text{ factors}}.$$

Look what happens when you multiply a^5 by a^3:

$$a^5 a^3 = (\underbrace{aaaaa}_{5})(\underbrace{aaa}_{3}) = \underbrace{aaaaaaaa}_{8 \text{ factors}} = a^8.$$

Thus $a^5 a^3 = a^8$. Similarly, if you multiply m factors a by n factors a, you get $m + n$ factors a. Therefore

$$a^m a^n = a^{m+n}$$

To multiply, add exponents.

Now look what happens when you divide a^5 by a^3:

$$\frac{a^5}{a^3} = \frac{aaaaa}{aaa} = \frac{aaa}{aaa}\frac{aa}{1} = aa = a^2.$$

Thus $a^5/a^3 = a^{5-3}$. You subtract exponents because the three factors in the denominator cancel three of the factors in the numerator. By the same reasoning

$$\frac{a^m}{a^n} = a^{m-n} \quad \text{when} \quad m > n.$$

It would be nice if we could also subtract exponents when the denominator contains as many or more factors than the numerator. That would require zero and negative exponents. Since $a^n/a^n = 1$, we should have $a^{n-n} = a^0 = 1$. So we define

$$\boxed{a^0 = 1 \quad (a \neq 0).}$$

We do this only for $a \neq 0$. The symbol 0^0 is undefined in mathematics.

Next we observe that

$$\frac{a^3}{a^5} = \frac{a \cdot a \cdot a}{a \cdot a \cdot a \cdot a \cdot a} = \frac{1}{a \cdot a} = \frac{1}{a^2}$$

so we should have $a^{3-5} = a^{-2} = 1/a^2$. In general, if n is a positive integer, we define

$$\boxed{a^{-n} = \frac{1}{a^n} \quad (a \neq 0).}$$

Note that a^{-1} means $1/a^1 = 1/a$. Thus $a^{-1} \cdot a = 1$. This is consistent with Section 1, where a^{-1} meant the multiplicative inverse of a. Also $a^{-n} \cdot a^n = 1$, so a^{-n} is the multiplicative inverse of a^n, that is, $(a^n)^{-1}$. Therefore a^{-n} can be interpreted in several ways, all equivalent:

$$\boxed{a^{-n} = (a^n)^{-1} = (a^{-1})^n = \frac{1}{a^n}.}$$

We have now defined a^n for every integer exponent n. Here a is any real number except that $a \neq 0$ when $n \leq 0$. Now let us state (without proofs) the basic rules for working with exponents.

> **Rules of Exponents**   If a and b are non-zero real numbers and if m and n are integers, then
>
> (1) $a^m a^n = a^{m+n}$ (2) $\dfrac{a^m}{a^n} = a^{m-n}$ (3) $(a^m)^n = a^{mn}$
>
> (4) $(ab)^n = a^n b^n$ (5) $\left(\dfrac{a}{b}\right)^n = \dfrac{a^n}{b^n}.$

We emphasize that these rules are valid for *all* integers m and n, not just positive ones.

The rules can be extended in various ways. For instance, by the associative law and repeated applications of (3) and (4) we prove

$$(a^m b^n c^p)^r = a^{mr} b^{nr} c^{pr}.$$

EXAMPLE 1   Use the rules of exponents to simplify

(a) $(xy)^2(x^2y^3)^{-1}$ (b) $(x^2y^{-3})^{-5}$.

Solution   (a) $(xy)^2(x^2y^3)^{-1} = (x^2y^2)\dfrac{1}{x^2y^3} = \dfrac{x^2}{x^2}\dfrac{y^2}{y^3} = \dfrac{1}{y} = y^{-1}$.

Alternatively, $(xy)^2(x^2y^3)^{-1} = (x^2y^2)(x^{-2}y^{-3}) = x^{2-2}y^{2-3} = x^0y^{-1} = y^{-1}$.

(b) $(x^2y^{-3})^{-5} = (x^2)^{-5}(y^{-3})^{-5} = x^{2(-5)}y^{(-3)(-5)} = x^{-10}y^{15}$.

Answer   (a) $y^{-1} = 1/y$ (b) $x^{-10}y^{15} = y^{15}/x^{10}$

EXAMPLE 2   Express $\dfrac{2^{-3} \cdot 8^5}{4^3 \cdot 16}$ as a power of 2.

Solution $\dfrac{2^{-3} \cdot 8^5}{4^3 \cdot 16} = \dfrac{2^{-3}(2^3)^5}{(2^2)^3(2^4)} = \dfrac{2^{-3} \cdot 2^{15}}{2^6 \cdot 2^4} = \dfrac{2^{12}}{2^{10}} = 2^2$.

Answer 2^2

EXAMPLE 3   Use the rules of exponents to compute $\dfrac{2^6 \cdot 5^7}{25 \cdot 10^4}$.

Solution $\dfrac{2^6 \cdot 5^7}{25 \cdot 10^4} = \dfrac{2^6 \cdot 5^7}{5^2 \cdot (2 \cdot 5)^4} = \dfrac{2^6 \cdot 5^7}{5^2 \cdot 2^4 \cdot 5^4} = \dfrac{2^6 \cdot 5^7}{2^4 \cdot 5^6} = 2^2 \cdot 5 = 20$.

Answer   20

Roots

Given a real number a, a number x is an n-th **root** of a if $x^n = a$. Here n is an integer, $n \geq 2$. Each positive real number a has a unique positive n-th root, which we denote by the symbol $\sqrt[n]{a}$. For example,

$$\sqrt[3]{125} = 5 \quad \text{because} \quad 5 > 0 \quad \text{and} \quad 5^3 = 125,$$
$$\sqrt[4]{81} = 3 \quad \text{because} \quad 3 > 0 \quad \text{and} \quad 3^4 = 81.$$

We also write $\sqrt[n]{0} = 0$ since $0^n = 0$. The symbol $\sqrt[n]{}$ is called a **radical.** Note that $\sqrt[2]{}$ is the same as $\sqrt{}$.

Remark   If n is even, then $-\sqrt[n]{a}$ is also an n-th root of the positive number a.

If n is odd, then negative numbers also have n-th roots. For instance -2 is a 5-th root of -32. Nevertheless, we shall stick to n-th roots of positive numbers only. The following rules for n-th roots are not 100% correct without qualifications if roots of negative numbers are allowed.

Rules for *n*-th roots   Let a and b be positive real numbers. Then

(1) $(\sqrt[n]{a})^n = a$ (2) $\sqrt[n]{a^n} = a$ (3) $\sqrt[n]{ab} = \sqrt[n]{a}\sqrt[n]{b}$

(4) $\sqrt[n]{\dfrac{a}{b}} = \dfrac{\sqrt[n]{a}}{\sqrt[n]{b}}$ (5) $\sqrt[m]{\sqrt[n]{a}} = \sqrt[mn]{a}$.

EXAMPLE 4   Simplify   (a) $\sqrt[3]{16x^3y^6}$ (b) $\sqrt[4]{\dfrac{48}{x^{12}y^4}}$ (c) $\sqrt[3]{\sqrt{27}}$.

Solution   (a) By rule (3),

$$\sqrt[3]{16x^3y^6} = \sqrt[3]{16}\sqrt[3]{x^3}\sqrt[3]{(y^2)^3} = \sqrt[3]{16}\,xy^2.$$

But $16 = 2^4 = 2^3 \cdot 2$, so $\sqrt[3]{16} = \sqrt[3]{2^3}\sqrt[3]{2} = 2\sqrt[3]{2}$. Hence the answer is $2xy^2\sqrt[3]{2}$.

(b) By rules (3) and (4),

$$\sqrt[4]{\frac{48}{x^{12}y^4}} = \frac{\sqrt[4]{48}}{\sqrt[4]{x^{12}y^4}} = \frac{\sqrt[4]{16 \cdot 3}}{\sqrt[4]{x^{12}}\sqrt[4]{y^4}} = \frac{\sqrt[4]{16}\sqrt[4]{3}}{x^3y} = \frac{2\sqrt[4]{3}}{x^3y}.$$

(c) $\sqrt[3]{\sqrt{27}} = \sqrt[6]{27} = \sqrt{\sqrt[3]{27}} = \sqrt{3}$.

Answer   (a) $2xy^2\sqrt[3]{2}$ (b) $\dfrac{2\sqrt[4]{3}}{x^3y}$ (c) $\sqrt{3}$

Remark   Note that we write the answer to (a) in the form $2xy^2\sqrt[3]{2}$ rather than $2\sqrt[3]{2}\,xy^2$. In general we prefer to write radical factors on the right because it is so easy to confuse symbols like $\sqrt[3]{2}x$ and $\sqrt[3]{2x}$. Just draw the bar a bit too long and you aren't sure which one is meant.

Rational Exponents

What meaning can we give to $a^{m/n}$, where m/n is a rational number not necessarily an integer? If the rules of exponents are to hold, we should have

$$(a^{m/n})^n = a^{(m/n)n} = a^m.$$

This means that $a^{m/n}$ must be the *n*-th root of a^m. Therefore we *define*

$$a^{m/n} = \sqrt[n]{a^m},$$

where m/n is a rational number with $n > 0$.

Examples

$$8^{2/3} = \sqrt[3]{8^2} = \sqrt[3]{64} = 4 \qquad 25^{3/2} = \sqrt{(25)^3} = \sqrt{(5^2)^3} = \sqrt{5^6} = 5^3 = 125$$

$$9^{-1/2} = \sqrt{9^{-1}} = \sqrt{\tfrac{1}{9}} = \tfrac{1}{3}.$$

According to the definition of $a^{m/n}$, if $m = 1$,

$$a^{1/n} = \sqrt[n]{a},$$

so $a^{1/n}$ is just the *n*-th root of a in new clothes.

Examples $36^{1/2} = 6$ $1000^{1/3} = 10$ $(\frac{1}{32})^{1/5} = \frac{1}{2}$.

Remark   There is a subtle point in the definition of $a^{m/n}$. The same rational number might be expressed in two ways. For instance, suppose $m/n = p/q$. The definition then gives two possibilities,

$$a^{m/n} = \sqrt[n]{a^m} \quad \text{and} \quad a^{p/q} = \sqrt[q]{a^p}.$$

These had better be the same, or the definition is plain nonsense. They are the same, as is verified by showing that their qn-th powers are equal.

Example $\frac{15}{6} = \frac{10}{4}$ $(=\frac{5}{2})$ so we claim that $7^{15/6} = 7^{10/4}$.

Indeed, $7^{15/6} = \sqrt[6]{7^{15}} = \sqrt[6]{7^{6 \cdot 2} \cdot 7^3} = 7^2 \cdot \sqrt[6]{7^3} = 7^2 \sqrt{7}$

and $7^{10/4} = \sqrt[4]{7^{10}} = \sqrt[4]{7^{4 \cdot 2} \cdot 7^2} = 7^2 \cdot \sqrt[4]{7^2} = 7^2 \sqrt{7}$.

Rules of Exponents

The definition of $a^{m/n}$ is a useful one because all the rules of exponents for integer exponents carry over to fractional exponents.

Rules of exponents   If m/n is rational with $n > 0$ and if $a > 0$, then

(1) $a^{m/n} = \sqrt[n]{a^m} = (\sqrt[n]{a})^m$.

If s and t are rational and if a and b are positive, then

(2) $a^s a^t = a^{s+t}$ (3) $\dfrac{a^s}{a^t} = a^{s-t}$

(4) $(a^s)^t = a^{st}$ (5) $a^s b^s = (ab)^s$.

EXAMPLE 5   Simplify using rules for exponents

(a) $(9u^4)^{-3/2}$ (b) $\left(\dfrac{x^3}{8y^{-6}}\right)^{1/3}$ (c) $(16x^4 y^8 z^{13})^{1/4}$.

Solution

(a) $(9u^4)^{-3/2} = 9^{-3/2}(u^4)^{-3/2} = (3^2)^{-3/2} u^{(4)(-3/2)} = 3^{-3} u^{-6} = \dfrac{1}{27 u^6}$.

(b) $\left(\dfrac{x^3}{8y^{-6}}\right)^{1/3} = \dfrac{(x^3)^{1/3}}{(8y^{-6})^{1/3}} = \dfrac{x}{8^{1/3} y^{-6/3}} = \dfrac{x}{2y^{-2}} = \dfrac{xy^2}{2}$.

(c) $(16x^4 y^8 z^{13})^{1/4} = (2^4 x^4 y^8 z^{12} z)^{1/4}$

$= (2^4)^{1/4}(x^4)^{1/4}(y^8)^{1/4}(z^{12})^{1/4} z^{1/4} = 2xy^2 z^3 \sqrt[4]{z}$.

Answer   (a) $\dfrac{1}{27 u^6}$ (b) $\dfrac{xy^2}{2}$ (c) $2xy^2 z^3 \sqrt[4]{z}$

EXAMPLE 6   Express using only one radical

(a) $\sqrt[3]{9}\sqrt{\frac{1}{3}}$ (b) $\dfrac{\sqrt{r^3 s^5}}{\sqrt[4]{r^2 s}}$.

Solution   (a) $\sqrt[3]{9}\sqrt[3]{\frac{1}{3}} = \sqrt[3]{3^2}\sqrt[3]{3^{-1}} = \sqrt[6]{3^4}\sqrt[6]{3^{-3}} = \sqrt[6]{3^4 \cdot 3^{-3}} = \sqrt[6]{3}$

(b) $\dfrac{\sqrt{r^3 s^5}}{\sqrt[4]{r^2 s}} = (r^3 s^5)^{1/2}(r^2 s)^{-1/4} = r^{3/2}s^{5/2}r^{-1/2}s^{-1/4}$

$= r^{3/2-1/2}s^{5/2-1/4} = rs^{9/4} = rs^{2+1/4} = rs^2 s^{1/4} = rs^2\sqrt[4]{s}.$

Answer   (a) $\sqrt[6]{3}$ (b) $rs^2\sqrt[4]{s}$

Scientific Notation

One important practical application of exponents is in computations. For scientific work, we need an efficient way of writing and computing with very large or very small numbers, such as

    32,000,000,000   1,876,000   0.00000 00000 006

Imagine multiplying such numbers as they are written!

The idea of scientific notation is to express each positive number as a small number times a power of 10, more precisely, in the form $c \times 10^n$, where $1 \le c < 10$ and n is an appropriate exponent.

Examples $140 = 1.4 \times 10^2$ $0.05 = 5 \times 10^{-2}$

$2550 = 2.55 \times 10^3$ $0.0031 = 3.1 \times 10^{-3}$

$1,876,000 = 1.876 \times 10^6$ $0.000988 = 9.88 \times 10^{-4}$

$32,000,000,000 = 3.2 \times 10^{10}$ $0.00000 00000 006 = 6 \times 10^{-13}$

The mass of a neutron is approximately

    $0.00000\ 00000\ 00000\ 00000\ 00016 = 1.6 \times 10^{-24}$ gram.

EXAMPLE 7   Multiply $(140)(32,000,000,000)(0.00000\ 00000\ 006)$.

Solution $(1.4 \times 10^2)(3.2 \times 10^{10})(6 \times 10^{-13})$

$= (1.4)(3.2)(6) \times 10^{2+10-13} = 26.88 \times 10^{-1}.$

Answer   2.688

EXAMPLE 8   Compute $\dfrac{(14000)(0.00003)(8800000)}{(1100)(0.000002)}.$

Solution

$\dfrac{(1.4 \times 10^4)(3 \times 10^{-5})(8.8 \times 10^6)}{(1.1 \times 10^3)(2 \times 10^{-6})} = \dfrac{(1.4)(3)(8.8)}{(1.1)(2)} \times 10^{4-5+6-3+6}$

$= 16.8 \times 10^8.$

Answer 1.68×10^9

Very large and very small numbers are displayed in scientific calculators by means of scientific notation. For example, the display

$$\boxed{\mathit{3.94016\ 67}}$$

means 3.94016×10^{67}. Your calculator has a key such as $\boxed{\text{EXP}}$ or $\boxed{\text{EE}}$ that allows you to enter a suitable power of 10 (usually up to 99 or down to -99). For example, to key in the number shown you would press

   3 . 9 4 0 1 6 $\boxed{\text{EXP}}$ 6 7 .

To key in 1.6×10^{-24}, you would press

   1.6 $\boxed{\text{EXP}}$ 2 4 $\boxed{+/-}$.

The key $\boxed{+/-}$ (or a similar key) changes the exponent from 24 to -24.
  Scientific notation on a calculator is also called **floating point** notation.

Calculating Powers and Roots

Note the key $\boxed{y^x}$ on your calculator. It is used for the (approximate) calculation of powers. To calculate y^x, you first key in y, then $\boxed{y^x}$, then the exponent x, and finally $\boxed{=}$.

Examples

(1) $(3.52)^7$ 3 . 5 2 $\boxed{y^x}$ 7 $\boxed{=}$ $\mathit{6695.74\ 10}$

(2) $(145.3)^{-12}$ 1 4 5 . 3 $\boxed{y^x}$ 1 2 $\boxed{+/-}$ $\boxed{=}$ $\mathit{1.1293086 \times 10^{-26}}$

(3) $(81)^{1/4}$ 8 1 $\boxed{y^x}$. 2 5 $\boxed{=}$ $\mathit{3}$

(4) $(27)^{2/3}$ 2 7 $\boxed{y^x}$. 6 6 6 6 6 6 7 $\boxed{=}$ $\mathit{9.000000\ 10}$

Note that the sequence

   2 7 $\boxed{y^x}$ 2 $\boxed{\div}$ 3 $\boxed{=}$

does not calculate $(27)^{2/3}$, it calculates $27^2/3 = 243$. Thus to estimate something like $5^{3/17}$ you proceed in two steps:

   3 $\boxed{\div}$ 1 7 $\boxed{=}$ $\mathit{0.1764706}$
   5 $\boxed{y^x}$. 1 7 6 4 7 0 6 $\boxed{=}$ $\mathit{1.3284575}$

If your calculator has parentheses, then use

   5 $\boxed{y^x}$ $\boxed{(}$ 3 $\boxed{\div}$ 1 7 $\boxed{)}$ $\boxed{=}$ $\mathit{1.3284575}$

If your (left-to-right) calculator has a $\boxed{\updownarrow}$ key, which exchanges x and y, then use

   3 $\boxed{\div}$ 1 7 $\boxed{y^x}$ 5 $\boxed{\updownarrow}$ $\boxed{=}$ $\mathit{1.3284575}$

Another possibility is

   3 $\boxed{\div}$ 1 7 $\boxed{=}$ $\boxed{\text{STO}}$ 5 $\boxed{y^x}$ $\boxed{\text{RCL}}$ $\boxed{=}$

for the same result.

  If your calculator has a key $\boxed{y^{1/x}}$, you can use it to estimate $\sqrt[x]{y}$ directly.

Example

$\sqrt[7]{45}$ 4 5 $\boxed{y^{1/x}}$ 7 $\boxed{=}$ 1.7225555

Without the key $\boxed{y^{1/x}}$ you can work indirectly, using both $\boxed{y^x}$ and $\boxed{1/x}$.

Example

$\sqrt[7]{45}$ 4 5 $\boxed{y^x}$ 7 $\boxed{1/x}$ $\boxed{=}$ 1.7225555

Check   1 . 7 2 2 5 5 5 5 $\boxed{y^x}$ 7 $\boxed{=}$ 45.000005

Exercises

[All letters in the exercises represent positive numbers.]

Express as simply as possible, with no negative exponents

1 $\dfrac{(xy)^6}{xy^2}$ **2** $\dfrac{1}{x^3}(x^2)^3 x^{-4}$ **3** $a^2(a^{-1} + a^{-3})$

4 $(aba^{-4})(a^3b^{-2})^0$ **5** $(8a^3b)^{-4}(2a/b)^{12}$ **6** $(-5x^2y^{-3})^{-20}$

7 $(-xy^2)^3(-2x^2y^2)^{-4}$ **8** $(xy)^{-5}(2xy^2)(3xy)^3$ **9** $(4x^3y^2z)^2(4x^3y^2z)^{-7}$

10 $(2x^2y^2)^3(-3x^4y)^{-2}$ **11** $\dfrac{x^{-2}}{y^{-2}} + \dfrac{y^2}{x^2}$ **12** $\dfrac{(2pqr)^2}{(p^3q)^2(pr)^{-1}}$

13 $\left(\dfrac{a}{b}\right)^{-2}\left(\dfrac{b}{a}\right)^0$ **14** $\left(\dfrac{ab^2c^3d^4}{a^4b^3c^2d}\right)^2.$

Simplify

15 $\sqrt{81}$ **16** $\sqrt{144}$ **17** $\sqrt{\dfrac{1}{9}}$ **18** $\sqrt{\dfrac{49}{25}}$

19 $\sqrt{9a^6}$ **20** $\sqrt{\tfrac{1}{4}y^2}$ **21** $\sqrt{50}$ **22** $\sqrt{128}$

23 $\sqrt{\dfrac{1}{12}}$ **24** $\sqrt{ab^2c^5}$ **25** $\sqrt{\dfrac{18x^3}{(y+z)^4}}$ **26** $\dfrac{\sqrt{6}}{\sqrt{3}}$

27 $\sqrt[3]{1000}$ **28** $\sqrt[3]{\dfrac{1}{8}}$ **29** $\sqrt[3]{\dfrac{27}{64}}$ **30** $\sqrt[3]{27 \cdot 64 \cdot 125}$

31 $\sqrt[4]{\dfrac{1}{16}}$ **32** $\sqrt[4]{81}$ **33** $\sqrt[4]{32}$ **34** $\sqrt[4]{162}$

35 $\sqrt[3]{8a^3b^9}$ **36** $\sqrt[4]{10,000u^2}$ **37** $\sqrt[5]{64u^5v^6}$ **38** $\sqrt[3]{\dfrac{a^7}{24b^9}}.$

Find an equivalent expression involving at most one radical sign

39 $\sqrt{x}\sqrt{x^6}$ **40** $\sqrt{3}(1 + \sqrt{3})$ **41** $\sqrt{2}\sqrt{xy}\sqrt{10yz}$ **42** $\sqrt[3]{\dfrac{16xy}{z}}\sqrt[3]{4y^2z^7}.$

Simplify

43 $(25x^4)^{-3/2}$ **44** $\left(\dfrac{8}{u^6}\right)^{2/3}$ **45** $\left(\dfrac{u^4}{v^{12}}\right)^{3/4}$

46 $\left(\dfrac{27a^3}{8b^3c^6}\right)^{-4/3}$ **47** $(x^4y^6z^{-8})^{5/2}$ **48** $(x^{-3/2}\sqrt{y})^{-2}$

49 $(x^{4/3}y^{-2/3})^3$ **50** $(xy^2)^{1/3}(x^2y)^{-2/3}$ **51** $(8x\sqrt{x})^{5/3}$

52 $(x^{1/2} + y^{3/2})^2$ **53** $u^{1/3}(2u^{2/3} - u^{-1/6})$ **54** $v^{1/3}(8v^6)^{-2/3}.$

Express in terms of integral exponents and at most a single radical

55 $\sqrt{2} \cdot \sqrt[3]{2}$ **56** $\sqrt[3]{\sqrt{x^{1/4}}}$ **57** $\sqrt[3]{4}/\sqrt[6]{16}$

58 $\dfrac{\sqrt{xy}}{\sqrt[4]{x^2 y}}$ **59** $\left(\dfrac{\sqrt{3a}}{\sqrt[3]{6a^2}}\right)^4$ **60** $\sqrt{b\sqrt{b}}$

61 $\sqrt{x} \cdot \sqrt{x^2} \cdot \sqrt{x^3}$ **62** $\sqrt{x} \cdot \sqrt[3]{x} \cdot \sqrt[4]{x}$ **63** $\sqrt{\sqrt{\sqrt{a}}}$

64 $\sqrt{a\sqrt{a\sqrt{a}}}$ **65** $\sqrt{3\sqrt[3]{2}}$ **66** $\sqrt[3]{27a^4 b^2}\sqrt{64ab}$.

Express without radicals, using only positive exponents; simplify as much as possible

67 $(\sqrt[3]{xy^2})^{-3/5}$ **68** $\sqrt[5]{(xy^2)^{-10/3}/(x^2 y)^{-15/7}}$

69 $(\sqrt[4]{x^{14}y^{-21/5}})^{-3/7}$ **70** $\sqrt[3]{x^{5/6}/x^{-5/6}}$

71 $\dfrac{\sqrt[3]{x}}{(x^{5/6})^{42}(x^{51})^{-2/3}}$ **72** $\left(\dfrac{x}{y}\right)^{1/5}\left(\dfrac{y}{z}\right)^{2/5}\left(\dfrac{z}{w}\right)^{3/5}\left(\dfrac{w}{x}\right)^{4/5}$.

Compute and express the result in scientific notation

73 $(180)(30{,}000{,}000)(0.00012)$ **74** $\dfrac{(20{,}100)(0.006)}{(0.00000\,02)(402{,}000)}$

75 $(0.002)^3(0.00004)(0.00000\,5)(6{,}000{,}000{,}000)$ **76** $\dfrac{1}{(800)(200{,}000)^2(0.00001)^4}$

77 $(200{,}000)^6$ **78** $(10)(200)(3000)(40{,}000)(500{,}000)$

79 $\dfrac{(0.00000\,02)^2}{(5000)^4}$ **80** $(8000)(30{,}000)^5(0.00000\,1)^6$

81 the number of inches in 100 miles

82 the number of cubic centimeters in a cubic kilometer

83 the total mass in grams of 3,000,000 neutrons

84 the length in kilometers of a light-year, the distance light travels in a year. Take the speed of light to be 300,000 km/sec and a year to be 365 days.

Estimate on a calculator

85 $(3.5)^{10}$ **86** $(1.723)^{35}$ **87** $(1.01)^{96}$

88 $(1.01)^{-120}$ **89** $(1.5708 \times 10^7)^{-9}$ **90** $(2.718 \times 10^{-12})^5$

91 $8^{3/4}$ **92** $10^{6/5}$ **93** $(2.79)^{4/7}$

94 $(6.207)^{13/21}$ **95** $(7.434 \times 10^{13})^{11/19}$ **96** $(1.043 \times 10^{-40})^{7/6}$.

97 $\sqrt{2991}$ **98** $\sqrt[3]{17}$ **99** $\sqrt[10]{10}$

100 $\sqrt[5]{371293}$ **101** $\sqrt[9]{1.4 \times 10^{30}}$ **102** $\sqrt{8.039 \times 10^{91}}$

103 $\sqrt[12]{5.62 \times 10^{-17}}$ **104** $\sqrt[14]{6.77 \times 10^{-551}}$ **105** $\sqrt[3]{8 + 17\sqrt{5}}$

106 $\sqrt[5]{16 + \sqrt[4]{78}}$ **107** $\sqrt{2 + \sqrt{2 + \sqrt{2}}}$ **108** $\sqrt[3]{4^4 + 5^5 + 6^6}$.

2 EXPONENTIAL FUNCTIONS

In Section 1, we discussed exponential expressions a^p, where p is an integer or a rational number, and $a > 0$. Now we introduce exponential *functions* $f(x) = a^x$, where x takes *all real values*.

A strict definition of a^x for all real values is technical. It requires defining such numbers as $a^{\sqrt{2}}$. Rather than attempt a definition, let us see what properties an exponential function ought to have. For example, suppose $f(x) = 2^x$ were defined for *all* real x. What would it be like?

If $x = n$, an integer, then 2^x should agree with our former definition of 2^n. We can tabulate some values of the function:

x	0	1	2	3	4	5	6	7	8	9	10
2^x	1	2	4	8	16	32	64	128	256	512	1024

The values increase rapidly; joining the plotted points by a smooth curve, we get a sketch of the graph of $y = 2^x$ for $x \geq 0$. See Fig. 1a.

If $x = -n$, where n is a positive integer, then 2^x should mean $2^{-n} = 1/2^n$. We tabulate some values, using 2-place accuracy:

x	-10	-9	-8	-7	-6	-5	-4	-3	-2	-1	0
$f(x) = 2^x$	0.00	0.00	0.00	0.01	0.02	0.03	0.06	0.12	0.25	0.50	1.00

The values decrease very rapidly. They suggest the graph shown in Fig. 1b. (Note that the x-axis is a horizontal asymptote.)

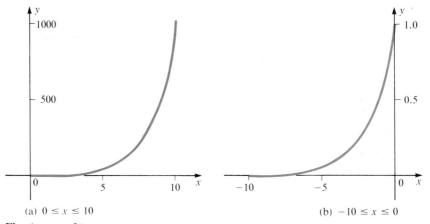

(a) $0 \leq x \leq 10$ (b) $-10 \leq x \leq 0$

Fig. 1 $y = 2^x$

Now let's plot $y = 2^x$ for "reasonable" values of x, say $-3 \leq x \leq 3$ (Fig. 2). We see that the graph always rises as x increases. It rises very fast as x increases through positive values, and it approaches the x-axis very fast as x decreases through negative values.

The graph of $y = a^x$, for any $a > 1$, is similar. Figure 3 shows some examples. The larger a is, the faster the curve zooms up as x increases, and the faster it decays as x decreases.

Properties of Exponential Functions

We leave the actual construction of exponential functions to more advanced courses. In this course, we simply accept their existence and list their properties, based on experimental evidence.

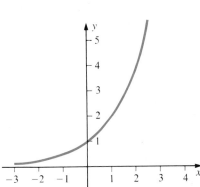

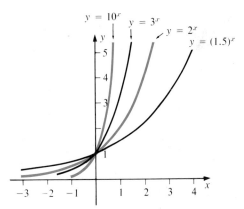

Fig. 2 $y = 2^x$

Fig. 3 $y = a^x$ for various
$a > 1$

For each number $a > 0$, there exists an exponential function a^x with the following properties:

(1) a^x is defined for all real x, and $a^x > 0$.

(2) a^x takes all positive real values.

(3) $a^n = \underbrace{a \cdot a \cdot a \cdots a}_{n \text{ factors}}$ for each positive integer n.

(4) If $a > 1$, then a^x is a strictly increasing function ($a^x < a^y$ whenever $x < y$). Also $a^x \longrightarrow \infty$ as $x \longrightarrow \infty$ and $a^x \longrightarrow 0+$ as $x \longrightarrow -\infty$.

(5) The rules of exponents hold:

$$a^{x+y} = a^x a^y \qquad a^{x-y} = \frac{a^x}{a^y} \qquad a^{-x} = \frac{1}{a^x} \qquad a^0 = 1$$

$$(a^x)^y = a^{xy} \qquad a^x b^x = (ab)^x \qquad 1^x = 1.$$

The number a is called the **base** of the exponential function a^x.

Graph of $y = a^x$ for $a < 1$

So far we have sketched the graphs of exponential functions $y = a^x$ only for $a > 1$. Now let us graph $y = a^x$ for $0 < a < 1$. To do so we write $a = 1/b$, where $b > 1$. By the rules of exponents

$$a^x = b^{-x} = \frac{1}{b^x}$$

Since b^x is an increasing function, a^x is a decreasing function. We can say even more: the graph of $y = a^x$ is the mirror image in the y-axis of the graph of $y = b^x$. Why? Because the height of $y = a^x$ at $-x$ is the height of $y = b^x$ at x. For example, the graph of $y = (\frac{1}{2})^x$ is the mirror image in the y-axis of the graph of $y = 2^x$. See Fig. 4, next page.

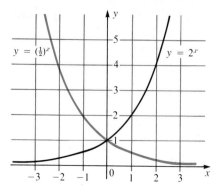

Fig. 4 $y = (\tfrac{1}{2})^x$

Use of Exponential Functions

Exponential functions occur frequently in applications concerning natural growth. We give a few examples now; others later when we have more technique.

EXAMPLE 1   A culture has 5000 bacteria now. In t hours it will have

$$N = 5000 \times 2^{t/k}$$

bacteria, where k is the time it takes the culture to double. If $k = 7$ hours, how many bacteria will there be in 16.5 hours?

Solution $N = 5000 \times 2^{16.5/7}$

Calculate:

$$1\ 6\ .\ 5\ \boxed{\div}\ 7\ \boxed{=}\ \boxed{\text{STO}}\ 2\ \boxed{y^x}\ \boxed{\text{RCL}}\ \boxed{\times}\ 5\ 0\ 0\ 0\ \boxed{=}\quad \mathit{25617.733}$$

Answer 25,618

EXAMPLE 2   A container of hot water is at temperature 90 °C at time $t = 0$. It is surrounded by air at constant temperature 20 °C. If stirred steadily, its temperature after t minutes will be

$$u(t) = 20 + 70 \times 1.071^{-t}\ °\text{C}.$$

Find its temperature after (a) 15 min. (b) 30 min. (c) 60 min.

Solution   A calculator routine for computing $u(t)$ is

$$1\ .\ 0\ 7\ 1\ \boxed{y^x}\ t\ \boxed{+/-}\ \boxed{\times}\ 7\ 0\ \boxed{+}\ 2\ 0\ \boxed{=}.$$

For $t = 15$, the calculation yields $u(15) \approx 45.02$. Similarly $u(30) \approx 28.94$ and $u(60) \approx 21.14$.

Answer   (a) 45.02 °C   (b) 28.94 °C   (c) 21.14 °C.

EXAMPLE 3 If $1000 is invested at $9\frac{1}{2}\%$ annual interest, compounded annually, for t years, it will be worth

$$(1000)(1 + 0.095)^t$$

dollars. Show that it will triple in approximately 12.1 years.

Solution In 12.1 years the value of the investment will be

$$1000 \times 1.095^{12.1}$$

By calculator this is approximately $2998.55, pretty close to $3000.

Exercises

Graph

1 $y = 3^x \quad -3 \le x \le 0$

2 $y = 3^x \quad 0 \le x \le 3$

3 $y = (1.5)^x \quad -3 \le x \le 3$

4 $y = 10^x \quad 0 \le x \le 6$

5 $y = 10^{-x} \quad 0 \le x \le 6$

6 $y = (0.4)^x \quad -3 \le x \le 3$

7 $y = 2^{x-1}$

8 $y = \frac{1}{2}(2^x + 2^{-x})$

9 $y = \frac{1}{2}(2^x - 2^{-x})$

10 $y = 2^{-x^2}$

11 $y = 2^{|x|}$

12 $y = 2^{-|x|}$

13 $y = 1 - 2^{-x}$

14 $y = 2^x - 3$

15 $y = 3^x - 2^x$

16 $y = 2^x - x$.

If you graph $y = f(x)$ and $y = g(x)$ on the same coordinate plane, each intersection of the graphs corresponds to a solution of $f(x) = g(x)$, and vice versa. Use this technique to determine the number of solutions of

17 $x = 5^{-x}$

18 $x = 2^x$

19 $10^x = x + 1$

20 $5^x = 2 - x^2$.

21 Find a positive function $f(x)$ for which $f(2x) = f(x)^2$.

22 Find a positive function $f(x)$ for which $f(x + 1) = 3f(x)$.

23 Suppose a certain culture of bacteria doubles in 4.5 hours. By how much does it multiply in 10 hours? [See Example 1.]

24 A culture of bacteria contains 8500 bacteria now, and we know that it doubles in 10 hours. How many bacteria were there 3 hours ago? [See Example 1.]

25 The temperature of a certain container of hot oil being cooled is

$$u(t) = 25 + 130 \times 1.208^{-t} \,°C$$

after t minutes. Find (a) its initial temperature, (b) its ultimate temperature (t very large), (c) its temperature after 12 minutes.

26 (cont.) Show that the temperature drops half way from 155 °C to 25 °C after approximately 3.67 minutes.

27 Show that money invested at 16% annual interest, compounded annually, will double in about 4 years 8 months. [See Example 3.]

28 Show that money invested at $11\frac{1}{2}\%$ annual interest, compounded annually, will triple in about 10 years 1 month. [See Example 3.]

29 Which is worth more now, $1000 invested at 4% eleven years ago, or $1000 invested at 8% six years ago? [See Example 3.]

30 How much invested at 5% twelve years ago would be worth today exactly as much as $2000 invested at 10% six years ago? [See Example 3.]

3 LOGARITHM FUNCTIONS

We next study the inverse of the exponential function $f(x) = a^x$. If $a > 1$, then $y = a^x$ is strictly increasing. Its domain is $\{\text{all } x\}$, and its range is $\{y > 0\}$. By the discussion of page 155, there exists an inverse function $x = g(y)$ with domain $\{y > 0\}$ and range $\{\text{all } x\}$. The function g is called the **logarithm function** to the **base** a. We write it as

$$x = \log_a y.$$

By the nature of inverse functions,

$$g[f(x)] = x \qquad f[g(y)] = y.$$

In this case, these relations mean

$$\log_a(a^x) = x \qquad a^{\log_a y} = y.$$

As usual, we prefer to interchange x and y. So let us consider the function

$$y = \log_a x.$$

Its graph is the reflection of the graph of $y = a^x$ in the line $y = x$. See Fig. 1. We see that $\log_a x$ is defined for all $x > 0$. It is a strictly increasing function taking each real value exactly once. In fact,

$$\log_a x \longrightarrow \infty \quad \text{as} \quad x \longrightarrow \infty.$$

The y-axis is a vertical asymptote of its graph,

$$\log_a x \longrightarrow -\infty \quad \text{as} \quad x \longrightarrow 0+.$$

Logarithm to base a

Let $a > 1$. There is a unique function $y = \log_a x$ defined for all $x > 0$ and satisfying

$$a^{\log_a x} = x \quad (x > 0) \qquad \log_a a^x = x \quad (\text{all } x).$$

The function $\log_a x$ is strictly increasing.

$$\log_a x \longrightarrow \infty \quad \text{as} \quad x \longrightarrow \infty$$

and

$$\log_a x \longrightarrow -\infty \quad \text{as} \quad x \longrightarrow 0+.$$

Since $a^0 = 1$, it follows that $\log_a 1 = 0$. The function $\log_a x$ is strictly increasing. Therefore if $0 < x < 1$, then $\log_a x < \log_a 1 = 0$. Similarly if $x > 1$, then $\log_a x > \log_a 1 = 0$.

Let $a > 1$. Then
$$\begin{cases} \log_a x < 0 & \text{for } 0 < x < 1 \\ \log_a 1 = 0 \\ \log_a x > 0 & \text{for } 1 < x. \end{cases}$$

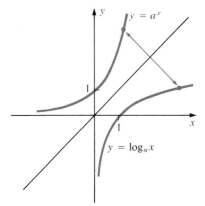

Fig. 1   Inverse functions: $y = a^x$ and $y = \log_a x$

Since $\log_a x_1 < \log_a x_2$ if $x_1 < x_2$, the function $\log_a x$ takes different values for different values of x. This fact is important in applications, especially in the following form:

Let $a > 1$. If $\log_a x_1 = \log_a x_2$, then $x_1 = x_2$.

Exponential and Logarithmic Relations

Because the functions a^x and $\log_a x$ are inverse to each other, each relation of the form $y = a^x$ is equivalent to the relation $x = \log_a y$.

Examples

Exponential Relation	Equivalent Logarithmic Relation
$3^0 = 1$	$\log_3 1 = 0$
$3^1 = 3$	$\log_3 3 = 1$
$3^2 = 9$	$\log_3 9 = 2$
$3^3 = 27$	$\log_3 27 = 3$
$3^{-1} = \frac{1}{3}$	$\log_3 \frac{1}{3} = -1$
$3^{-2} = \frac{1}{9}$	$\log_3 \frac{1}{9} = -2$
$3^{1/2} = \sqrt{3}$	$\log_3 \sqrt{3} = \frac{1}{2}$

To find $x = \log_a y$, it suffices to find the number x that satisfies $a^x = y$.

EXAMPLE 1   Find

(a) $\log_2 32$ (b) $\log_5 \frac{1}{25}$ (c) $\log_4 8$ (d) $7^{\log_7 12}$

Solution   (a)   Set $x = \log_2 32$. Then x must satisfy $2^x = 32$. But $2^5 = 32$, therefore $x = 5$.

(b)   Set $x = \log_5 \frac{1}{25}$. Then x must satisfy $5^x = \frac{1}{25}$. But $5^{-2} = \frac{1}{25}$, therefore $x = -2$.

(c)   Set $x = \log_4 8$. Then x must satisfy $4^x = 8$. But $4^{3/2} = 2^3 = 8$, therefore $x = \frac{3}{2}$.

(d)   Set $x = \log_7 12$. Then x must satisfy $7^x = 12$. Therefore $7^{\log_7 12} = 12$.

Answer   (a)   5   (b) -2 (c) $\frac{3}{2}$ (d)   12

EXAMPLE 2   Solve for x

(a) $\log_2(4x - 1) = -3$ (b) $\log_5(x^2 - 6) = \log_5 x$.

Solution   (a)   Convert the equation to an equivalent exponential form:

$$4x - 1 = 2^{-3} = \tfrac{1}{8}.$$

Solving for x yields $x = \frac{9}{32}$.

(b)   The equation $\log_5(x^2 - 6) = \log_5 x$ holds only if

$$x^2 - 6 = x.$$

Solve this quadratic:

$$x^2 - x - 6 = 0 \qquad (x - 3)(x + 2) = 0 \qquad x = -2, 3.$$

The solution $x = -2$ is unacceptable because $\log_5(-2)$ is undefined. The other solution is OK because if $x = 3$, then $x^2 - 6 = 3$ so $\log_5(x^2 - 6) = \log_5 3$.

Answer   (a) $\frac{9}{32}$ (b)   3

Properties of Logarithm Functions

Logarithms satisfy certain rules (algebraic properties) of great importance for both theory and computation:

Rules for Logarithm Functions

(1) $\log_a(x_1 x_2) = \log_a x_1 + \log_a x_2$

(2) $\log_a\left(\dfrac{x_1}{x_2}\right) = \log_a x_1 - \log_a x_2$

(3) $\log_a x^r = r \cdot \log_a x$.

These properties are practically restatements of corresponding properties of a^x. Take (1), for instance. Suppose $y_1 = \log_a x_1$ and $y_2 = \log_a x_2$, that is, $x_1 = a^{y_1}$ and $x_2 = a^{y_2}$. Then

$$x_1 x_2 = a^{y_1} a^{y_2} = a^{y_1 + y_2},$$

so

$$\log_a(x_1 x_2) = y_1 + y_2 = \log_a x_1 + \log_a x_2.$$

There are similar verifications of the other two properties; they are left as exercises.

EXAMPLE 3   Given

$$\log_7 2 \approx 0.3562 \qquad \log_7 3 \approx 0.5646 \qquad \log_7 5 \approx 0.8271,$$

estimate   (a) $\log_7 15$ (b) $\log_7 \frac{32}{9}$ (c) $\log_7 \sqrt[3]{0.4}$ (d) $\log_7 35$.

Solution   (a) By rule (1),

$$\log_7 15 = \log_7(3 \times 5) = \log_7 3 + \log_7 5$$
$$\approx 0.5646 + 0.8271 = 1.3917.$$

(b) By rules (2) and (3),

$$\log_7 \tfrac{32}{9} = \log_7 32 - \log_7 9 = \log_7(2^5) - \log_7(3^2)$$
$$= 5 \log_7 2 - 2 \log_7 3 \approx 5(0.3562) - 2(0.5646) = 0.6518.$$

(c) By rules (3) and (2),

$$\log_7 \sqrt[3]{0.4} = \log_7(0.4)^{1/3} = \tfrac{1}{3} \log_7 0.4 = \tfrac{1}{3} \log_7 \tfrac{2}{5}$$
$$= \tfrac{1}{3}(\log_7 2 - \log_7 5) \approx \tfrac{1}{3}(0.3562 - 0.8271) \approx -0.1570.$$

(d) By rule (1),

$$\log_7 35 = \log_7(7 \times 5) = \log_7 7 + \log_7 5$$
$$= 1 + \log_7 5 \approx 1 + 0.8271 = 1.8271$$

Answer   (a) 1.3917   (b) 0.6518   (c) -0.1570 (d) 1.8271

EXAMPLE 4   Write as $\log_a x$

(a) $-3(\log_a 4 + \log_a 5)$ (b) $2 \log_a 7 + \log_a 3$

(c) $\tfrac{1}{2} \log_a 2 + \tfrac{1}{2} \log_a 3 - 2 \log_a 5$.

Solution

(a) $-3(\log_a 4 + \log_a 5) = -3 \log_a(4 \times 5)$
$$= -3 \log_a 20 = \log_a 20^{-3} = \log_a \tfrac{1}{8000}.$$

(b) $2 \log_a 7 + \log_a 3 = \log_a 7^2 + \log_a 3 = \log_a(7^2 \times 3) = \log_a 147.$

(c) $\tfrac{1}{2} \log_a 2 + \tfrac{1}{2} \log_a 3 - 2 \log_a 5 = \log_a 2^{1/2} + \log_a 3^{1/2} - \log_a 5^2$
$$= \log_a \sqrt{2} + \log_a \sqrt{3} - \log_a 25 = \log_a(\sqrt{2} \times \sqrt{3}) - \log_a 25$$
$$= \log_a \sqrt{6} - \log_a 25 = \log_a \left(\frac{\sqrt{6}}{25} \right).$$

Answer   (a) $\log_a \tfrac{1}{8000}$ (b) $\log_a 147$ (c) $\log_a \tfrac{1}{25} \sqrt{6}$.

EXAMPLE 5   Solve $\log_2 x - \log_2(x - 2) = 4$.

Solution   By the rules of logarithms,

$$\log_2 x - \log_2(x - 2) = \log_2 \frac{x}{x - 2} = 4.$$

Hence $\dfrac{x}{x-2} = 2^4 = 16$.

Solve for x:

$$x = 16(x-2)   15x = 32   x = \tfrac{32}{15} .$$

This answer checks. Both $\log_2 x$ and $\log_2 (x-2)$ are defined for $x = \tfrac{32}{15}$.

Answer $\tfrac{32}{15}$

Exercises

Graph

1 $y = \log_2 x$ **2** $y = \log_3 x$ **3** $y = \log_{10} x$
4 $y = \log_2 |x|$ **5** $y = \log_2 x^2$ **6** $y = \log_2 (x^2 + 1)$.

The graph of $y = \log_{10} x$ rises very slowly.

7 Where does it reach the level $y = 1$? 2? 10?

8 How much does it rise as x increases from $1{,}000{,}000$ to $10{,}000{,}000$?

Express as a statement about logarithms

9 $2^{10} = 1024$ **10** $5^4 = 625$
11 $8^{-2/3} = \tfrac{1}{4}$ **12** $\sqrt{961} = 31$

Find (without a calculator)

13 $\log_2 128$ **14** $\log_2 1/(2\sqrt{2})$ **15** $\log_2 \sqrt[5]{16}$
16 $\log_3 81$ **17** $\log_3 9\sqrt{3}$ **18** $\log_4 4$
19 $\log_4 2$ **20** $\log_4 1024$ **21** $\log_5 5^{12}$
22 $\log_5 0.008$ **23** $\log_{10} 0.0001$ **24** $\log_8 64^{15}$
25 $\log_{24} 1$ **26** $10^{\log_{10} 17}$ **27** $\log_2 0.03125$
28 $\log_{16} \tfrac{1}{2}$ **29** $2^{3 \log_2 5}$ **30** $\log_{a^2} a$.

Solve for x

31 $\log_{12} x = 0$ **32** $\log_5 x = 2$
33 $\log_3 (x + 5) = 4$ **34** $\log_3 (x + 7) = -1$
35 $\log_4 (2x + 3) = \tfrac{1}{2}$ **36** $\log_{27} x^2 = \tfrac{1}{3}$
37 $\log_9 (x^2 - 10) = 1$ **38** $\log_5 (x^2 - 7x + 7) = 0$
39 $\log_3 3^{2x+1} = 15$ **40** $\log_3 27^x = 4.5$
41 $\log_x 8 = 3$ **42** $\log_x 8 = -\tfrac{3}{2}$

Use the approximations $\log_{11} 2 \approx 0.2891$ $\log_{11} 3 \approx 0.4582$ $\log_{11} 5 \approx 0.6712$
to estimate

43 $\log_{11} 6$ **44** $\log_{11} 48$ **45** $\log_{11} \tfrac{9}{16}$
46 $\log_{11} \sqrt{12}$ **47** $\log_{11} 45$ **48** $\log_{11} 225$
49 $\log_{11} (\tfrac{1}{96} \sqrt{5})$ **50** $\log_{11} \sqrt[3]{\tfrac{36}{5}}$ **51** $\log_{11} \tfrac{3}{25}$
    52 $\log_{11} (\tfrac{6}{125})^{1/5}$.

Write as $\log_a x$

53 $\log_a 6 + 3 \log_a 2$ **54** $\log_a \sqrt{2} - \tfrac{5}{2} \log_a 4$
55 $\log_a 2 \cdot 3 - \log_a 3 \cdot 4 + \log_a 4 \cdot 5 - \log_a 5 \cdot 6$
56 $2 \log_a 2 + 4 \log_a 4 - 8 \log_a 8$.

Solve for x

57 $\log_6 x + \log_6 (x - 1) = 1$

58 $\log_6 (2x + 1) - \log_6 (2x - 1) = 1$

59 $\log_2 \sqrt{x} + \log_2 \dfrac{1}{x} = 3$

60 $\frac{1}{2} \log_3 x - \frac{1}{3} \log_3 x^2 = 1$

61 $\log_2 x^2 - \log_2 (3x + 8) = 1$

62 $\log_{10} (\log_{10} x) = 1$.

63 Prove Rule (2) on page 276.

64 Prove Rule (3) on page 276.

65 Given an example showing that $\log_a (x_1 + x_2) \neq \log_a x_1 + \log_a x_2$.

66 Find a non-zero function $f(x)$ satisfying $f(x_1 x_2) = f(x_1) + f(x_2)$.

67 Find all x such that $-2 < \log_{10} x < -1$.

68 Find the domain of $\log_2 \log_2 x$.

4   COMPUTING LOGARITHM FUNCTIONS

For practical computations, the most useful base of logarithms is 10. This is because we write numbers in decimal notation. A key feature of decimal notation is the way we multiply a number by powers of 10. We merely shift the decimal point. For logarithms to the base 10, there is a corresponding feature:

$$\log_{10} (10^n x) = \log_{10} 10^n + \log_{10} x = n + \log_{10} x.$$

Thus multiplying x by a power of 10 (shifting the decimal point) just adds an integer to $\log_{10} x$. For instance,

$$\log_{10} (1057.398) = 3 + \log_{10} (1.057398),$$

$$\log_{10} (0.000092) = -5 + \log_{10} 9.2.$$

The logarithm of x to the base 10 is called the **common logarithm** of x and is written $\log x$. When you see the expression $\log x$ without an explicit base, then base 10 is understood.

Until about 1975, computations with logarithms were done using tables of common logs. Now hand-held calculators have made log table calculations obsolete. Most scientific calculators have keys for $\log x$ and 10^x. To find $\log x$ on your calculator, just key in x and follow by $\boxed{\log}$.

Examples

(1) $\log(1057.398)$ $1 \ 0 \ 5 \ 7 \ . \ 3 \ 9 \ 8 \ \boxed{\log}$ 3.024238

(2) $\log(0.000092)$ $. \ 0 \ 0 \ 0 \ 0 \ 9 \ 2 \ \boxed{\log}$ -4.036212

Exponential Equations

Common logarithms are useful in equations involving exponents. The calculator is a magnificent tool for finding numerical solutions.

> *EXAMPLE 1*   Solve for x $7^x = 54$.
>
> **Solution**   Since 54 is a little larger than 7^2, the answer should be a little more than 2. Now the exponential statement $7^x = 54$ is equivalent to the logarithmic statement
>
> $$x = \log_7 54.$$

Theoretically at least, this solves the problem. But this theoretical solution does not give the numerical value of x.

To estimate the number x, apply the function log to both sides of the original equation $7^x = 54$:

$$\log 7^x = \log 54 \qquad x \log 7 = \log 54 \qquad x = \frac{\log 54}{\log 7}.$$

Now find $\log 54$ and $\log 7$, and divide. This can all be done on your calculator by a short sequence of steps:

5 4 $\boxed{\log}$ $\boxed{\div}$ 7 $\boxed{\log}$ $\boxed{=}$ *2.049932*

The result is slightly more than 2, as predicted.

Answer   2.049932

EXAMPLE 2   Estimate the solution of

(a) $11^x = 5000$ (b) $5^{-x} = 0.038$.

Solution   (a)   Apply the function log to both sides:

$$x \log 11 = \log 5000 \qquad x = \frac{\log 5000}{\log 11}.$$

Now estimate x by the key sequence

5 0 0 0 $\boxed{\log}$ $\boxed{\div}$ 1 1 $\boxed{\log}$ $\boxed{=}$ *3.551945*

(b)   Apply log to both sides:

$$-x \log 5 = \log 0.038 \qquad x = -\frac{\log 0.038}{\log 5}.$$

Now use the sequence

.0 3 8 $\boxed{\log}$ $\boxed{\div}$ 5 $\boxed{\log}$ $\boxed{=}$ $\boxed{+/-}$ *2.031870*

Answer   (a)   3.551945   (b)   2.031870

EXAMPLE 3   Estimate the solution of $5^x = 9 \times 4^x$.

Solution 1   Apply log to both sides and use the rules for logarithms:

$$\log 5^x = \log(9 \times 4^x) = \log 9 + \log 4^x$$

$$x \log 5 = \log 9 + x \log 4$$

$$x(\log 5 - \log 4) = \log 9.$$

Therefore

$$x = \frac{\log 9}{\log 5 - \log 4} = \left(\frac{1}{\log 5 - \log 4}\right)(\log 9).$$

Now calculate x by the sequence

5 $\boxed{\log}$ $\boxed{-}$ 4 $\boxed{\log}$ $\boxed{=}$ $\boxed{1/x}$ $\boxed{\times}$ 9 $\boxed{\log}$ $\boxed{=}$ *9.846686*

Solution 2   Divide both sides by 4^x and use laws of exponents:

$$\frac{5^x}{4^x} = 9 \qquad (\tfrac{5}{4})^x = 9 \qquad (1.25)^x = 9.$$

By the usual technique

$$x \log 1.25 = \log 9 \qquad x = \frac{\log 9}{\log 1.25} \approx 9.846686$$

The answer is the same as in Solution 1 because

$$\log 1.25 = \log \tfrac{5}{4} = \log 5 - \log 4.$$

Answer   9.846686

Inverse Logarithms (Antilogs)

Suppose you are given $\log x = 1.816$. How do you find x? Remember that $\log x$ and 10^x are inverse functions. Therefore

$$x = 10^{\log x} = 10^{1.816}$$

If your calculator has a $\boxed{10^x}$ key, use the sequence

1.816 $\boxed{10^x}$ 65.463617

If it doesn't have a $\boxed{10^x}$ key, then the sequence $\boxed{\text{INV}}$ $\boxed{\log}$ will do the same thing. Another route is

10 $\boxed{y^x}$ 1.816 $\boxed{=}$ 65.463617

Logs to Other Bases

For practical computations with logarithms, the most useful base is 10 because it goes so well with decimals. However, for theoretical work the most natural base is a number called e. To 10 places,

$$e = 2.71828\ 18285.$$

A strict definition of e requires calculus. There is a nice geometric interpretation of e, however. The graph of each exponential function passes through the point $(0, 1)$. Among all of these, only the graph of $y = e^x$ has slope 1 at that point (Fig. 1).

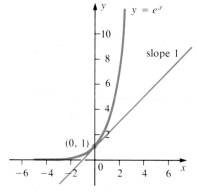

Fig. 1 $y = e^x$

The function $\log_e x$ is called the **natural logarithm** function. It is written $\ln x$. You can easily obtain $\ln x$ and e^x on any scientific calculator. Many have keys for each of these functions. Others may not have a key for e^x, but you can calculate e^x as the inverse of $\ln x$. For example, you can find $e^{3.86}$ by the sequence

3.8 6 | INV | | ln | *47.46535*

Logarithms to Any Base

Suppose we need a numerical value for $\log_a x$ and a is not 10 or e. There is no calculator key for this, so what can we do? Actually we have already run into this problem. Back in Example 1, we found that the *theoretical* solution to $7^x = 54$ was $\log_7 54$. For a *numerical* solution, we applied log to both sides of the equation and found $x = (\log 54)/(\log 7)$. The answers must be the same. Therefore

$$\log_7 54 = \frac{\log 54}{\log 7}.$$

There is a general principle here. To find $y = \log_a x$, write the equivalent exponential relation,

$$a^y = x.$$

Apply log to both sides:

$$\log a^y = \log x$$

$$y \cdot \log a = \log x.$$

Now solve for y.

$$\boxed{\log_a x = \frac{\log x}{\log a} = \left(\frac{1}{\log a}\right) \log x.}$$

This formula shows that each function $\log_a x$ is just a constant multiple of $\log x$. For example

$$\log_2 x \approx (3.321928) \log x$$

because $1/\log 2 \approx 3.321928$.

Suggestion Don't memorize the formula in the box. It's better to remember the way it was derived: by applying log to both sides of $a^y = x$.

EXAMPLE 4 Let $1 < a < b$. Show that $(\log_a b)(\log_b a) = 1$.

Solution 1 Apply the boxed formula twice, once to $\log_a$ with $x = b$ and once to $\log_b$ with $x = a$:

$$\log_a b = \frac{\log b}{\log a} \qquad \log_b a = \frac{\log a}{\log b}.$$

Clearly the product $(\log_a b)(\log_b a)$ equals 1.

Solution 2 Set $y = \log_a b$ so $a^y = b$. Apply the function $\log_b$ to both sides:

$$\log_b a^y = \log_b b = 1 \qquad y \cdot \log_b a = 1.$$

But $y = \log_a b$, so we are done.

EXAMPLE 5 Estimate an x for which $\log_3 x = \log x + 1$.

Solution Replace $\log_3 x$ by an equivalent expression in terms of common logs:

$$\frac{\log x}{\log 3} = \log x + 1.$$

Solve for $\log x$:

$$\left(\frac{1}{\log 3} - 1\right) \log x = 1 \qquad \log x = \frac{1}{\dfrac{1}{\log 3} - 1}.$$

Calculate the number on the right, then press $\boxed{10^x}$ (or $\boxed{\text{INV}}$ $\boxed{\log}$) to obtain x:

$$3 \ \boxed{\log} \ \boxed{1/x} \ \boxed{-} \ 1 \ \boxed{=} \ \boxed{1/x} \ \boxed{10^x}$$

$\mathit{8.175029}$

Answer 8.175029

Exercises

Calculate to 6 places

1 $\log 2.385$	**2** $\log 0.01105$	**3** $\log (2.385 \times 10^{26})$
4 $\log (4.7 \times 10^{-13})$	**5** $\ln 3.14159$	**6** $\ln 178{,}000$.

Estimate the solution to 6 significant figures

7 $2^x = 10$	**8** $3^x = 77$	**9** $3^{-x} = 0.002$
10 $10^{-x} = 1.7$	**11** $1066^x = 1492$	**12** $1492^x = 1066$
13 $(3.5)^x = 4$	**14** $(9.1)^x = 5.8$	**15** $(1.001)^x = 10$
16 $(0.999)^x = 0.1$	**17** $(0.99)^x = 1.01$	**18** $(1.001)^x = 0.999$
19 $10^x = e$	**20** $e^{-x} = 3.82$	**21** $\log \log x = 1.013$
22 $\log_x 10 = 3$	**23** $\log_x 10 = -1.54$	**24** $8^x = 9^x$.

Estimate to 6 places

25 $\log_7 5$	**26** $\log_{11} 9$	**27** $\log_8 2$
28 $\log_{13} 17$	**29** $\log_{1.48} 2.2$	**30** $\log_{6.331} 1.049$.

Estimate the solution to 6 significant figures

31 $\log_4 x = 2 + \log_5 x$	**32** $\ln x - \log x = 1$
33 $(\log x)^2 = \ln x$	**34** $\log_2 x + \log_3 x + \log_4 x = 3$.

Estimate the base a to 5 places

35 $\log_a x = \frac{3}{4} \log x$ all $x > 0$	**36** $\log_a x = \frac{1}{2} \log x$ all $x > 0$.

5 APPLICATIONS

Natural Growth

We shall discuss a growth law that occurs frequently in nature. To illustrate the idea, let us imagine an experiment. A group of laboratory mice are split randomly into a small group of 100 and a large group of 500. After two months, the first group has increased to 150. What is a reasonable guess for the size of the second group?

Let $S(t)$ denote the population of the smaller group after t months, and $L(t)$ the same for the larger group. We are given that

$$S(0) = 100 \qquad L(0) = 500 \qquad \text{and that} \quad S(2) = 150.$$

We want an estimate for $L(2)$.

One "solution" is to assume that both populations grow at the *same constant rate*. Then

$$S(t) = 100 + kt \qquad \text{and} \qquad L(t) = 500 + kt,$$

where k is the rate of increase. But this is nonsense: in t months, each population would increase by the same amount, kt. Obviously the larger group should have a larger increase.

Let's think. If a population of 100 increases to 150 in two months, that is a 50% increase. Hence a population of 500 also should increase by 50% in the same time, so $L(2) = 750$. This is the key. The populations do not grow at the same rate, but they do increase by the same percentage of themselves in a fixed time. So, regardless of its actual size, the population should increase by 50% in any period of two months. If so, then

$$S(2) = S(0) \times 1.5$$

$$S(4) = S(2) \times 1.5 = [S(0) \times 1.5] \times 1.5 = S(0) \times (1.5)^2$$

$$S(6) = S(4) \times 1.5 = [S(0) \times (1.5)^2] \times 1.5 = S(0) \times (1.5)^3$$

and in general,

$$S(2n) = S(0) \times (1.5)^n = S(0) \times (1.5)^{2n/2}.$$

Thus we obtain an exponential function on the domain $\{2, 4, 6, \cdots, 2n, \cdots\}$. It seems reasonable to interpolate and guess that

$$S(t) = S(0) \times (1.5)^{t/2} = 100 \times (1.5)^{t/2}$$

for any time t. For instance

$$S(3) = 100 \times (1.5)^{3/2} \approx 184$$

$$S(5.5) = 100 \times (1.5)^{5.5/2} \approx 305.$$

The function $L(t)$ has the same growth law (50% increase in population for 2 months) except for a different initial value, so

$$L(t) = L(0) \times (1.5)^{t/2} = 500 \times (1.5)^{t/2}.$$

For instance

$$L(2) = 500 \times 1.5 = 750$$

$$L(3) = 500 \times (1.5)^{3/2} \approx 919 .$$

We have been led to consider a growth law of the form

$$F(t) = c \cdot a^t \qquad c \quad \text{a constant} .$$

For such a law, the percentage increase over any time interval of duration d depends only on d, not on the time or the size of $F(t)$. For

$$F(t + d) = c \cdot a^{t+d} = c \cdot a^t \cdot a^d = F(t) \cdot a^d .$$

Thus the population grows by a fixed factor a^d, that is, by a fixed percentage of itself.

For very short time intervals, we can almost imagine "instantaneous rate of growth." This idea is made precise in calculus. It is proved that at each instant, the growth rate of $F(t) = c \cdot a^t$ is proportional to $F(t)$ itself. Conversely, if $F(t)$ is a function growing this way, then $F(t) = c \cdot a^t$.

We now state what we shall call the Natural Growth Law.

Natural Growth Law Let $F(t)$ be a positive function with domain $\{\text{all } t\}$. Suppose the rate of increase of $F(t)$ at any instant of time t is proportional to $F(t)$. Then

$$F(t) = F_0 \cdot a^t$$

where a is a positive constant and $F_0 = F(0)$.

We say that $F(t)$ obeys the **natural growth law.**

EXAMPLE 1 A function $F(t)$ obeys the natural growth law. Suppose its value triples in 5 minutes. (a) Find the function. (b) By what factor does the function increase in 12 minutes? (c) How long does it take the function to multiply six-fold?

Solution (a) We are given

$$F(t) = F_0 \cdot a^t \quad \text{and} \quad F(5) = 3F_0 .$$

Therefore

$$F_0 \cdot a^5 = 3F_0 \qquad a^5 = 3 \qquad a = 3^{1/5}$$

$$F(t) = F_0 \cdot a^t = F_0 \cdot (3^{1/5})^t = F_0 \cdot 3^{t/5} .$$

(b) $F(12) = F_0 \cdot 3^{12/5}$

so the function increases to $3^{12/5}$ times its initial value in the first 12 minutes.

(c) We must solve $F(t) = 6F_0$ for t:

$$F_0 \cdot 3^{t/5} = 6F_0 \qquad 3^{t/5} = 6 \qquad \frac{t}{5} \cdot \log 3 = \log 6$$

$$t = 5 \frac{\log 6}{\log 3} \approx 8.155 \text{ minutes} .$$

Answer   (a) $F(t) = F_0 \cdot 3^{t/5}$ (b) $3^{12/5} \approx 13.97$

(c) $5\dfrac{\log 6}{\log 3} \approx 8.155$ minutes

Population Growth

Under certain conditions the rate of growth of a population is proportional to the current population. Therefore the population obeys the natural growth law.

EXAMPLE 2   A yeast culture has initial weight 10 g. The weight doubles in 2 hours. Assuming the growth rate is proportional to current weight, find

(a)   the growth law   (b)   the weight after $3\frac{1}{2}$ hours

(c)   when the weight reaches 24 g.

Solution   (a)   Let $W(t)$ denote the weight at t hours. Then

$$W(t) = W_0 \cdot a^t = 10 \times a^t \quad \text{and} \quad W(2) = 20.$$

Therefore

$$10 \times a^2 = 20 \qquad a^2 = 2 \qquad a = 2^{1/2}$$

$$W(t) = 10 \times (2^{1/2})^t = 10 \times 2^{t/2}.$$

(b) $W(3.5) = 10 \times 2^{3.5/2} = 10 \times 2^{1.75} \approx 33.64$ g

(c)   Solve $W(t) = 24$ for t:

$$10 \times 2^{t/2} = 24 \qquad 2^{t/2} = 2.4 \qquad \frac{t}{2}\log 2 = \log 2.4$$

$$t = 2\,\frac{\log 2.4}{\log 2} \approx 2.526 \text{ hr}$$

Answer   (a) $W(t) = 10 \times 2^{t/2}$ (b)   33.64 g   (c)   2.526 hr

---

EXAMPLE 3   On 1 Jan. 1980 the population of city A was 1,772,000 and the population of city B was 2,091,000. The population of A is increasing at the rate of 3.13% per year and the population of B is increasing at the rate of 1.66% per year. When will A catch up to B?

Solution   The population $A(t)$ of the first city adds $3.13\% = 0.0313$ times its value to itself each year, that is, it *multiplies* by the factor $1 + 0.0313 = 1.0313$ each year. By the natural growth law,

$$A(t) = 1,772,000 \times (1.0313)^t.$$

Similarly for the second city

$$B(t) = 2,091,000 \times (1.0166)^t.$$

We must solve $A(t) = B(t)$ for t:

$$1,772,000 \times (1.0313)^t = 2,091,000 \times (1.0166)^t$$

$$\frac{(1.0313)^t}{(1.0166)^t} = \frac{2091}{1772} \qquad \left(\frac{1.0313}{1.0166}\right)^t = \frac{2091}{1772} \qquad (1.01446)^t \approx 1.18002$$

$$t \log(1.01446) \approx \log(1.18002) \qquad t \approx \frac{\log(1.18002)}{\log(1.01446)} \approx 11.530 \text{ years}.$$

Now 0.530 years = 0.530×12 months = 6.36 months, and 0.36 months $= 0.36 \times 31$ days ≈ 11 days, so the populations will be equal approximately 11 yr 6 mo 11 days after 1 Jan. 1980.

Answer   mid-July 1991.

Natural Decay

If a positive function $F(t)$ *decreases* (decays) at a rate proportional to the value of the function, then

$$F(t) = F_0 \cdot b^t$$

where $0 < b < 1$. This is a particular case of natural growth. Since we are generally more comfortable with bases greater than one, we can replace b by $1/a = a^{-1}$, where $a > 1$. Thus

$$F(t) = F_0 \cdot a^{-t} \qquad a > 1$$

is the law of natural decay.

For instance, a poor investment might decrease in value at the rate of 6% per year. Assuming the natural decay law, its value after t years would be

$$V(t) = V_0 \cdot (1 - 0.06)^t = V_0 \cdot (0.94)^t.$$

Since $1/0.94 \approx 1.063830$, we can also write

$$V(t) = V_0 \cdot (1.063830)^{-t}.$$

The law of natural decay applies to the disintegration of radioactive elements. With each such element is associated an important constant, its **half-life.** This is the time in which a sample decays to half of its mass.

EXAMPLE 4   A 1 kg sample of Thorium X decays to 1 g in 36.2755 days. Find its half-life to two places.

Solution   The decay law is of the form

$$X(t) = 1000 \cdot a^{-t} \text{ g},$$

where t is in days. To apply it, we need the value of a. We are given

$$X(36.2755) = 1,$$

hence

$$1000 \cdot a^{-36.2755} = 1 \qquad a^{36.2755} = 1000$$

$$a = (1000)^{1/36.2755} \approx 1.20976.$$

Therefore the decay law is

$$X(t) = 1000 \times (1.20976)^{-t}.$$

Now the half-life is the time in which the sample decays to 500 g. So we solve $X(t) = 500$ for t:

$$1000 \times (1.20976)^{-t} = 500 \qquad (1.20976)^t = 2$$

$$t = \frac{\log 2}{\log(1.20976)} \approx 3.64 \text{ days}$$

Answer   3.64 days

The long half-life of ^{14}C (carbon-14) has been applied to determine the age of very old objects. The next example is a typical application of carbon-14 dating.

EXAMPLE 5   All living wood contains the same concentration of ^{14}C. When wood dies, the ^{14}C decays at a rate proportional to its mass; its half-life is 5568 years.

In 1950, archeologists excavated wood from a city built in the time of King Hammurabi, the law giver. Analysis showed that this wood contained 61.28% of the normal concentration of ^{14}C. Find an approximate date for King Hammurabi.

Solution   Let $m(t)$ be the mass of ^{14}C at time t years from King H. Then

$$m(t) = m_0 \cdot a^{-t}.$$

Since $m(5568) = \frac{1}{2}m_0$, we have

$$m_0 \cdot a^{-5568} = \frac{1}{2}m_0 \qquad a^{-5568} = \frac{1}{2} \qquad a^{5568} = 2.$$

Therefore $a = 2^{1/5568}$ and

$$m(t) = m_0 \cdot 2^{-t/5568}.$$

We must solve $m(t) = (0.6128)m_0$ for t:

$$m_0 \cdot 2^{-t/5568} = (0.6128)m_0 \qquad 2^{t/5568} = \frac{1}{0.6128}$$

$$\frac{t}{5568} \log 2 = \log\left(\frac{1}{0.6128}\right) = -\log(0.6128)$$

$$t = -5568 \frac{\log(0.6128)}{\log 2} \approx 3934 \text{ yr}.$$

The decay started about 3934 years before 1950, that is, in 1985 BC. (There is no year 0; the year before 1 AD is 1 BC.)

Answer   1985 BC.

Exercises

In exercises concerning growth (decay), always assume the natural growth (decay) law unless otherwise stated.

1 A bacteria colony doubles in 3 hours. How long does it take to increase 5-fold? 10-fold?

2 (cont.) After 7 hours the population of the colony is 10^7. What was its original population?

3 There are 10^5 bacteria in a culture at the start of an experiment and 10^6 after 5 hours. Find the growth law.

4 A bacteria colony grows by 4% in 10 minutes. How long will it take to double?

5 See Example 3. When will the population of city A be 20% greater than the population of city B?

6 The 1970 population of San Antonio, Texas was 707,500. At what rate must the city grow annually in order to have one million inhabitants by the year 2000?

7 Find an interpretation for k in the growth law $N(t) = N_0 3^{t/k}$.

8 Cities C and D presently have equal populations. The population of C has been growing at the rate of 3.1% per year and the population of D has been growing at the rate of 1.7% per year. When was C's population 15% lower than D's population?

9 The estimated 1970 world population was 3.63×10^9. Assuming a steady growth rate of doubling in 35 years, estimate the world population in the years 2000, 2100, 2200.

10 (cont.) If this growth rate continues too long we will run out of space to hold all these people. When will there be one person for each square foot of land? The land area of the earth is about 1.6×10^{15} ft^2.

Exercise 10 shows that the basic growth law is unrealistic over a long period. A more sophisticated model takes into consideration the struggle for food and space as the population grows. It proposes the formula

$$p(t) = \frac{ak}{bk + e^{-at}},$$

where a and b are constants and

$$k = \frac{p_0}{a - bp_0} \qquad p_0 = \text{population at time } t = 0.$$

11 Taking $a = 0.0290$ and $b = 2.94 \times 10^{-12}$, find k. Then estimate world population in the years 2000, 2100, 2200 based upon $p_0 = 3.63 \times 10^9$ in 1970. What is the ultimate world population?

12 (cont.) The U.S. population in 1790 was 3.93×10^6. Taking $a = 3.054 \times 10^{-2}$ and $b = 1.189 \times 10^{-10}$, find k. Then find what population the formula predicts for 1950 and for 1970. What does it predict as the ultimate U.S. population?

13 Thorium-X has a half-life of 3.64 days. How long will it take for a sample to decay to $\frac{1}{10}$ of its original mass?

14 A certain radioactive substance loses $\frac{1}{5}$ of its original mass in 3 days. What is its decay law?

15 A substance decays according to the law $M(t) = M_0 e^{-t}$, where t is in seconds. What is its half-life?

16 Under ideal conditions the rate of *decrease* of atmospheric pressure with increasing altitude above sea level is proportional to the pressure. If the barometer reads 30 in. at sea level and 25 in. at 4000 ft., find the pressure at 20,000 ft.

17 A 5-lb. sample of radioactive material contains 2 lb. of radium-F, half-life 138.3 days, and 3 lb. of thorium-X, half-life 3.64 days. When will the sample contain equal amounts of radium-F and thorium-X?

18 Wood from the Lascaux caves in France was found in 1950 to contain 14.52% of the normal concentration of carbon-14. Estimate the date of the famous Lascaux cave paintings. [See Example 5, p. 288.]

6 FURTHER APPLICATIONS

Cooling

Suppose that an object has been heated to temperature T_0. At time $t = 0$ it is plunged into a cooling bath that is kept at the constant temperature B. Let $T(t)$ denote the temperature of the object at time t. It is known that the difference

$$T(t) - B$$

obeys the decay law; it decreases at a rate proportional to itself. Hence

$$T(t) - B = (T_0 - B)a^{-t} \qquad a > 1.$$

EXAMPLE 1   During heat treatment, a piece of steel at 400 °C is plunged into an oil bath kept at 25 °C. The steel cools to 200 °C in 68 sec. (a) Find the cooling law for the steel. (b) Find the time needed to cool to 50 °C.

Solution   (a)   By the decay law

$$T(t) - 25 = (400 - 25)a^{-t} = 375 \cdot a^{-t}.$$

Since $T(68) = 200$ we have

$$375 \cdot a^{-68} = T(68) - 25 = 200 - 25 = 175$$

$$a^{68} = \frac{375}{175} \approx 2.1429 \qquad a \approx (2.1429)^{1/68} \approx 1.0113.$$

Therefore

$$T(t) = 25 + 375 \times (1.0113)^{-t}.$$

(b)   We must solve $T(t) = 50$ for t:

$$25 + 375 \times (1.0113)^{-t} = 50 \qquad (1.0113)^{-t} = \frac{50 - 25}{375} \approx 0.066667$$

$$t \approx -\frac{\log(0.066667)}{\log(1.0113)} \approx 241 \text{ sec}.$$

Answer   (a) $T(t) = 25 + 375 \times (1.0113)^{-t}$ (b)   241 sec

Magnitudes of Stars

The magnitude of a star is a measure of its brightness (intensity). Bright stars have magnitudes around 0; the faintest star that can be seen with the naked eye has magnitude about 6. Thus magnitude is a decreasing function of brightness. Precisely, the **magnitude** of a star is

$$m = -2.5 \log \frac{B}{B_0},$$

where B is the brightness of the star and B_0 is a reference point chosen so that a star of brightness B_0 has magnitude 0. The constant 2.5 is chosen so that a star of brightness $\frac{1}{100}B_0$ has magnitude 5. (Check this.)

EXAMPLE 2   The brightest star in the sky is Sirius, with magnitude -1.5. How many times brighter is Sirius than a star (Eri A) of magnitude 4.4?

Solution   We want the ratio B_S/B, where B_S is the brightness of Sirius and B is the brightness of the other star. We are given

$$-2.5 \log \frac{B_S}{B_0} = -1.5 \quad \text{and} \quad -2.5 \log \frac{B}{B_0} = 4.4.$$

To find the relation between B_S and B, we must eliminate B_0 from these equations. To do so, rewrite the equations as

$$-2.5(\log B_S - \log B_0) = -1.5 \quad \text{and} \quad -2.5(\log B - \log B_0) = 4.4$$

and subtract:

$$-2.5(\log B_S - \log B) = -1.5 - 4.4 = -5.9$$

$$-2.5 \log \frac{B_S}{B} = -5.9 \qquad \log \frac{B_S}{B} = \frac{5.9}{2.5} = 2.36.$$

Therefore $\dfrac{B_S}{B} = 10^{2.36} \approx 229$.

Answer   229 times brighter

Semi-log and Log-log Paper

These are graph papers that use logarithmic scales on one or both axes. On semi-log paper (Fig. 1a) the horizontal axis is scaled as usual. But the vertical axis is scaled logarithmically. This means that the vertical distance from the horizontal axis to the level marked y is not y, but $\log y$.

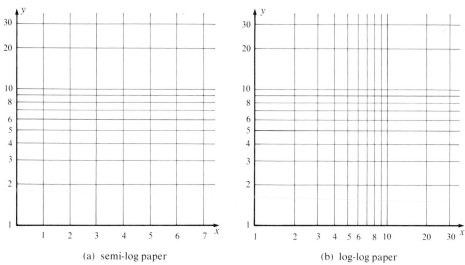

(a) semi-log paper   (b) log-log paper

Fig. 1

Imagine a rectangular grid superimposed on the semi-log paper, with both sets of axes coinciding and the same scale on the horizontal axes. Then a given point will have two sets of coordinates: (x, y) on the semi-log grid and (u, v) on the rectangular grid. These coordinates are related by

$$u = x \qquad v = \log y.$$

In log-log paper (Fig. 1b) both axes have logarithmic scales. The coordinates (x, y) are related to rectangular coordinates (u, v) by

$$u = \log x \qquad v = \log y.$$

EXAMPLE 3   Find the equation of the general straight line plotted on semi-log paper.

Solution   The rectangular equation of the line is

$$v = au + b.$$

But $u = x$ and $v = \log y$. So the equivalent equation in semi-log coordinates is

$$\log y = ax + b.$$

It can be written as

$$y = 10^{ax+b},$$

which is equivalent to

$$y = 10^b \cdot 10^{ax} = kc^x,$$

where $k = 10^b$ and $c = 10^a$ are positive constants.

Answer $y = kc^x$ where $k > 0$ and $c > 0$

Remark   Example 3 illustrates the convenience of semi-log paper for dealing with exponential functions. For example, suppose a scientist suspects an exponential relation between two quantities in an experiment. He can plot data points on semi-log paper. If they show an approximate straight line pattern, that is strong evidence for his suspicion.

Overflow

The largest number the calculator holds is $9.999 \cdots \times 10^{99}$. A calculation that results in a larger number at any stage puts the calculator into an error state called **overflow.** Similarly, numbers smaller than 10^{-99} yield **underflow** and the calculator probably displays 0. Numbers beyond the capacity of the calculator can sometimes be handled indirectly by using logarithms.

EXAMPLE 4   Estimate

(a) 2^{3000} (b) 3^{-5000}

Solution   (a)   The number is so large that

$2 \; \boxed{y^x} \; 3 \; 0 \; 0 \; 0 \; \boxed{=}$

results in **Error.** However, we can calculate

$$\log 2^{3000} = 3000 \cdot \log 2$$

by

2 | log | × | 3 0 0 0 | = |

resulting in 903.089988. Therefore

$$2^{3000} \approx 10^{903.089988} = 10^{0.089988} \times 10^{903}.$$

The factor $10^{0.089988}$ is within the capacity of the calculator; its value is approximately 1.2302, hence

$$2^{3000} \approx 1.23 \times 10^{903}.$$

(The last digits are doubtful so we round.) The following routine does the whole calculation:

2 | log | × | 3 0 0 0 | − | 9 0 3 | = | | 10ˣ | .

You must pause at | − | to read 903, the integer part of the logarithm, then subtract this integer part. Also you must multiply the result by 10^{903}.

(b)   Similarly, first we do

3 | log | × | 5 0 0 0 | +/− | | + | ,

read the integer part −2385, and finish with

2 3 8 6 | = | | 10ˣ | ,

to obtain 2.4759×10^{-2386}. (Why do we add 2386 rather than 2385; what would have happened had we added 2385 instead?)

Answer   (a) 1.23×10^{903} (b) 2.48×10^{-2386}

Equations Involving Exponentials or Logs

We apply the methods of Chapter 8, Sections 2 and 3, to equations involving exponential and/or logarithm functions.

EXAMPLE 5   Estimate to 5 places the solution of

$$e^x = 2 - x^3.$$

Solution   We start by graphing

$$y = e^x \quad \text{and} \quad y = 2 - x^3$$

on the same grid (Fig. 2, next page). Even a fairly crude figure indicates that there is only one solution, around $x = 0.6$.
   The equation is equivalent to $f(x) = 0$, where

$$f(x) = e^x + x^3 - 2.$$

(Note that $f(x)$ can be evaluated by

x | STO | | yˣ | 3 | − | 2 | + | | RCL | | eˣ | | = | .)

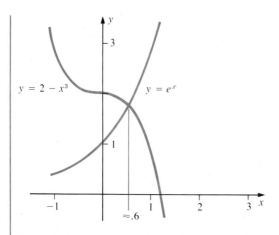

Fig. 2   To solve: $e^x = 2 - x^3$

We tabulate $f(x)$ to three places, starting at $x = 0.5$.

x	0.5	0.6
$f(x)$	-0.226	0.038

There is a solution between 0.5 and 0.6, close to 0.6. We estimate it graphically by linear interpolation (Fig. 3a). A good guess is $x \approx 0.586$, so let us start again at $x = 0.585$ and work forward by increments of 0.001:

x	0.585	0.586	0.587
$f(x)$	-0.00481	-0.00198	0.00085

The solution lies between 0.586 and 0.587. We interpolate again (Fig. 3b).

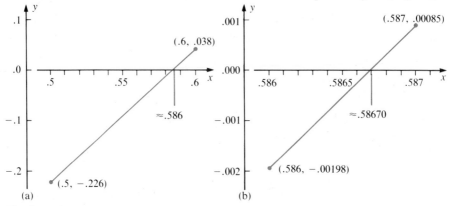

(a)   (b)

Fig. 3

The graph indicates $x \approx 0.58670$. We calculate values:

x	0.58669	0.58670	0.58671
$f(x)$	-3.12×10^{-5}	-2.89×10^{-6}	2.54×10^{-5}

The solution is between 0.58670 and 0.58671, closer to 0.58670.

Answer   0.58670

Exercises

1 A metal object in air at 20 °C cools from 85 °C to 40 °C in 13 minutes. How much longer will it take to cool to 21 °C?

2 In a superconductivity experiment, a piece of metal at 25 °C must be cooled to -268 °C. The metal is immersed in liquified gas held at -271 °C, and cools to -100 °C in 6.2 min. What is the total time required to reach -268 °C?

3 A teapot is cooling in a room at 68 °F. Its temperature decreases from 200 °F to 175 °F in the first minute. How much longer will it take to reach the desirable drinking temperature of 130 °F?

4 Molten lava at 900 °C is discharged from a volcano into air at 25 °C. If it cools to 780 °C in three hours, find its temperature after 35 hours.

5 How many times brighter is the sun, magnitude -26.8, than Sirius, magnitude -1.5?

6 Show that the difference in magnitude between two stars depends only on the *ratio* of their brightness.

7 Loudness of sounds is measured in decibels. The loudness in decibels of a sound of intensity I is

$$10 \log \frac{I}{I_0},$$

where I_0 is a standard very faint sound. The loudness of normal speech is 60 decibels. How many times more intense is the noise of a jet landing at 118 decibels?

8 If two sounds differ by one decibel, by what percentage does the intensity of the louder sound exceed the intensity of the softer one?

9 (a) Show that $y = 3x^2$ is a straight line in log-log coordinates.
 (b) What is the equation of the general straight line in log-log coordinates? See Fig. 1b.

10* The sequence $1, 1, 2, 3, 5, 8, 13, 21, 34. \cdots$ is the sequence of **Fibonacci numbers.** Each is the sum of the preceding two. If we denote the sequence by $F_1, F_2, F_3, \cdots$ then $F_1 = F_2 = 1$ and $F_{n+2} = F_{n+1} + F_n$. It is known that

$$F_n = \frac{1}{\sqrt{5}}(\alpha^n - \beta^n) \quad \text{where} \quad \alpha = \tfrac{1}{2}(1 + \sqrt{5}), \quad \beta = \tfrac{1}{2}(1 - \sqrt{5}).$$

Find the n for which F_n is closest to 10^{12}.

11* In calculus it is shown that for large values of n,

$$1 + \tfrac{1}{2} + \tfrac{1}{3} + \tfrac{1}{4} + \cdots + \frac{1}{n} \approx \ln n + \gamma,$$

where $\gamma \approx 0.5772$. Estimate how many terms are needed until the sum $1 + \tfrac{1}{2} + \tfrac{1}{3} + \cdots$ exceeds 10.

12* (cont.) Estimate the sum

$$\tfrac{1}{1001} + \tfrac{1}{1002} + \tfrac{1}{1003} + \cdots + \tfrac{1}{2000}.$$

13 In 1971 a computation proved that $2^{19937} - 1$ is a prime number (the largest prime known at that time). How large is it?

14 In 1978 it was announced that $2^{21701} - 1$ is a prime number. How large is it? The same question for $2^{23209} - 1$, the largest known prime as of April, 1979, and for $2^{44497} - 1$, the largest one as of June, 1979.

Estimate to 3 significant figures.

15 $(0.99)^{1,000,000}$

16 $\left(\dfrac{1}{\sqrt{2}}\right)^{10,000}$

Solve approximately to 6 places

17 $10^x = x + 1$ $x < 0$

18 $x \cdot 10^x = 100$

19 $\log x = -x$

20 $2 - \log x = x$

21 $e^x = 10x$ $x > 2$

22 $2^{-x} = x$

23 $2^x + x^3 = 0$

24 $2^x + 3^x = 15$.

25 $e^\theta = \sin 3\theta$ $-\frac{1}{2}\pi < \theta < 0$

26 $e^{\sin \theta} = \theta^2$.

Solve to 4 places by the method of iteration (page 181)

27 $x = 3 \ln x$ $x \approx 5$

28* $x = 3 \ln x$ $x \approx 2$.

7 RATES OF GROWTH [Optional]

In some applications it is important to understand the growth rate as $x \longrightarrow \infty$ of a^x, and the decay rate of a^{-x}, where $a > 1$. Our standard of comparison is powers of 10. This is psychological; we feel in our hearts that a billion (10^9) is pretty large, but we are not so sure at first sight if, say, 2^{25} is or isn't so large.

We'll first make some rough estimates of the growth rate of 2^x as $x \longrightarrow \infty$, starting with

$$2^{10} = 1024 > 10^3.$$

From this inequality follow, for instance,

$$2^{20} = (2^{10})^2 > (10^3)^2 = 10^6$$

$$2^{30} = (2^{10})^3 > (10^3)^3 = 10^9$$

$$2^{100} = (2^{10})^{10} > (10^3)^{10} = 10^{30} \qquad \text{etc.}$$

In general,

$$2^{10n} > 10^{3n}.$$

This gives a pretty good idea of the extremely rapid growth of 2^x. Actually, we can also express the inequality in the form $2^x > 10^{0.3x}$, another way of looking at it.

We are convinced that 2^x grows rapidly. But so do x^5 and x^{10}. How does 2^x stack up against these functions for rapid growth? Just for fun let's compare 2^x and x^{10} for some large values of x, using our estimates for 2^x. First we'll try $x = 100$:

$$2^{100} > 10^{30} \quad \text{and} \quad 100^{10} = (10^2)^{10} = 10^{20},$$

$$\text{hence} \quad \frac{2^{100}}{100^{10}} > \frac{10^{30}}{10^{20}} = 10^{10}.$$

Thus if $x = 100$, then 2^x is more than 10^{10} times as large as x^{10}. Next, we'll try $x = 1000$:

$$2^{1000} = 2^{10 \times 100} > 10^{3 \times 100} = 10^{300} \quad \text{and} \quad 1000^{10} = (10^3)^{10} = 10^{30},$$

$$\text{hence} \quad \frac{2^{1000}}{1000^{10}} > \frac{10^{300}}{10^{30}} = 10^{270}.$$

Thus if $x = 1000$, then 2^x is more than 10^{270} times as large as x^{10}. Wow! Even though $2^x < x^{10}$ for small values of x (up to about 58.77), 2^x is overwhelmingly larger than x^{10} as $x \longrightarrow \infty$.

Similar reasoning applies to any power x^n, not just x^{10}, and to any exponential function a^x with $a > 1$, not just 2^x.

Growth of a^x Let $a > 1$ and let n be any whole number. Then

$$\frac{a^x}{x^n} \longrightarrow \infty \quad \text{as} \quad x \longrightarrow \infty.$$

In words, the exponential function a^x grows more rapidly than any power function x^n.

Example Again just for fun, let's try to find an x such that $(1.01)^x > x^{100}$. Since $(1.01)^x$ doesn't grow very rapidly at first:

$$(1.01)^5 \approx 1.051 \qquad (1.01)^{10} \approx 1.105 \qquad (1.01)^{25} \approx 1.282$$

this seems difficult. It isn't actually once you get a start. Play around on your calculator; you soon will find an x such that $(1.01)^x > 10$. For instance, $(1.01)^{250} \approx 12.0 > 10$. This is the start we need. Now we have for instance

$$(1.01)^{250,000} > 10^{1000}$$

But

$$(250,000)^{100} < (1,000,000)^{100} = (10^6)^{100} = 10^{600},$$

so

$$\frac{(1.01)^{250,000}}{(250,000)^{100}} > \frac{10^{1000 \cdot}}{10^{600}} = 10^{400}.$$

Therefore for $x = 250,000$, the number $(1.01)^x$ is more than 10^{400} times as large as x^{100}. Convinced?

Note Let us try briefly to see *why* $2^x/x^{10} \longrightarrow \infty$ as $x \longrightarrow \infty$. To do so, we analyze what happens when x increases to $x + 1$. The numerator of $2^x/x^{10}$ increases from 2^x to $2^{x+1} = 2 \cdot 2^x$, that is, it doubles. The denominator increases from x^{10} to $(x + 1)^{10}$. But

$$\frac{(x + 1)^{10}}{x^{10}} = \left(1 + \frac{1}{x}\right)^{10}$$

is little larger than 1 when x is large. For instance $(\frac{1001}{1000})^{10} \approx 1.01$. So while the numerator of $2^x/x^{10}$ is doubling, the denominator is increasing by a tiny percentage; therefore the fraction itself is practically doubling.

Now let's look at the decay of a^{-x} as $x \longrightarrow \infty$. If $a > 1$, then $a^{-x} \longrightarrow 0+$ as $x \longrightarrow \infty$ because

$$a^{-x} = \frac{1}{a^x} \quad \text{and} \quad a^x \longrightarrow \infty.$$

How rapidly does a^{-x} decay to 0? More rapidly than any function x^{-n}, where $n > 0$.

Decay of a^{-x}   Let $a > 1$ and $n > 0$. Then

$$\frac{a^{-x}}{x^{-n}} = x^n a^{-x} \longrightarrow 0+ \quad \text{as} \quad x \longrightarrow \infty.$$

As an example let us compare 2^{-x} and x^{-10} for $x = 1000$:

$$\frac{2^{-1000}}{(1000)^{-10}} = \frac{(1000)^{10}}{2^{1000}} < \frac{1}{10^{270}} = 10^{-270}.$$

The ratio is indeed small.

Instead of using a^{-x} where $a > 1$, we can use b^x where $0 < b < 1$; we just take $b = a^{-1}$.

EXAMPLE 1   Estimate an x such that

(a) $(1.001)^x > 10$ (b) $(0.999)^x < 0.1$.

Solution   (a)   By calculator, $(1.001)^{10} > 1.01$. But $(1.01)^{250} > 10$ was observed earlier, hence

$$(1.001)^{2500} = [(1.001)^{10}]^{250} > (1.01)^{250} > 10.$$

(b) $\dfrac{1}{0.999} = 1.001001001 \cdots > 1.001$,

therefore $\left(\dfrac{1}{0.999}\right)^{2500} > (1.001)^{2500} > 10$. It follows that

$$(0.999)^{2500} < 0.1.$$

Answer   (a)   2500 [actually $(1.001)^{2304} > 10 > (1.001)^{2303}$]

(b)   2500 [actually $(0.999)^{2302} < 0.1 < (0.999)^{2301}$]

EXAMPLE 2   Estimate an x such that

(a) $(1.001)^x > x^{100}$ (b) $(0.999)^x < x^{-1000}$.

Solution   (a)   From Example 1, $(1.001)^{2500} > 10$. It follows that

$$(1.001)^{2,500,000} > 10^{1000}.$$

But $(2,500,000)^{100} < (10^7)^{100} = 10^{700} < 10^{1000}$,

so $x = 2.5 \times 10^6$ is large enough.

(b)   From Example 1, $(0.999)^{2500} < 0.1 = 10^{-1}$. It follows that

$$(0.999)^{25,000,000} < 10^{-10,000}.$$

But $(25,000,000)^{-1000} > (10^8)^{-1000} = 10^{-8000} > 10^{-10,000}$,

so $x = 2.5 \times 10^7$ is large enough.

Answer   (a) 2.5×10^6 (b) 2.5×10^7

Remark We have emphasized the rapid growth as $x \longrightarrow \infty$ of a^x, where $a > 1$. Does anything grow even more rapidly? Sure, b^x, where $b > a$, because the ratio

$$\frac{b^x}{a^x} = \left(\frac{b}{a}\right)^x$$

is itself an exponential function with base $b/a > 1$, hence grows rapidly.

Exercises

1 Compare the values of 2^x and x^{10} for $x = 10, 20, 30, \cdots, 100$ by computing their ratio.

2 Compare the values of 3^x and x^3 for $x = 1, 2, 3, \cdots, 10$ by computing their ratio.

3 Compare the values of 2^{-x} and x^{-2} for $x = 1, 2, 3, \cdots, 10$ by computing their ratio.

4 Compare the values of 3^{-x} and x^{-4} for $x = 1, 2, 3, \cdots, 10$ by computing their ratio.

5 Find a value of n for which $2^n > 10^{50}$.

6 Find a value of x for which $2^x > x^{100}$.

7 Find a value of x for which $(1.01)^x > x^{10}$.

8 Find a value of x for which $(1.0001)^x > x^5$.

9 Compare 2^x and x^{100} for $x = 10^3$ and $x = 10^6$ by examining their ratio.

10 Compare $(1.1)^x$ and x^{10} for $x = 10^3$ and $x = 10^6$ by examining their ratio. Use $2 < (1.1)^{10}$.

11 Give convincing numerical evidence that

$$\frac{\log_{10} x}{x} \longrightarrow 0 \quad \text{as} \quad x \longrightarrow \infty.$$

12 $\log_2 x$ grows slowly as $x \longrightarrow \infty$. So does $\sqrt[100]{x}$. Which grows more slowly in the long run?

13 $\log_{10} x \longrightarrow -\infty$ as $x \longrightarrow 0+$. How about $x \log x$ as $x \longrightarrow 0+$? [As usual, $\log = \log_{10}$.]

14 (cont.) The same question for $\sqrt[10]{x} \log x$.

15 Which is larger for $x > 1$, $\log_2 x$ or $\log_3 x$? [*Hint* Compare the graphs of the corresponding exponential functions.]

16 (cont.) The same question for $0 < x < 1$.

17 Find the smallest number x on your calculator so that $x \boxed{\log}\,\boxed{\log}\,\boxed{\log}\,\boxed{\log}$ does not result in **Error.**

18 Find the largest number x on your calculator so that $x \boxed{e^x}\,\boxed{e^x}\,\boxed{e^x}\,\boxed{e^x}$ does not result in **Error.**

REVIEW EXERCISES

1 Express $[3^5(\frac{1}{3})^2]^2/[27\cdot(\frac{1}{9})^2]$ as a power of 3.

2 Express $(50,000,000,000)(0.00000\ 00000\ 002)$ in scientific notation.

3 Estimate $(356.187 \times 10^5)^{11}$ on a calculator.

4 Express $3^{4/5}\sqrt[7]{9}/\sqrt[4]{(27)^6}$ as a power of 3.

Graph

5 $y = (0.8)^x$

6 $y = (1.2)^x$.

Find the inverse function: give its domain and range

7 $y = 1 + 2^x$

8* $y = \frac{1}{2}(2^x - 2^{-x})$.

Graph

9 $y = \log_{10}(100x)$

10 $y = \log_2(x - 1)$

11 $y = \log_{10}\left(1 + \dfrac{1}{x}\right)$

12 $y = \log_{10}\left(\dfrac{1}{x} - 1\right)$.

Calculate

13 $\log 8 + \log(0.125)$

14 $(\log 3.000)/(\ln 3.000)$.

Solve to 5 significant figures

15 $(1.1432)^x = 17.611$

16 $\log(5^x + 2) = 2$.

How accurate are the approximations

17 $\log \ln 2 \approx -\dfrac{1}{2\pi}$

18 $\dfrac{4}{\sqrt{t}} \approx \pi$ where $t = \frac{1}{2}(1 + \sqrt{5})$?

19 How long will it take a bacteria colony to triple if it doubles in 7 hr 35 min?

20 A radioactive chemical loses $\frac{1}{10}$ of its original mass in $5\frac{1}{2}$ hours. Give its decay law.

Which grows faster as $x \longrightarrow \infty$

21 $2^{x+\log_2 x}$ or 3^x

22 $\log_3(x + 3^x)$ or x?

Solve to 6 places

23 $3^x - 2^x = 2$

24 $\cos x = e^{2x}$ $-\pi < x < 0$.

APPENDIX

SAMPLE PROGRAMS FOR SOLVING TRIANGLES

In this appendix we present a selection of triangle solving programs in three common languages: the "machine" language of the Texas Instruments TI-58/59 series, BASIC, and PASCAL.* We choose these languages because (1) the TI-58 and TI-59 are probably the most widely used programmable calculators; (2) BASIC is almost a universal language for microcomputers; and (3) PASCAL is rapidly becoming the first language taught in introductory courses on programming and data structures. Also, the popular APPLE microcomputers have PASCAL available, and other makes will undoubtedly follow this trend.

Here is a rundown of what follows:

A. Programs for SSS (See p. 124 of the text.)

A1. A TI-58/59 program. There are three columns in the program printout. The first column is the line number. The third column is an abbreviated step, and the second is a number that tells what key to press for the step (by its coordinates).

Following the program is a flowchart explaining the program line-by-line. Next are instructions for running the program, followed by examples.

A2. BASIC program for SSS. This program has a nicely formatted output and includes tests for error conditions. The REM's (remarks) explain what is going on. The computations at lines 350, 410, and 460 are each split into two steps to avoid ugly line breaks in equations (the APPLE we used has a *very* short print line in BASIC).

Note in this and the following PASCAL program the necessity to develop a routine for the arc cosine function. Most languages do not have arc sin and arc cos built in, only arc tan. The necessary relations are

$$(\text{arc cos } x)^\circ = \begin{cases} 180^\circ + \dfrac{180^\circ}{\pi} \text{ arc tan}\left(\dfrac{\sqrt{1 - x^2}}{x}\right) \\ 90^\circ \\ \dfrac{180^\circ}{\pi} \text{ arc tan}\left(\dfrac{\sqrt{1 - x^2}}{x}\right) \end{cases} \quad \text{if} \begin{cases} -1 \leq x < 0 \\ x = 0 \\ 0 < x \leq 1 \end{cases}$$

See lines 350 to 480 in the program.

Examples of the BASIC program follow.

---

* The name should really be written "Pascal" rather than "PASCAL" since it, unlike BASIC, FORTRAN, etc., is not an acronym.

A3. PASCAL program for SSS. Note carefully how the program is structured into functions and procedures.

The boolean variable FLAG, initially set TRUE, becomes FALSE whenever the program discovers an error condition in the data, and it controls the output.

Examples of the PASCAL program follow.

B. A program for SAA

This TI-58/59 program is very long because it includes a fancy output routine, showing off the alpha-numeric capability of the TI printer.

C. Programs for SSA

These programs are closely modeled on the flowcharts on pp. 131 and 135 of the text.

C1. TI-58/59 program and examples.

C2. PASCAL program and examples.

Remark 1   The BASIC and PASCAL programs were run on an APPLE II PLUS with PAPER TIGER printer. The PASCAL programs had to be modified slightly to direct output to the printer rather than the screen.

Remark 2   The TI-58/59 Master Library module includes two programs (ML-11 and ML-12) for triangle solving, so our programs are unnecessary in a way. However, it is worthwhile writing your own program in each case to get the input and output features you like. Also, you can shorten the running time by carefully programming around the (bad) feature of the TI's that they search for labels from the *beginning* of the program. Finally, the program in ML-11 for SSA overlooks the possibility of two solutions (ambiguous case)!

Remark 3   APPLE PASCAL uses ATAN instead of the standard PASCAL identifier ARCTAN.

Remark 4   The statement USES TRANSCEND is an APPLE PASCAL call to a library package containing the standard transcendental functions SQRT, SIN, ATAN, etc. It is not part of standard or UCSD PASCAL.

A1 TEXAS INSTRUMENTS TI-58/59 PROGRAM FOR SSS

Input Procedure

000	76	LBL
001	11	A
002	42	STO
003	01	01
004	92	RTN
005	76	LBL
006	12	B
007	42	STO
008	02	02
009	92	RTN
010	76	LBL
011	13	C
012	42	STO
013	03	03
014	92	RTN

Set-up for Computation

015	76	LBL
016	17	B'
017	43	RCL
018	03	03
019	42	STO
020	09	09
021	32	X⧧T
022	43	RCL
023	01	01
024	42	STO
025	07	07
026	42	STO
027	03	03
028	43	RCL
029	02	02
030	42	STO
031	08	08
032	42	STO
033	01	01
034	32	X⧧T
035	42	STO
036	02	02
037	71	SBR
038	18	C'
039	92	RTN

Law of Cosines Subroutine

040	76	LBL
041	18	C'
042	53	(
043	43	RCL
044	08	08
045	33	X²
046	85	+
047	43	RCL
048	09	09
049	33	X²
050	75	−
051	43	RCL
052	07	07
053	33	X²
054	54	)
055	55	÷
056	53	(
057	02	2
058	65	×
059	43	RCL
060	08	08
061	65	×
062	43	RCL
063	09	09
064	54	)
065	95	=
066	22	INV
067	39	COS
068	92	RTN

Main Program and Output

069	76	LBL
070	15	E
071	22	INV
072	58	FIX
073	43	RCL
074	01	01
075	99	PRT
076	43	RCL
077	02	02
078	99	PRT
079	43	RCL
080	03	03
081	99	PRT
082	98	ADV
083	22	INV
084	52	EE
085	58	FIX
086	02	02
087	71	SBR
088	17	B'
089	42	STO
090	04	04
091	99	PRT
092	71	SBR
093	17	B'
094	42	STO
095	05	05
096	99	PRT
097	71	SBR
098	17	B'
099	42	STO
100	06	06
101	99	PRT
102	98	ADV
103	98	ADV
104	91	R/S

Examples of the TI-58/59 program for SSS

a	4.578	
b	5.822	
c	5.197	
α	48.69	
β	72.80	
γ	58.51	

a	5.	
b	12.	
c	13.	
α	22.62	
β	67.38	
γ	90.00	

a	7.5486	11
b	6.5883	11
c	8.0539	11
α	61.10	
β	49.83	
γ	69.08	

Flowchart for the Program

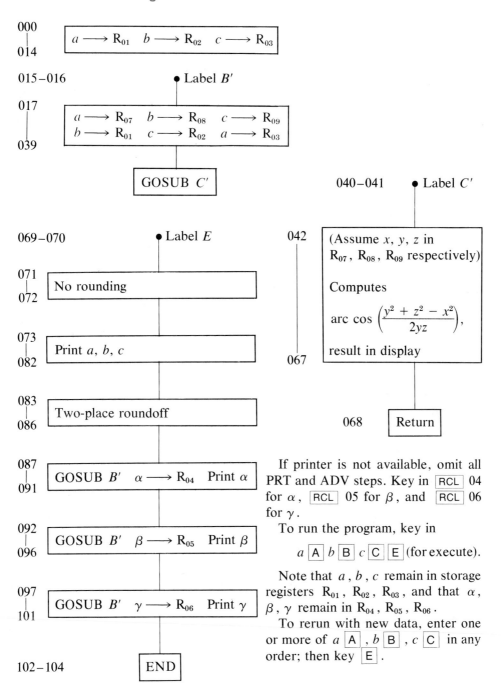

000

014

$$a \longrightarrow R_{01} \quad b \longrightarrow R_{02} \quad c \longrightarrow R_{03}$$

015–016 • Label B'

017

039

$$a \longrightarrow R_{07} \quad b \longrightarrow R_{08} \quad c \longrightarrow R_{09}$$
$$b \longrightarrow R_{01} \quad c \longrightarrow R_{02} \quad a \longrightarrow R_{03}$$

GOSUB C'

040–041 • Label C'

069–070 • Label E

042

(Assume x, y, z in R_{07}, R_{08}, R_{09} respectively)

Computes

$$\text{arc cos} \left(\frac{y^2 + z^2 - x^2}{2yz} \right),$$

067 result in display

071

072

No rounding

073

082

Print a, b, c

068 Return

083

086

Two-place roundoff

087

091

GOSUB B' $\alpha \longrightarrow R_{04}$ Print α

092

096

GOSUB B' $\beta \longrightarrow R_{05}$ Print β

097

101

GOSUB B' $\gamma \longrightarrow R_{06}$ Print γ

102–104 END

If printer is not available, omit all PRT and ADV steps. Key in $\boxed{\text{RCL}}$ 04 for α, $\boxed{\text{RCL}}$ 05 for β, and $\boxed{\text{RCL}}$ 06 for γ.

To run the program, key in

$a \boxed{A} \ b \boxed{B} \ c \boxed{C} \boxed{E}$ (for execute).

Note that a, b, c remain in storage registers R_{01}, R_{02}, R_{03}, and that α, β, γ remain in R_{04}, R_{05}, R_{06}.

To rerun with new data, enter one or more of $a \boxed{A}$, $b \boxed{B}$, $c \boxed{C}$ in any order; then key $\boxed{E}$.

A2 BASIC PROGRAM FOR SSS

```
100  REM   PROGRAM TO SOLVE A
110  REM    TRIANGLE GIVEN ITS
120  REM    SIDES A, B, AND C.
130  PRINT
140  PRINT "ENTER SIDES A, B, AND
       C"
150  INPUT A,B,C
160  IF A >  = B + C THEN 580
170  IF C + A <  = B THEN 600
180  IF C >  = A + B THEN 620
190  GOSUB 350
200  PRINT "ALPHA = "T" DEG"
210  GOSUB 510
220  GOSUB 350
230  PRINT " BETA = "T" DEG"
240  GOSUB 510
250  GOSUB 350
260  PRINT "GAMMA = "T" DEG"
270  PRINT
280  PRINT "PRESS 1 TO CONTINUE"
290  PRINT "PRESS 0 TO QUIT"
300  INPUT Q
310  IF Q = 1 THEN 130
320  GOTO 640
330  REM   SUB TO COMPUTE
340  REM    ONE ANGLE T(HETA:
350  X = B * B + C * C - A * A
360  X = X / (2 * B * C)
370  K = 45 /  ATN (1)
380  IF X < 0 THEN 410
390  IF X = 0 THEN 440
400  IF X > 0 THEN 460
410  W =  SQR (1 - X * X)
420  T = 180 + K *  ATN (W / X)
430  GOTO 480
440  T = 90
450  GOTO 480
460  W =  SQR (1 - X * X)
470  T = K *  ATN (W / X)
480  RETURN
490  REM   SUB TO PERMUTE:
500  REM    B-->A   C-->B   A-->C
510  D = A
520  A = B
530  B = C
540  C = D
550  RETURN
560  REM   ERROR SUBS IF TRIANGLE
570  REM   INEQUALITY FAILS:
580  PRINT "A TOO LARGE"
590  GOTO 130
600  PRINT "B TOO LARGE"
610  GOTO 130
620  PRINT "C TOO LARGE"
630  GOTO 130
640  END
```

Examples of the BASIC Program for SSS

```
ENTER SIDES A, B, AND C
?4.578
??5.822
??5.197
ALPHA = 48.6909817 DEG
 BETA = 72.7996012 DEG
GAMMA = 58.5094172 DEG

PRESS 1 TO CONTINUE
PRESS 0 TO QUIT
?1

ENTER SIDES A, B, AND C
?5
??12
??13
ALPHA = 22.619865 DEG
 BETA = 67.3801351 DEG
GAMMA = 90 DEG

PRESS 1 TO CONTINUE
PRESS 0 TO QUIT
?1


ENTER SIDES A, B, AND C
??7.5486E11
??6.5883E11
??8.0539E11
ALPHA = 61.0986095 DEG
 BETA = 49.825348 DEG
GAMMA = 69.0760426 DEG

PRESS 1 TO CONTINUE
PRESS 0 TO QUIT
? 1
```

```
ENTER SIDES A, B, AND C
?2
??4
??6
C TOO LARGE


ENTER SIDES A, B, AND C
?7
??4
??2
A TOO LARGE

ENTER SIDES A, B, AND C
?6
??8
??1
B TOO LARGE

ENTER SIDES A, B, AND C
?10
??10
??19.9999
ALPHA = .18119016 DEG
 BETA = .18119016 DEG
GAMMA = 179.637626 DEG

PRESS 1 TO CONTINUE
PRESS 0 TO QUIT
?0
```

A3 PASCAL PROGRAM FOR SSS

```
PROGRAM SSS(INPUT, OUTPUT);
(* INPUT: SIDES A,B,C; OUTPUT ANGLES ALPHA, BETA, GAMMA *)
USES TRANSCEN;

VAR A,B,C,ALPHA,BETA,GAMMA: REAL;
    FLAG: BOOLEAN;

FUNCTION F(X,Y,Z: REAL): REAL;
BEGIN F:=(Y*Y+Z*Z-X*X)/(2*Y*Z) END;

FUNCTION ACOS(X: REAL): REAL;
VAR Y,PI,K: REAL;
BEGIN
   PI:=4*ATAN(1); K:=180/PI;
   IF X=0 THEN ACOS:=90 ELSE Y:=K*ATAN(SQRT(1-X*X)/X);
   IF X>0 THEN ACOS:=Y;
   IF X<0 THEN ACOS:=Y+180
END;

PROCEDURE COMPUTE(A,B,C: REAL; VAR ANG: REAL);
VAR W: REAL;
BEGIN
   W:=F(A,B,C);
   IF W<=-1 THEN
   BEGIN
     WRITELN('BAD DATA: ',A,'>=',B,'+',C);
     FLAG:=FALSE
   END
   ELSE
   IF W>=1 THEN FLAG:=FALSE
   ELSE ANG:=ACOS(W)
END;

PROCEDURE PRINTOUT;
BEGIN
WRITELN('A= ',A:10,'   ALPHA=',ALPHA:7:2,' DEG');
WRITELN('B= ',B:10,'   BETA =',BETA:7:2,' DEG');
WRITELN('C= ',C:10,'   GAMMA=',GAMMA:7:2,' DEG');
WRITELN('CHECK: ',ALPHA:7:2,'+',BETA:7:2,'+',
          GAMMA:7:2,'=', ALPHA+BETA+GAMMA:7:2);
END;

BEGIN
WRITELN('GIVEN THREE SIDES, TO FIND THE ANGLES:');
WHILE NOT EOF DO
   BEGIN
   FLAG:=TRUE;
   WRITE('ENTER A '); READ(A); WRITELN;
   WRITE('ENTER B '); READ(B); WRITELN;
   WRITE('ENTER C '); READ(C); WRITELN;
   IF (A<=0) OR (B<=0) OR (C<=0) THEN
     BEGIN WRITELN('BAD DATA'); FLAG:=FALSE END
     ELSE
       BEGIN
         COMPUTE(A,B,C,ALPHA);
         COMPUTE(B,C,A,BETA);
         COMPUTE(C,A,B,GAMMA);
         IF FLAG THEN PRINTOUT
       END;
   WRITELN('PRESS <CR> TO CONTINUE OR <ETX> TO QUIT');
   READLN
   END (*WHILE*)
END.
```

Examples of the PASCAL Program for SSS

```
GIVEN THREE SIDES, TO FIND THE ANGLES:
ENTER A
ENTER B
ENTER C
A=     1.00000    ALPHA=   60.00 DEG
B=     1.00000    BETA =   60.00 DEG
C=     1.00000    GAMMA=   60.00 DEG
CHECK:   60.00+  60.00+   60.00= 180.00
PRESS <CR> TO CONTINUE OR <ETX> TO QUIT


ENTER A
ENTER B
ENTER C
A=    4.57800 B=    5.82200 C=    5.19700

A=     4.57800    ALPHA=   48.69 DEG
B=     5.82200    BETA =   72.80 DEG
C=     5.19700    GAMMA=   58.51 DEG
CHECK:    48.69+  72.80+   58.51= 180.00
PRESS <CR> TO CONTINUE OR <ETX> TO QUIT


ENTER A
ENTER B
ENTER C
A=    5.00000 B= 1.20000E1 C= 1.30000E1

A=     5.00000    ALPHA=   22.62 DEG
B=  1.20000E1     BETA =   67.38 DEG
C=  1.30000E1     GAMMA=   90.00 DEG
CHECK:   22.62+  67.38+   90.00= 180.00
PRESS <CR> TO CONTINUE OR <ETX> TO QUIT


ENTER A
ENTER B
ENTER C
A= 7.54860E11 B= 6.58830E11 C= 8.05390E11

A=  7.54860E11    ALPHA=   61.10 DEG
B=  6.58830E11    BETA =   49.83 DEG
C=  8.05390E11    GAMMA=   69.08 DEG
CHECK:   61.10+  49.83+   69.08= 180.00
PRESS <CR> TO CONTINUE OR <ETX> TO QUIT
```

B TEXAS INSTRUMENTS TI-58/59 PROGRAM FOR SAA

a Input

000	42	STO
001	01	01
002	43	RCL
003	07	07
004	22	INV
005	58	FIX
006	69	OP
007	04	04
008	43	RCL
009	01	01
010	58	FIX
011	03	03
012	69	OP
013	06	06
014	43	RCL
015	10	10
016	22	INV
017	58	FIX
018	69	OP
019	04	04
020	91	R/S

α Input

021	42	STO
022	04	04
023	22	INV
024	52	EE
025	58	FIX
026	02	02
027	69	OP
028	06	06
029	43	RCL
030	11	11
031	22	INV
032	58	FIX
033	69	OP
034	04	04
035	91	R/S

β Input

036	42	STO
037	05	05
038	22	INV
039	52	EE
040	58	FIX
041	02	02
042	69	OP
043	06	06
044	43	RCL
045	08	08
046	22	INV
047	58	FIX
048	69	OP
049	04	04
050	58	FIX
051	03	03

b Comp. and Output

052	43	RCL
053	01	01
054	55	÷
055	43	RCL
056	04	04
057	38	SIN
058	95	=
059	42	STO
060	00	00
061	65	×
062	43	RCL
063	05	05
064	38	SIN
065	95	=
066	42	STO
067	02	02
068	98	ADV
069	69	OP
070	06	06

c Comp. and Output

071	01	1
072	08	8
073	00	0
074	75	−
075	43	RCL
076	04	04
077	75	−
078	43	RCL
079	05	05
080	95	=
081	42	STO
082	06	06
083	38	SIN
084	65	×
085	43	RCL
086	00	00
087	95	=
088	42	STO
089	03	03
090	43	RCL
091	09	09
092	22	INV
093	58	FIX
094	69	OP
095	04	04
096	43	RCL
097	03	03
098	58	FIX
099	03	03
100	69	OP
101	06	06

Program continued on p. 311

γ Comp. and Output

102	43	RCL
103	12	12
104	22	INV
105	58	FIX
106	69	OP
107	04	04
108	22	INV
109	52	EE
110	58	FIX
111	02	02
112	43	RCL
113	06	06
114	69	OP
115	06	06
116	98	ADV
117	91	R/S
118	81	RST

Fancy Output Initialization

119	76	LBL
120	11	A
121	22	INV
122	58	FIX
123	01	1
124	03	3
125	00	0
126	00	0
127	00	0
128	00	0
129	00	0
130	00	0
131	42	STO
132	07	07
133	01	1
134	04	4
135	00	0
136	00	0
137	00	0
138	00	0
139	00	0
140	00	0
141	42	STO
142	08	08
143	01	1
144	05	5
145	00	0
146	00	0
147	00	0
148	00	0
149	00	0
150	00	0
151	42	STO
152	09	09

153	01	1
154	03	3
155	02	2
156	07	7
157	03	3
158	03	3
159	02	2
160	03	3
161	42	STO
162	10	10
163	01	1
164	04	4
165	01	1
166	07	7
167	03	3
168	07	7
169	01	1
170	03	3
171	42	STO
172	11	11
173	02	2
174	02	2
175	01	1
176	03	3
177	03	3
178	00	0
179	03	3
180	00	0
181	42	STO
182	12	12
183	92	RTN
184	81	RST

To run the program, key $\boxed{A}$ (the first time only) to initialize the output format. Then key

(*) a $\boxed{R/S}$ α $\boxed{R/S}$ β $\boxed{R/S}$.

To run with new data, repeat (*), all three steps.

Examples of the TI-58/59 Program for SAA

1.000	A
30.00	ALPH
40.00	BETA
1.286	B
1.879	C
110.00	GAMM

4.257	A
32.56	ALPH
127.48	BETA
6.277	B
2.700	C
19.96	GAMM

7.500	20	A
35.00		ALPH
45.00		BETA
9.246	20	B
1.288	21	C
100.00		GAMM

C1 TEXAS INSTRUMENTS TI-58/59 PROGRAM FOR SSA

Data Input

000	42	STO
001	01	01
002	99	PRT
003	91	R/S
004	42	STO
005	03	03
006	99	PRT
007	22	INV
008	52	EE
009	91	R/S
010	42	STO
011	06	06
012	99	PRT
013	98	ADV

First Sol. Output

052	99	PRT
053	22	INV
054	52	EE
055	58	FIX
056	02	02
057	43	RCL
058	04	04
059	99	PRT
060	43	RCL
061	05	05
062	99	PRT
063	22	INV
064	58	FIX
065	98	ADV

Second Sol. Output

092	99	PRT
093	22	INV
094	52	EE
095	58	FIX
096	02	02
097	43	RCL
098	04	04
099	99	PRT
100	43	RCL
101	05	05
102	99	PRT
103	22	INV
104	58	FIX
105	98	ADV
106	98	ADV
107	91	R/S
108	81	RST

First Sol. Computation

014	43	RCL
015	03	03
016	55	÷
017	43	RCL
018	06	06
019	38	SIN
020	95	=
021	42	STO
022	00	00
023	43	RCL
024	01	01
025	55	÷
026	43	RCL
027	00	00
028	95	=
029	22	INV
030	38	SIN
031	42	STO
032	04	04
033	01	1
034	08	8
035	00	0
036	75	−
037	43	RCL
038	04	04
039	75	−
040	43	RCL
041	06	06
042	95	=
043	42	STO
044	05	05
045	38	SIN
046	65	×
047	43	RCL
048	00	00
049	95	=
050	42	STO
051	02	02

Second Sol. Computation

066	43	RCL
067	04	04
068	75	−
069	43	RCL
070	06	06
071	95	=
072	42	STO
073	05	05
074	01	1
075	08	8
076	00	0
077	75	−
078	43	RCL
079	04	04
080	95	=
081	42	STO
082	04	04
083	43	RCL
084	05	05
085	38	SIN
086	65	×
087	43	RCL
088	00	00
089	95	=
090	42	STO
091	02	02

To run program, key

a R/S c R/S γ R/S .

Printout:

a

c

γ

b_1

α_1

β_1

b_2

α_2

β_2

To rerun, enter all new a, c, γ.

Examples of the TI-58/59 Programs for SSA

4.783	a
3.244	c
22.68	γ

7.08190456	b_1
34.65	α_1
122.67	β_1

1.74438287	b_2
145.35	α_2
11.97	β_2

7.829 11	a
4.578 11	c
18.56	γ

1.1262167 12	b_1
32.98	α_1
128.46	β_1

3.5814739 11	b_2
147.02	α_2
14.42	β_2

1.	a
1.	c
45.	γ

1.414213562	b
45.00	α
90.00	β

$$-9.8730732-13$$ $$135.00$$ $$0.00$$ } no second solution

8.	a
3.	c
75.	γ

3.033094701	?	
2.58	?	
102.42	?	no
-2.960843132	?	solution
177.42	?	
-72.42	?	

7.	a
9.	c
54.	γ

11.10942948	b
38.99	α
87.01	β

$$-2.880435945$$ $$141.01$$ $$-15.01$$ } no second solution

Memory allocation:

$a \longrightarrow R_{01}$		$\alpha_1 \longrightarrow R_{04}$	
$b_1 \longrightarrow R_{02}$		$\beta_1 \longrightarrow R_{05}$	
$c \longrightarrow R_{03}$		$\gamma \longrightarrow R_{06}$	

Second solution:

$b_2 \longrightarrow R_{02}$	
$\alpha_2 \longrightarrow R_{04}$	
$\beta_2 \longrightarrow R_{05}$	

C2 PASCAL PROGRAM FOR SSA

```
PROGRAM SSA(INPUT, OUTPUT);
(* INPUT: SIDES A,C AND ANGLE GAMMA; OUTPUT: REMAINING 3 PARTS *)

USES TRANSCEND;

VAR A,B,C,H,P,K,TA,TB,TC: REAL;

FUNCTION ASIN(T: REAL): REAL; (* ARC SIN IN DEGREES *)
  BEGIN ASIN:=ATAN(T/SQRT(1-T*T))/K END;

PROCEDURE PRINTOUT;
  BEGIN
  WRITELN('A=',A:10,'    ALPHA=',TA:7:2,'  DEG');
  WRITELN('B=',B:10,'     BETA=',TB:7:2,'  DEG');
  WRITELN('C=',C:10,'    GAMMA=',TC:7:2,'  DEG');
  END;

BEGIN (* MAIN *)
K:=ATAN(1)/45;
WHILE NOT EOF DO
  BEGIN
  WRITELN;
  WRITELN('ENTER A,C,GAMMA');
  READLN(A,C,TC);
  WRITELN('A=',A:10,' C=',C:10,' GAMMA=',TC:6:2);
  H:=A*SIN(TC*K);
  IF C<H THEN WRITELN('NO SOLUTIONS')
    ELSE IF C=H THEN
    BEGIN
    TA:=90; TB:=90-TC; B:=A*COS(TC*K);
    PRINTOUT
    END
    ELSE (* C>H *) IF C>=A THEN
      BEGIN
      P:=C/SIN(TC*K); TA:=ASIN(A/P);
      TB:=180-TC-TA; B:=P*SIN(TB*K);
      WRITELN('ONE SOLUTION; OBLIQUE TRIANGLE:');
      PRINTOUT
      END
    ELSE (* A>C>H ; AMBIGUOUS CASE *)
      BEGIN
      P:=C/SIN(TC*K); TA:=ASIN(A/P);
      TB:=180-TC-TA; B:=P*SIN(TB*K);
      WRITELN('TWO SOLUTIONS');
      WRITELN('FIRST SOLUTION:');
      PRINTOUT;
      TB:=TA-TC; TA:=180-TA; B:=P*SIN(TB*K);
      WRITELN('SECOND SOLUTION:');
      PRINTOUT
      END;
  WRITELN('TYPE <CR> TO CONTINUE OR <ETX> TO QUIT');
  WRITELN;
  READLN
  END (* WHILE *)
END.
```

Examples of PASCAL Program SSA

```
ENTER A,C,GAMMA
A=    4.78300 C=     3.24400 GAMMA= 22.68
TWO SOLUTIONS
FIRST SOLUTION:
A=    4.78300    ALPHA=   34.65   DEG
B=    7.08190     BETA=  122.67   DEG
C=    3.24400    GAMMA=   22.68   DEG
SECOND SOLUTION:
A=    4.78300    ALPHA=  145.35   DEG
B=    1.74438     BETA=   11.97   DEG
C=    3.24400    GAMMA=   22.68   DEG
TYPE <CR> TO CONTINUE OR <ETX> TO QUIT


ENTER A,C,GAMMA
A= 7.82900E11 C= 4.57800E11 GAMMA= 18.56
TWO SOLUTIONS
FIRST SOLUTION:
A= 7.82900E11    ALPHA=   32.98   DEG
B= 1.12622E12     BETA=  128.46   DEG
C= 4.57800E11    GAMMA=   18.56   DEG
SECOND SOLUTION:
A= 7.82900E11    ALPHA=  147.02   DEG
B= 3.58147E11     BETA=   14.42   DEG
C= 4.57800E11    GAMMA=   18.56   DEG
TYPE <CR> TO CONTINUE OR <ETX> TO QUIT


ENTER A,C,GAMMA
A=    1.00000 C=     1.00000 GAMMA= 45.00
ONE SOLUTION; OBLIQUE TRIANGLE:
A=    1.00000    ALPHA=   45.00   DEG
B=    1.41421     BETA=   90.00   DEG
C=    1.00000    GAMMA=   45.00   DEG
TYPE <CR> TO CONTINUE OR <ETX> TO QUIT


ENTER A,C,GAMMA
A=    8.00000 C=     3.00000 GAMMA= 75.00
NO SOLUTIONS
TYPE <CR> TO CONTINUE OR <ETX> TO QUIT


ENTER A,C,GAMMA
A=    7.00000 C=     9.00000 GAMMA= 54.00
ONE SOLUTION; OBLIQUE TRIANGLE:
A=    7.00000    ALPHA=   38.99   DEG
B= 1.11094E1      BETA=   87.01   DEG
C=    9.00000    GAMMA=   54.00   DEG
TYPE <CR> TO CONTINUE OR <ETX> TO QUIT
```

ANSWERS TO ODD-NUMBERED EXERCISES

CHAPTER 1

Section 1   page 5

1
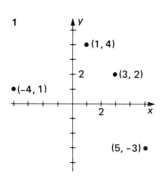

3
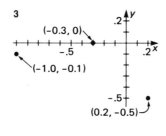

5
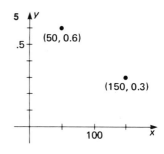

7
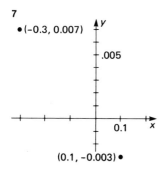

9
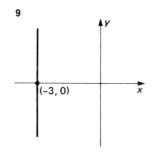

11
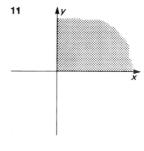

13
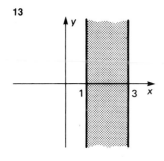

15
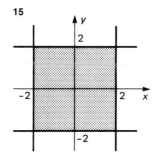

17
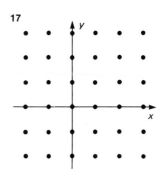

19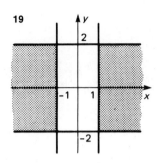

21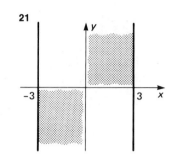

23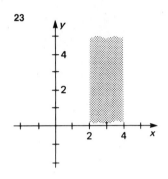

25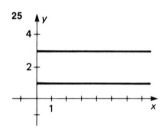

27 $(1,1)$ $(1,-1)$ $(-1,1)$ $(-1,-1)$ **29** $(0,0),(0,9)$, and $(12,0)$ **31** 10
33 $2\sqrt{17}$ **35** $\sqrt{53}$ **37** Call the points A, B, C respectively. Then
$\overline{AB}^2 = 130 = 32 + 98 = \overline{AC}^2 + \overline{BC}^2$.   **39** Call the points A, B, C, D. Then
$\overline{AB} = \overline{BC} = \overline{CD} = \overline{DA} = 5$. Also $\overline{AB} \perp \overline{DA}$ since $\overline{BD}^2 = 50 = 5^2 + 5^2 = \overline{AB}^2 + \overline{DA}^2$.
41 The distance between each pair of successive points is 2, and each of the points is 2 units
from $(0,0)$.   **43** No. The distance between their centers is 25, greater than the sum of
their radii, 24.   **45** $2x + 3y = 13$ **47** $(-1,2)$ only   **49** $(\pm\sqrt{3},1)$ $(1,\pm\sqrt{3})$
51 The vertical distance from $(6,9)$ to $y = -1$ is $9 + 1 = 10 =$ distance from $(6,9)$
to $(0,1)$.

Section 2   page 10

1 (a) 2   (b) -4 (c) -0.55 (d) -37 **3** (a) $\frac{1}{5}$ (b) 1000   (c) x (d) $\dfrac{2}{\sqrt{x}}$

5 (a) and (b)   **7** (a) and (b)   **9** $f(x_1 x_2) = (x_1 x_2)^n = x_1^n x_2^n = f(x_1) f(x_2)$.
If $f(x) = x + 1$ for example, then $f(x_1 x_2) = x_1 x_2 + 1 \neq (x_1 + 1)(x_2 + 1) =$
$f(x_1) f(x_2)$.   **11** 0 for all x **13** $f(2x) = f(x)$ **15** $x f(x) = \begin{cases} x \text{ if } x \geq 0 \\ -x \text{ if } x < 0 \end{cases} = |x|$
17 all real x **19** all real x **21** $x \neq \frac{3}{2}$ **23** $x \neq \frac{5}{3}$ **25** $x \geq 6$ **27** $|x| \leq \frac{2}{3}$
29 $x \geq \frac{3}{2}$ **31** $|x| \leq \frac{1}{2}$ **33** $x \geq 4$ or $x \leq 1$ **35** $x \leq -3$ or $x \geq 4$
37 $f(x) = 3600x$ **39** $f(x) = (\sqrt{2})x$ **41** $f(x) = 1.06x + 1.09(10,000 - x)$

Section 3   page 16

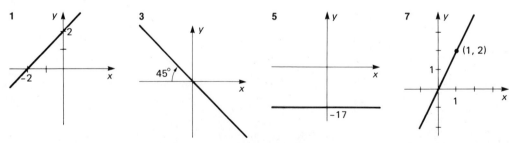

1   **3**   **5**   **7**

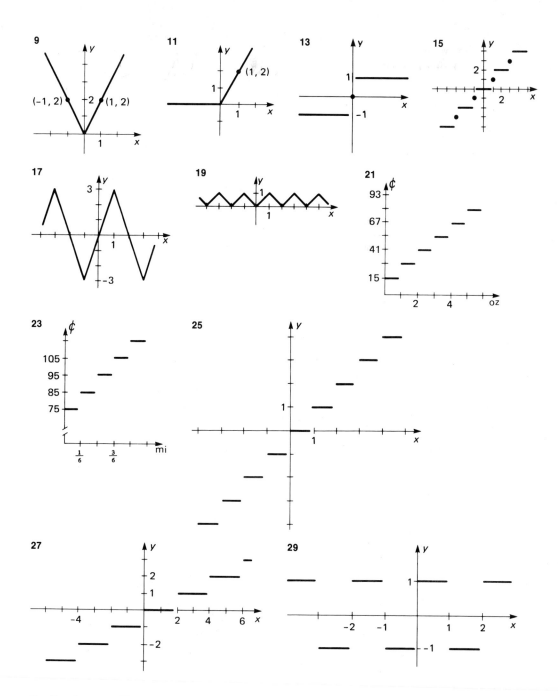

Section 4   page 19

1 $4x - 3$ $3x^2 - 11x - 4$ all x **3** $x^2 - 2x + 1$ $-2x^3 + x^2$ all x

5 $x + \dfrac{1}{x}$ 1 $x \neq 0$ **7** Both are undefined since the domains of f and g have no points

in common.   **9** $3x - 5$ $3x - 1$ all x **11** $2(x + 1)^2$ $-2x^2 - 1$ all x

13 $-4x$ $-4x$ all x **15** 9 3 all x **17** $\sqrt{x^2 - 9}$ $x \leq -3$ or $x \geq 3$;

$x - 9$ $x \geq 0$ **19** $\dfrac{1}{x^2 + 2}$ all x; $\dfrac{1}{x^2} + 2$ $x \neq 0$ **21** x x all x **23** $g(x)$

25 x **27** No; $\sqrt{-3 - 2x^2}$ is undefined for all real x.   **29** $3 \leq x \leq 5$

31 $0 \leq x \leq 1$ **33** If $f(x_1) < f(x_2)$ and $g(x_1) < g(x_2)$ for $x_1 < x_2$, then

$$[f + g](x_1) = f(x_1) + g(x_1) < f(x_2) + g(x_2) = [f + g](x_2).$$

35 $g(x) = x - 2$ **37** Let $f(x) = \sqrt{x}$ and $g(x) = x^2$. The first sequence computes $[f \circ g](x) = \sqrt{x^2} = |x|$ for all x. The second computes $[g \circ f](x) = (\sqrt{x})^2 = x$, but only for $x \geq 0$. (They agree for $x \geq 0$.)   **39** $y = \frac{80}{21}x$ **41** $y = 4x$

Review Exercises   page 21

1

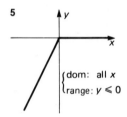

Two sides, $y = \pm 3$, are horizontal; two sides, $x = \pm 2$, are vertical; hence a rectangle.

3 (a) $f(0) = 3 \cdot 0 + 1 = 1$ $f(-2) = 3 \cdot (-2) + 1 = -5$
$f[f(x)] = 3(3x + 1) + 1 = 9x + 4$
(b) $f(a + b) = 3(a + b) + 1 = 3a + 3b + 1 = (3a + 1) + (3b + 1) - 1 =$
$f(a) + f(b) - 1$

5

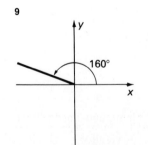

$\begin{cases} \text{dom: all } x \\ \text{range: } y \leqslant 0 \end{cases}$

7 $h(x) = 6x^2 + 2$
9 $y = -x + 6$

11

CHAPTER 2

Section 1   page 25

1, 3, 5, 7

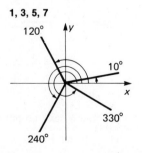

9

11

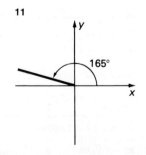

Section 2   page 30

1 $\frac{3}{5}$ $\frac{4}{5}$ **3** $-\frac{9}{41}$ $\frac{40}{41}$ **5** $-\frac{12}{13}$ $-\frac{5}{13}$ **7** $-\frac{60}{61}$ $-\frac{11}{61}$

9 $\frac{24}{25}$ $\frac{7}{25}$ **11** $\sqrt{\frac{2}{3}}$ $\sqrt{\frac{1}{3}}$ **13** $\frac{1}{3}\sqrt{7}$ $\frac{1}{3}\sqrt{2}$ **15** $-\sqrt{\frac{5}{8}}$ $\sqrt{\frac{3}{8}}$

17 0.3025   **19** 0.9959   **21** 0.9829   **23** -0.0654

25 -0.4410 **27** -0.9454 **29** -0.9876 **31** 0.1805

Section 3   page 34

1 $\frac{1}{2}\sqrt{3}$ **3** $-\frac{1}{2}$ **5** $-\frac{1}{2}\sqrt{3}$ **7** $-\frac{1}{2}\sqrt{3}$ **9** $-\frac{1}{2}$ **11** $\sqrt{1-K^2}$ **13** $-K$

15 $\sqrt{1-K^2}$ **17** 70.53°   **19** 26.05°   **21** 101.54°   **23** 102.59°

25 19.47° 160.53°   **27** 67.56° 112.44°

Section 4   page 39

1 18°   **3** 108°   **5** 40°   **7** 195°   **9** 285°   **11** $\frac{1}{18}\pi$ **13** $\frac{1}{5}\pi$ **15** $\frac{3}{10}\pi$

17 $\frac{119}{60}\pi$ **19** $\frac{19}{15}\pi$ **21** 0.2243   **23** 3.0178   **25** 30.94°   **27** 1.81°   **29** 13.02

31 1.0167   **33** 0 -1 **35** 0 -1 **37** $\frac{1}{2}\sqrt{2}$ $-\frac{1}{2}\sqrt{2}$ **39** $-\frac{1}{2}\sqrt{2}$ $-\frac{1}{2}\sqrt{2}$

41 $-\frac{1}{2}\sqrt{3}$ $-\frac{1}{2}$ **43** $\frac{1}{2}\sqrt{3}$ $-\frac{1}{2}$ **45** $-\frac{1}{6}\pi$ $-\frac{5}{6}\pi$ **47** $\pm\frac{1}{4}\pi$ **49** $-\frac{1}{2}\pi$

51 $\pm\frac{1}{6}\pi$ **53** $\theta = n\pi$ **55** $\theta = (2n + \frac{1}{2})\pi$

57 Because he calculated $\cos(\frac{1}{4}\pi)°$ rather than $\cos\frac{1}{4}\pi$.   **59** radian mode

Section 5   page 44

1 $-\sin\theta$ **3** $-\sin\theta$ **5** $\cos\theta$ **7** $-y$ **9** $-\sqrt{1-y^2}$ **11** $-x$

13 $-x$ **15** $0 \le \theta \le \frac{1}{2}\pi$ or $\frac{3}{2}\pi \le \theta \le 2\pi$ **17** $\frac{1}{4}\pi \le \theta \le \frac{5}{4}\pi$ **19** $0 \le \theta \le \frac{1}{2}\pi$

21 $\cos(\theta + \frac{1}{2}\pi) + \cos(\theta - \frac{1}{2}\pi) = -\sin\theta + \cos(\frac{1}{2}\pi - \theta) = -\sin\theta + \sin\theta = 0$

23 To prove $\sin(\frac{9}{2}\pi + \theta) = \sin(\frac{9}{2}\pi - \theta)$: $\sin(\frac{9}{2}\pi + \theta) = \sin(\theta - \frac{9}{2}\pi + 9\pi) = \sin(\theta - \frac{9}{2}\pi + \pi) = \sin[\pi - (\frac{9}{2}\pi - \theta)] = \sin(\frac{9}{2}\pi - \theta)$.

Review Exercises   page 45

1 $\sqrt{\frac{7}{12}}$ $\sqrt{\frac{5}{12}}$ **3** 0.54106   **5** $-S$ **7** $77\frac{1}{7}°$ **9** $\frac{1}{4}\pi$ $\frac{5}{4}\pi$

CHAPTER 3

Section 1   page 51

1 0   **3** $\frac{1}{3}\sqrt{3}$ **5** $-\sqrt{3}$ **7** $\frac{1}{3}\sqrt{3}$ **9** $\sqrt{3}$ **11** -1

13 $-\sqrt{1-K^2}/K$ **15** $1/T$ **17** $1/\sqrt{1+T^2}$ **19** $1/T$ **21** $-T/\sqrt{1+T^2}$

23 0.3174   **25** -5.3383 **27** 0   **29** -1 **31** $-\sqrt{3}$ **33** $\sqrt{3}$ **35** $\sqrt{3}$

37 $n\pi$ **39** $\frac{5}{6}\pi + n\pi$ or $-\frac{1}{6}\pi + n\pi$ **41** $n\pi < \theta < (n + \frac{1}{2})\pi$ for any integer n

43 $\tan(\pi - \theta) = \dfrac{\sin(\pi - \theta)}{\cos(\pi - \theta)} = \dfrac{\sin\theta}{-\cos\theta} = -\tan\theta$

45 $(1 + \tan^2\theta)\cos^2\theta = \left(1 + \dfrac{\sin^2\theta}{\cos^2\theta}\right)\cos^2\theta = \left(\dfrac{\cos^2\theta + \sin^2\theta}{\cos^2\theta}\right)\cos^2\theta =$

$\cos^2\theta + \sin^2\theta = 1$ **47** $\pm\frac{3}{4}$ **49** $\pm\sqrt{3}$ **51** 1.15782   **53** 0.691831

55 -1581.67

Section 2   page 55

1

$10^3\,\theta$	2	4	6	8	10
$10^3 \sin\theta$	1.9999987	3.9999893	5.9999640	7.9999147	9.9998333

3 ≈ 0.14 **5** $1° = \dfrac{\pi}{180}$, so $\sin 1° = \sin\dfrac{\pi}{180} \approx \dfrac{\pi}{180} \approx 0.01745$

7 $\tan(90° - \theta°) = \tan(\tfrac{1}{2}\pi - \dfrac{\pi}{180}\theta) \approx \dfrac{1}{\pi\theta/180} = \dfrac{180}{\pi\theta}$

9 $\tan(\theta - \tfrac{1}{2}\pi) = -\tan(\tfrac{1}{2}\pi - \theta) \approx -1/\theta$

11 $\tfrac{1}{2}(\sin\theta + \tan\theta) \approx \tfrac{1}{2}(\theta + \theta) = \theta$. Set $f(\theta) = \tfrac{1}{2}\sin\theta + \tfrac{1}{2}\tan\theta$. Then:

θ	0.1	0.05	0.01	0.005
$\sin\theta$	.0998334166	.0499791693	.0099998333	.0049999792
$\tan\theta$	.1003346721	.0500417084	.0100003333	.0050000417
$f(\theta)$	.1000840444	.0500104388	.0100000833	.0050000104

Section 3   page 59

1 1 $\sqrt{2}$ $\sqrt{2}$ **3** $-\tfrac{1}{3}\sqrt{3}$ 2 $-\tfrac{2}{3}\sqrt{3}$ **5** $-\tfrac{1}{3}\sqrt{3}$ 2 $-\tfrac{2}{3}\sqrt{3}$ **7** 1 $-\sqrt{2}$ $-\sqrt{2}$

9 $-\tfrac{1}{3}\sqrt{3}$ 2 $-\tfrac{2}{3}\sqrt{3}$ **11** $\pm\tfrac{1}{3}\pi + 2\pi n$ **13** $\tfrac{1}{6}\pi \pm \pi n$ **15** $(2n + 1)\pi$

17 3.10789   **19** -1.24822 **21** 3.73819   **23** $\cot\theta = 1/\tan\theta \approx 1/\theta$

25 $1/\theta - \cot\theta \approx \tfrac{1}{3}\theta$. Set $g(\theta) = 1/\theta - \cot\theta - \tfrac{1}{3}\theta$:

θ	0.10	0.08	0.06	0.04	0.02	0.01	0.005
$g(\theta)$	2.2×10^{-5}	1.1×10^{-5}	4.8×10^{-6}	1.4×10^{-6}	1.8×10^{-7}	2.2×10^{-8}	2.8×10^{-9}

27 $\csc\theta = 1/\sin\theta \approx 1/\theta$

θ	0.1	0.05	0.01	0.005	0.001
$\csc\theta$	10.0167	20.0083	100.0017	200.00083	1000.00017
$1/\theta$	10	20	100	200	1000

29 $(n + \tfrac{1}{4})\pi \le \theta < (n + \tfrac{1}{2})\pi$ or $(n + \tfrac{3}{4})\pi \le \theta < (n + 1)\pi$ **31** $\tan\theta$

33 $\sec\theta$ **35** $\cos\theta$ **37** θ

Section 4   page 63

1

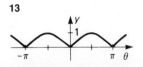

3

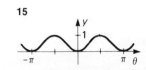

5 20π **7** 1   **9** π **11** π

13 $y = 0$ for $\theta = 0.1n$. The points are too far apart; 0.1 is the period of y.

Section 5   page 67

1

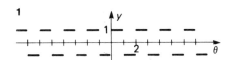

3

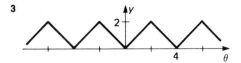

5

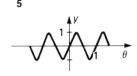

7

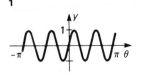

9

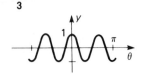

11

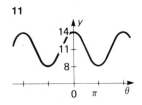

13

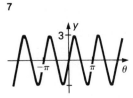

15

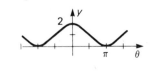

17

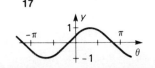

19

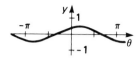

21

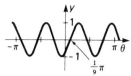

23

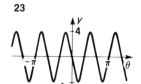

25 $x = 3 \cos[2\pi(t + \frac{1}{2})]$
       $= -3 \cos(2\pi t)$

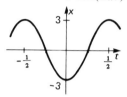

27 $x = 1 + \cos(20\pi t)$

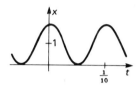

29   odd, per $= 2\pi$

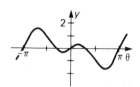

31   shifted sine, per $= 2\pi$

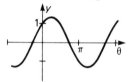

33   per $= \pi$, odd

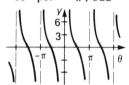

35   odd, per $= \pi$

37   per $= 2\pi$

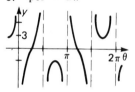

39   odd

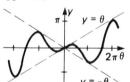

41

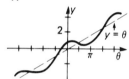

43   odd

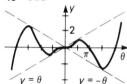

45

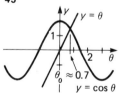

47

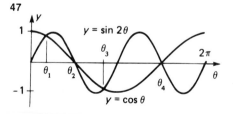

$$\begin{cases} \theta_1 \approx 0.5 \\ \theta_2 = \frac{1}{2}\pi \\ \theta_3 \approx 2.6 \\ \theta_4 = \frac{3}{2}\pi \end{cases}$$

Review Exercises   page 68

1   1   **3**   1.000000   **5** $\pm\frac{1}{6}\pi + 2\pi n$

7

9 $\approx 4.46095 \times 10^{-5}$ on a Sharp EL-5806S

CHAPTER 4

Section 1 page 73

1 $\sin \theta \sec \theta = \sin \theta (1/\cos \theta) = \sin \theta/\cos \theta = \tan \theta$

3 $\tan \theta \csc \theta = (\sin \theta/\cos \theta)(1/\sin \theta) = 1/\cos \theta = \sec \theta$

5 $\sin^2 \theta/(1 - \sin^2 \theta) = \sin^2 \theta/\cos^2 \theta = (\sin \theta/\cos \theta)^2 = \tan^2 \theta$

7 $\sec^2 \theta/(\sec^2 \theta - 1) = \sec^2 \theta/\tan^2 \theta = (1/\cos \theta)^2/(\sin \theta/\cos \theta)^2 = 1/\sin^2 \theta$

9 $\sin \theta/(1 - \sin^2 \theta) = \sin \theta/\cos^2 \theta = (\sin \theta/\cos \theta)(1/\cos \theta) = \tan \theta \sec \theta$

11 $\sec^4 \theta - \sec^2 \theta = \sec^2 \theta(\sec^2 \theta - 1) = \sec^2 \theta \tan^2 \theta$

13 $1 - (\cos \theta - \sin \theta)^2 = 1 - (\cos^2 \theta - 2 \cos \theta \sin \theta + \sin^2 \theta) = 1 - (1 - 2 \sin \theta \cos \theta) = 2 \sin \theta \cos \theta$　　**15** $\cos^2 \theta/(1 + \sin \theta) = (1 - \sin^2 \theta)/(1 + \sin \theta) = 1 - \sin \theta$

	$\sin \theta$	$\cos \theta$	$\tan \theta$	$\cot \theta$	$\sec \theta$	$\csc \theta$
17	$-\frac{3}{5}$		$\frac{3}{4}$	$\frac{4}{3}$	$-\frac{5}{4}$	$-\frac{5}{3}$
19	$3/\sqrt{10}$	$1/\sqrt{10}$		$\frac{1}{3}$	$\sqrt{10}$	$\frac{1}{3}\sqrt{10}$
21	$\frac{1}{2}\sqrt{3}$	$-\frac{1}{2}$	$-\sqrt{3}$	$-1/\sqrt{3}$		$2/\sqrt{3}$
23	$-2/\sqrt{13}$	$3/\sqrt{13}$	$-\frac{2}{3}$		$\frac{1}{3}\sqrt{13}$	$-\frac{1}{2}\sqrt{13}$

17　　　　**19**　　　　**21**　　　　**23**

25 $1/\sqrt{1 + \cot^2 \theta}$　　**27** $\sqrt{1 + \cot^2 \theta}$　　**29** $-\sqrt{1 + \cot^2 \theta}/\cot \theta$

31 $\sin \theta/\sqrt{1 - \sin^2 \theta}$　　**33** $1/\sqrt{1 - \cos^2 \theta}$

Section 2 page 76

1 $s^4 - c^4 = (s^2 + c^2)(s^2 - c^2) = s^2 - c^2 = \sin^2 \theta - \cos^2 \theta$

3 $(1 - c^2)/s = s^2/s = s = \sin \theta$　　**5** $(\sec^2 \theta - \tan^2 \theta)^3 = 1^3 = 1$

7 $\cos^2 \theta + \cos^2(\frac{1}{2}\pi - \theta) = \cos^2 \theta + \sin^2 \theta = 1$　　**9** $\tan \theta + \cot \theta = s/c + c/s = (s^2 + c^2)/cs = 1/cs = \sec \theta \csc \theta$　　**11** $(1 - \sin \theta)(1 + \csc \theta) = (1 - s)(1 + 1/s) = 1 - s + 1/s - 1 = 1/s - s = (1 - s^2)/s = c^2/s = c(c/s) = \cos \theta \cot \theta$　　**13** $\sec \theta - \cos \theta = 1/c - c = (1 - c^2)/c = s^2/c = s(s/c) = \sin \theta \tan \theta$　　**15** $\sec^2 \theta - \csc^2 \theta = (1 + \tan^2 \theta) - (1 + \cot^2 \theta) = \tan^2 \theta - \cot^2 \theta$　　**17** $\sec^2 \theta + \csc^2 \theta = 1/c^2 + 1/s^2 = (s^2 + c^2)/s^2 c^2 = 1/s^2 c^2 = \sec^2 \theta \csc^2 \theta$　　**19** $\sec^4 \theta - \tan^4 \theta = (\sec^2 \theta + \tan^2 \theta)(\sec^2 \theta - \tan^2 \theta) = \sec^2 \theta + \tan^2 \theta = (1 + \tan^2 \theta) + \tan^2 \theta = 1 + 2 \tan^2 \theta$　　**21** $\cot^4 \theta + \cot^2 \theta = \cot^2 \theta(\cot^2 \theta + 1) = \cot^2 \theta \csc^2 \theta = (\csc^2 \theta - 1)\csc^2 \theta = \csc^4 \theta - \csc^2 \theta$

23 LHS $= (c^2 - 2cs + s^2) + (c^2 + 2cs + s^2) = 2(c^2 + s^2) = 2$

25 LHS $= \dfrac{1 - s/c}{1 + s/c} = \dfrac{c - s}{c + s}$, RHS $= \dfrac{c/s - 1}{c/s + 1} = \dfrac{c - s}{c + s}$　　**27** equivalent relation: $\tan^2 \theta = \sec \theta(\sec \theta - \cos \theta) = \sec^2 \theta - 1$, which is an identity

29 RHS $= \left(\dfrac{s}{c} + \dfrac{c}{s}\right)(c + s) = \left(\dfrac{s^2 + c^2}{cs}\right)(c + s) = \dfrac{1}{cs}(c + s) = \dfrac{1}{s} + \dfrac{1}{c} =$ LHS

31 LHS $= \dfrac{s/c - sc}{1/c} = s - sc^2 = s(1 - c^2) = ss^2 = s^3 =$ RHS

33 $c^4 - c^2 = c^2(c^2 - 1) = (1 - s^2)(-s^2) = s^4 - s^2$　　**35** LHS $= 2 - \tan \theta - \cot \theta =$

$$2 - \frac{s}{c} - \frac{c}{s} = 2 - \frac{s^2 + c^2}{sc} = 2 - \frac{1}{sc} = \text{RHS} \qquad \textbf{37} \quad \text{LHS} = \frac{c}{1 - s/c} + \frac{s}{1 - c/s} =$$

$$\frac{c^2}{c - s} + \frac{s^2}{s - c} = \frac{c^2 - s^2}{c - s} = c + s = \text{RHS}$$

39 Equivalent relations:

$$(\cos \alpha - \sin \beta)(\cos \alpha + \sin \beta) = (\cos \beta + \sin \alpha)(\cos \beta - \sin \alpha),$$
$$\cos^2 \alpha - \sin^2 \beta = \cos^2 \beta - \sin^2 \alpha,$$
$$\cos^2 \alpha + \sin^2 \alpha = \cos^2 \beta + \sin^2 \beta.$$

The last is an identity, hence so is the given relation.

Section 3   page 80

1 $\cos(\theta + 2\pi) = \cos \theta \cos 2\pi - \sin \theta \sin 2\pi = \cos \theta$
$\sin(\theta + 2\pi) = \sin \theta \cos 2\pi + \cos \theta \sin 2\pi = \sin \theta$
3 $\cos(\theta + \frac{1}{2}\pi) = \cos \theta \cos \frac{1}{2}\pi - \sin \theta \sin \frac{1}{2}\pi = -\sin \theta$
$\sin(\theta + \frac{1}{2}\pi) = \sin \theta \cos \frac{1}{2}\pi + \cos \theta \sin \frac{1}{2}\pi = \cos \theta$
5 $\frac{1}{2}\sqrt{2}(\cos \theta - \sin \theta)$ $\frac{1}{2}\sqrt{2}(\cos \theta + \sin \theta)$
7 $(\cot \alpha \cot \beta - 1)/(\cot \alpha + \cot \beta)$ **9** $\frac{1}{4}(\sqrt{2} + \sqrt{6})$ **11** $\frac{1}{4}(\sqrt{2} + \sqrt{6})$
13 $\frac{1}{2}$ **15** $\frac{1}{2}\sqrt{2}$ **17** $\frac{5}{4}\pi$ **19** $\frac{1}{20} + \frac{3}{10}\sqrt{10}$ **21** $-\frac{4}{9}\sqrt{2}$ $-\frac{7}{9}$ **23** $-\frac{15}{17}$ $-\frac{8}{15}$
25 $4 \cos^3 \theta - 3 \cos \theta$ **27** $\csc^3 \theta/(3 \csc^2 \theta - 4)$ **29** $\sin 4\theta = 2 \sin 2\theta \cos 2\theta =$
$2(2 \sin \theta \cos \theta)(1 - 2 \sin^2 \theta) = 4 \sin \theta \cos \theta - 8 \sin^3 \theta \cos \theta$ **31** $\cot 2\theta + \csc 2\theta =$
$\dfrac{\cos 2\theta}{\sin 2\theta} + \dfrac{1}{\sin 2\theta} = \dfrac{\cos 2\theta + 1}{\sin 2\theta} = \dfrac{2 \cos^2 \theta}{2 \sin \theta \cos \theta} = \dfrac{\cos \theta}{\sin \theta} = \cot \theta$ **33** LHS $=$
$\dfrac{\sin \theta}{\cos \theta} + \dfrac{\cos \theta}{\sin \theta} = \dfrac{\sin^2 \theta + \cos^2 \theta}{\sin \theta \cos \theta} = \dfrac{1}{\sin \theta \cos \theta} = \dfrac{2}{2 \sin \theta \cos \theta} = \dfrac{2}{\sin 2\theta} = 2 \csc 2\theta$
35 $\tan 2\theta$ **37** $\cot 2\theta = \dfrac{1}{\tan 2\theta} = \dfrac{1 - \tan^2 \theta}{2 \tan \theta} = \dfrac{\cot^2 \theta - 1}{2 \cot \theta}$ **39** $\cot(\alpha + \beta) =$
$\dfrac{1}{\tan(\alpha + \beta)} = \dfrac{1 - \tan \alpha \tan \beta}{\tan \alpha + \tan \beta} = \text{RHS}$ **41** $\dfrac{1 + \cos 2\theta}{1 - \cos 2\theta} = \dfrac{2 \cos^2 \theta}{2 \sin^2 \theta} = \cot^2 \theta$
43 LHS $= \dfrac{1/c - 1/s}{1/c + 1/s} = \dfrac{s - c}{s + c} = \dfrac{(s - c)^2}{s^2 - c^2} = \dfrac{s^2 - 2sc + c^2}{-\cos 2\theta} =$
$\dfrac{1 - \sin 2\theta}{-\cos 2\theta} = \tan 2\theta - \sec 2\theta$

Section 4   page 86

1 $\sin 22.5° = \sqrt{\frac{1}{2}(1 - \cos 45°)} = \sqrt{\frac{1}{2}(1 - \frac{1}{2}\sqrt{2})} = \frac{1}{2}\sqrt{2 - \sqrt{2}}$

3 $\tan 67.5° = \dfrac{\sin 135°}{1 + \cos 135°} = \dfrac{\frac{1}{2}\sqrt{2}}{1 - \frac{1}{2}\sqrt{2}} = \dfrac{\frac{1}{2}\sqrt{2}(1 + \frac{1}{2}\sqrt{2})}{1 - \frac{1}{2}} = \sqrt{2} + 1$

5 $\tan \frac{1}{2}\theta = \dfrac{\sin \theta}{1 + \cos \theta} = \dfrac{\sin \theta(1 - \cos \theta)}{1 - \cos^2 \theta} = \dfrac{\sin \theta(1 - \cos \theta)}{\sin^2 \theta} = \dfrac{1 - \cos \theta}{\sin \theta}$

7 $5 \cos(\theta - \theta_0)$ $\theta_0 = \arcsin \frac{3}{5} \approx 0.6435$ **9** $\sqrt{5} \cos(\theta + \theta_0)$
$\theta_0 = \arcsin(2/\sqrt{5}) \approx 1.1071$ **11** $\sqrt{13} \cos(\theta - \theta_0)$ $\theta_0 = \arcsin(3/\sqrt{13}) \approx 0.9828$
13 $\sqrt{10} \cos(\theta + \theta_0)$ $\theta_0 = \arcsin(1/\sqrt{10}) \approx 0.3218$ **15** $|\sin \theta + \cos \theta| =$
$|\sqrt{2} \cos(\theta - \frac{1}{4}\pi)| \leq \sqrt{2}$ **17** $\sqrt{10}$ **19** $\frac{1}{2}(\cos \theta - \cos 3\theta)$
21 $\frac{1}{2}(\sin 7\theta - \sin \theta)$ **23** $2 \sin \frac{3}{2}\theta \cos \frac{1}{2}\theta$ **25** $2 \cos \frac{3}{2}\theta \sin \frac{1}{2}\theta$
27 $-2 \sin \frac{11}{2}\theta \sin \frac{1}{2}\theta$ **29** LHS $= 2 \sin 45° \cos 30° = 2(\frac{1}{2}\sqrt{2})(\frac{1}{2}\sqrt{3}) = \frac{1}{2}\sqrt{6}$
31 LHS $= 2 \cos 60° \cos 45° = \frac{1}{2}\sqrt{2}$ **33** $\frac{1}{4}(\sqrt{6} + \sqrt{2})$ $\frac{1}{4}(\sqrt{6} - \sqrt{2})$
35 $\sin(\alpha + \beta)/\cos \alpha \cos \beta = (\sin \alpha \cos \beta + \cos \alpha \sin \beta)/\cos \alpha \cos \beta =$
$(\sin \alpha/\cos \alpha) + (\sin \beta/\cos \beta) = \tan \alpha + \tan \beta$ **37** $\sin^4 \theta = [\frac{1}{2}(1 - \cos 2\theta)]^2 =$
$\frac{1}{4}(1 - 2 \cos 2\theta + \cos^2 2\theta) = \frac{1}{4}[1 - 2 \cos 2\theta + \frac{1}{2}(1 + \cos 4\theta)] = \text{RHS}$
39 $\cos(\alpha + \beta)\cos(\alpha - \beta) = (\cos \alpha \cos \beta - \sin \alpha \sin \beta)(\cos \alpha \cos \beta + \sin \alpha \sin \beta) =$
$\cos^2 \alpha \cos^2 \beta - \sin^2 \alpha \sin^2\beta = \cos^2 \alpha(1 - \sin^2 \beta) - (1 - \cos^2 \alpha)\sin^2 \beta =$
$\cos^2 \alpha - \sin^2 \beta$ **41** LHS $= \sin \theta(\sin 2\theta \cos 2\theta) = \sin \theta(\frac{1}{2} \sin 4\theta) =$
$\frac{1}{2} \sin \theta \sin 4\theta = \frac{1}{4}(\cos 3\theta - \cos 5\theta)$ **43** $4t(1 - t^2)/(1 + t^2)^2$

45 $\dfrac{\sin(\beta - \alpha)}{\sin\alpha\sin\beta} = \dfrac{\sin\beta\cos\alpha - \sin\alpha\cos\beta}{\sin\alpha\sin\beta} = \dfrac{\cos\alpha}{\sin\alpha} - \dfrac{\text{co }3}{\sin\beta} = \cot\alpha - \cot\beta$

47 LHS $= \dfrac{2\cos\frac{1}{2}(\alpha + \beta)\sin\frac{1}{2}(\alpha - \beta)}{2\cos\frac{1}{2}(\alpha + \beta)\cos\frac{1}{2}(\alpha - \beta)} = \dfrac{\sin\frac{1}{2}(\alpha - \beta)}{\cos\frac{1}{2}(\alpha - \beta)} = \tan\frac{1}{2}(\alpha - \beta)$

49 LHS $= \dfrac{2\sin\frac{1}{2}(\alpha + \beta)\cos\frac{1}{2}(\alpha - \beta)}{2\cos\frac{1}{2}(\alpha + \beta)\cos\frac{1}{2}(\alpha - \beta)} = \dfrac{\sin\frac{1}{2}(\alpha + \beta)}{\cos\frac{1}{2}(\alpha + \beta)} = \tan\frac{1}{2}(\alpha + \beta)$

51 LHS $\approx 0.642788 + 0.342020 = 0.984808$ RHS ≈ 0.984808

53 $\sin\frac{1}{6}\pi < \frac{1}{6}\pi < \tan\frac{1}{6}\pi$ $\frac{1}{2} < \frac{1}{6}\pi < \frac{1}{3}\sqrt{3}$ $3 < \pi < 2\sqrt{3} < 3.47$

55 $3.132 < \pi < 3.160$

Review Exercises   page 92

1 $-\frac{2}{3}\sqrt{2}$ **3** $\tan^2\theta - \sin^2\theta\tan^2\theta = \tan^2\theta(1 - \sin^2\theta) = \tan^2\theta\cos^2\theta = \sin^2\theta$, hence $\tan^2\theta - \sin^2\theta = \sin^2\theta\tan^2\theta$ **5** LHS $= \dfrac{1}{1/c + s/c} = \dfrac{c}{1 + s} = \dfrac{c(1 - s)}{1 - s^2} = \dfrac{c(1 - s)}{c^2} = \dfrac{1 - s}{c} =$ RHS   **7** $4\cos\frac{1}{2}\theta\cos\frac{1}{4}\theta\sin\frac{1}{4}\theta = 2\cos\frac{1}{2}\theta(2\cos\frac{1}{4}\theta\sin\frac{1}{4}\theta) = 2\cos\frac{1}{2}\theta\sin\frac{1}{2}\theta = \sin\theta$ **9** $\sin 105° = \sin 60°\cos 45° + \cos 60°\sin 45° = \frac{1}{2}\sqrt{3}\cdot\frac{1}{2}\sqrt{2} + \frac{1}{2}\cdot\frac{1}{2}\sqrt{2} = \frac{1}{4}(\sqrt{6} + \sqrt{2})$ **11** $13\cos(\theta - \theta_0)$ where $\cos\theta_0 = -\frac{5}{13}$ and $\sin\theta_0 = \frac{12}{13}$ **13** $2\sin 2\theta\sin\theta$

CHAPTER 5

Section 1   page 97

1   0.44   0.31   0.11   0.26   **3**   0.000   0.000   16.244   3.786   **5**   (a) 5.102   5.10   (b) $1.26 + 0.40 + 2.12 + 1.34 = 5.12$ **7** 1.05×10^3 55.5   10.0   **9**   0.40

11   2.3   **13**   12, 0   **15**   4, 5   **17** $\dfrac{ab - c}{d} + e$, $ab - \dfrac{c}{d} + e$

19 x $\boxed{+}$ 3 $\boxed{\div}$ y $\boxed{=}$, x $\boxed{+}$ 3 $\boxed{=}$ $\boxed{\div}$ y $\boxed{=}$ **21** x $\boxed{\div}$ y $\boxed{\div}$ z $\boxed{=}$, either logic

23 a $\boxed{\div}$ b $\boxed{+}$ c $\boxed{\div}$ 2 $\boxed{=}$, a $\boxed{\div}$ b $\boxed{+}$ c $\boxed{=}$ $\boxed{\div}$ 2 $\boxed{=}$ **25** -2 **27**   945   **29**   1

Section 2   page 103

1 $a + \sqrt{b} + \sqrt{c}$ **3** $\dfrac{a}{b} + \dfrac{1}{c}$ **5** $\left(\dfrac{ab}{c} + a\right)\dfrac{c}{b}$ **7** $(a + 10^b + c)^2 - a$

9 $\{[(a + 1)^{-1} + 1]^{-1} + 1\}^{-1}$ **11** $\dfrac{ab}{c} + \dfrac{ac}{b}$ **13** $a + b^c d$

15 $(a - b)c - (b - c)a$ **17** $(a + b^{a-b})(a - b)$ **19** $a + \dfrac{b}{a}$

21 a $\boxed{\text{STO}}$ $\boxed{+}$ 3 $\boxed{=}$ $\boxed{\log}$ $\boxed{\times}$ $\boxed{\text{RCL}}$ $\boxed{=}$ **23** a $\boxed{\text{STO}}$ $\boxed{y^x}$ $\boxed{\text{RCL}}$ $\boxed{1/x}$ $\boxed{-}$ $\boxed{\text{RCL}}$ $\boxed{=}$ $\boxed{+/-}$

25 1 $\boxed{\text{STO}}$ a $\boxed{+/-}$ $\boxed{\text{M}+}$ $\boxed{+/-}$ $\boxed{+}$ 1 $\boxed{\div}$ $\boxed{\text{RCL}}$ $\boxed{=}$ or 1 $\boxed{-}$ a $\boxed{\div}$ 2 $\boxed{=}$ $\boxed{1/x}$ $\boxed{-}$ 1 $\boxed{=}$

Note $\dfrac{1 + a}{1 - a} = \dfrac{2}{1 - a} - 1$.

27 a $\boxed{\text{STO}}$ $\boxed{x^2}$ $\boxed{x^2}$ $\boxed{\times}$ 5 $\boxed{+}$ 1 $\boxed{=}$ $\boxed{\sqrt{}}$ $\boxed{\sqrt{}}$ $\boxed{-}$ $\boxed{\text{RCL}}$ $\boxed{=}$ $\boxed{y^x}$ $\boxed{\text{RCL}}$ $\boxed{-}$ $\boxed{\text{RCL}}$ $\boxed{=}$

29 3 $\boxed{\text{STO}}$ a $\boxed{+/-}$ $\boxed{\text{M}+}$ $\boxed{+/-}$ 2 $\boxed{=}$ $\boxed{\div}$ $\boxed{\text{RCL}}$ $\boxed{=}$ or 3 $\boxed{-}$ a $\boxed{=}$ $\boxed{1/x}$ $\boxed{\times}$ 5 $\boxed{-}$ 1 $\boxed{=}$

Note $\dfrac{2 + a}{3 - a} = \dfrac{5}{3 - a} - 1$

31 1 $\boxed{+}$ a $\boxed{\text{STO}}$ $\boxed{y^x}$ $\boxed{\text{RCL}}$ $\boxed{1/x}$ $\boxed{=}$ $\boxed{\times}$ $\boxed{(}$ 1 $\boxed{+}$ $\boxed{\text{RCL}}$ $\boxed{1/x}$ $\boxed{y^x}$ $\boxed{\text{RCL}}$ $\boxed{)}$ $\boxed{=}$

33 a $\boxed{\text{STO}}$ $\boxed{x^2}$ $\boxed{x^2}$ $\boxed{\times}$ 3 $\boxed{-}$ 2 $\boxed{\times}$ $\boxed{\text{RCL}}$ $\boxed{\times}$ $\boxed{\text{RCL}}$ $\boxed{x^2}$ $\boxed{+}$ 5 $\boxed{\div}$ $\boxed{\text{RCL}}$ $\boxed{=}$

35   3.94482 37   **37**   0.30888 640   **39**   1.73205 08075 68

41 $\boxed{\text{M}+}$ $\boxed{-}$ $\boxed{\text{RCL}}$ $\boxed{=}$ $\boxed{\text{M}+}$ $\boxed{+/-}$:

dis.	a	a	$-b$	$-b$	b
mem.	b	$a + b$	$a + b$	a	a

Section 3   page 108

	a	b	c	α	β
1		5		67.4°	22.6°
3		138.9		37.69°	52.31°
5	16.1			47.0°	43.0°
7	48.56			46.67°	43.33°
9			6.4	38.7°	51.3°
11			28.68	71.04°	18.96°
13	7.51	29.0			75.5°
15		31.3	34.7		64.4°
17	194.8	132.4		55.8°	
19	58.22		80.80	46.1°	
21	0.22078	1.0017			77.57°
23		35.0636	35.2019		84.92°

25 0.000278°   **27** 0.016389°   **29** 0.203333°   **31** 0.5°   **33** 17.060833°
35 73.006389°   **37** 71.964167°   **39** 2.616667°   **41** 19° 6′ 29.5″
43 72° 56′ 59.6″   **45** 1′ 12″   **47** 1.72″   **49** 20° 35′ 11″   **51** 69° 24′ 49″
53 72° 8′ 50″   **55** 1″   **57** 0.26228   **59** 0.27797   **61** 1.33495
63 4.8481 × 10⁻⁶   **65** −0.68924   **67** −0.96109   **69** about −2.5 × 10⁻¹¹
71 1.7 × 10⁻⁷   **73** 4.8 × 10⁻⁶   **75** $\frac{1}{30}(D + \frac{1}{60}M + \frac{1}{3600}S)$ hours
77 $\frac{1}{6}(D + \frac{1}{60}M + \frac{1}{3600}S)$ sec

Section 4   page 113

1 77.8 ft   **3** 14.1 ft   **5** 0.39 km   **7** 2.2 km   **9** 0.885
11 Refer to Fig. 4(b) of the text, with 6 replaced by R_0, 3 by $R_1 - R_0$, and 20 by D.
Then $\theta = \text{arc cos}[(R_1 - R_0)/D]$, $L_0 = 2R_0\theta$, $L_1 = R_1(2\pi - 2\theta)$, and
$L_2 = \sqrt{D^2 - (R_1 - R_0)^2}$. Finally $L = L_0 + L_1 + 2L_2$.
13 When $R_0 = R_1$, the formula yields $L = 2D + 2\pi R_1$, D
which checks with the figure.

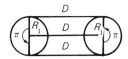

Section 5   page 117

15 $\sqrt{D^2 - 4R_1^2} + 4R_1(\pi - \theta)$ **17** 50.41 m   5.53 m   **19** 113.3 m   11.2 m
21 22.1 m   1.6 m   **23** 84.52 m   3.96 m   **25** 1.9[1/sin(360/28)° − 1] ≈ 6.64 cm
27 1.9[1/sin(360/14)° − 1] ≈ 6.28 cm

Review Exercises   page 120

1 54.78°   43.87°   67.39°   **3** $3r\sqrt{3}$ **5** $8r\sqrt{2 - \sqrt{2}}$ **7** $s = 2a \text{ arc tan}(y/a)$
9 $A = \frac{1}{4}b^2/\tan\frac{1}{2}\beta$ **11** $2r^2\sqrt{2}$ **13** 30°
15 $b = \sqrt{h^2 + \frac{1}{2}a^2}$ $\theta = 2 \text{ arc tan}(a/2\sqrt{h^2 + \frac{1}{4}a^2})$ **17** 5°26′19″   **19** 24°51′54″
21 2.4060 cm   **23** 2.3196 cm   **25** 34′22.6″   **27** 1°30′54.9″
29 12°42′50″   87.592   **31** 39.37 m   136.93 m

1 90   **3** 3.60813   **5** $\beta = 68.86°$ $c \approx 35.96$ $b \approx 33.54$
7 81.0 ft   **9** 13.51

CHAPTER 6

Section 1   page 127

	α	β	γ			α	β	γ
1	28.96°	46.57°	104.48°	**3**		50.35°	31.50°	98.16°
5	42.61°	39.80°	97.59°	**7**		21.56°	146.76°	11.68°
9	14.47°	17.62°	147.92°	**11**		no such triangle		

	c	α	β			a	β	γ
13	3.269	105.72°	35.28°	**17**		144.8	24.65°	2.98°
15	6.590	4.69°	164.17°	**19**		35050	33° 14′ 50″	107° 27′ 58″

	b	γ	α			b	γ	α
21	88.80	176.57°	0.37°	**23**		3.4231	28° 47′ 8″	151° 12′ 18″

25 3.9324 cm   12.1258 cm   **27** 149.5 mi   **29** 50.44° 56.71° 72.86°
31 $d_1^2 + d_2^2 = [a^2 + b^2 - 2ab \cos \theta] + [a^2 + b^2 - 2ab \cos(\pi - \theta)]$
$= 2(a^2 + b^2) - 2ab \cos \theta + 2ab \cos \theta = 2(a^2 + b^2)$
33 283.6 ft   **35** 3.522   **37** 14.979

Section 2   page 136

1 $b \approx 3.833$ $c \approx 4.115$ $\gamma = 72.17°$ **3** $b \approx 3.278$ $a \approx 14.08$ $\gamma = 63.84°$
5 $c \approx 0.5428$ $a \approx 1.108$ $\alpha = 135.91°$ **7** $b \approx 0.3839$ $c \approx 0.3194$ $\alpha = 135.91°$
9 $b \approx 0.7839$ $a \approx 1.332$ $\alpha = 135.91°$ **11** no sol.: $a \sin \gamma > c$
13 no sol.: $c \sin \alpha > a$ **15** $b = 3$ $\beta = 45°$ $\alpha = 90°$
17 $\alpha \approx 42.09°$ $\beta \approx 106.67°$ $b \approx 7.132$; $\alpha \approx 137.91°$ $\beta \approx 10.85°$ $b \approx 1.401$
19 $\alpha \approx 78.76°$ $\beta \approx 27.30°$ $b \approx 1455$; $\alpha \approx 101.24°$ $\beta \approx 4.82°$ $b \approx 266.8$
21 $\alpha \approx 115.43°$ $\beta \approx 37.26°$ $a \approx 1.006$; $\alpha \approx 9.95°$ $\beta \approx 142.74°$ $a \approx 0.1924$
23 $\beta \approx 68.63°$ $\gamma \approx 56.38°$ $c \approx 42.22$; $\beta \approx 111.37°$ $\gamma \approx 13.64°$ $c \approx 11.96$
25 $\alpha \approx 38.21°$ $\gamma \approx 123.44°$ $c \approx 55080$; $\alpha \approx 141.79°$ $\gamma \approx 19.86°$ $c \approx 22430$
27 $\alpha \approx 22.80°$ $\beta \approx 74.08°$ $b \approx 37.75$ **29** $\alpha \approx 6.83°$ $\gamma \approx 163.79°$ $c \approx 1232$
31 $\alpha \approx 37.81°$ $\beta \approx 3.58°$ $b \approx 6.836$ **33** $\beta \approx 36.87°$ $\gamma \approx 44.96°$ $b \approx 1847$
35 $\beta \approx 58.89°$ $\gamma \approx 2.81°$ $c \approx 0.4015$

Section 3   page 141

1 5.3 mi from A, 8.0 mi from B, 4.2 mi from the line   **3** 4.335 m   **5** 2.992
7 21.9 ft   **9** $\sqrt{43} \approx 6.557$ **11** 2.4718   0.94356   **13** 2.571150   2.694593
15 $\frac{15}{4}$ **17** 0.9473   **19** 7620.   **21** 23.05   **23** 8.162   **25** 104.6
27 1047×10^2 **29** two triangles: 148.0   27.05   **31** 104.5   **33** 2168×10^2
35 The second figure is inaccurate; there is a skinny diamond missing from the center of
the figure. For instance, the sum of the two angles at the lower left corner is
$\theta = \arctan \frac{3}{8} + \arctan \frac{5}{2} < 90°$; in fact,

$$\tan \theta = \frac{\frac{3}{8} + \frac{5}{2}}{1 - \frac{3}{8} \cdot \frac{5}{2}} = \frac{\frac{23}{8}}{\frac{1}{16}} = 46 \qquad \theta \approx 88.75°$$

Section 4   page 148

1 $206\frac{1}{4}$ m   **3** 2.063×10^5 **5** $1 \approx d \tan \alpha \approx d\alpha$ so $1/\alpha \approx d$

	parallax (″)	pc	ly	km
7		1.32	4.29	4.06×10^{13}
9	0.431		7.57	7.16×10^{13}
11	0.377	2.65		8.18×10^{13}
13	0.345	2.90	9.45	

15 $75 + 50 + 24.5 + 2.0 + 1.47 + 1.006 = 153.976$ mm   **17** $90° - 9° + 1° = 82°$
19 $41° - 27° + 1° - 27′ + 9′ - 3′ - 9″ = 14° 38′ 51″$ **21** 5; one part in 1000
23 6.02224; 1.6 parts in 1000   **25** $\frac{2}{10}$ of 1%   **27** 0.68 ft

Review Exercises   page 149

1 $127.2°$ **3** 2.122
5 $b \approx 8.081$ $\beta \approx 145.11°$ $\gamma \approx 20.69°$; $b \approx 1.596$ $\beta \approx 6.49°$ $\gamma \approx 159.31°$
7 3.69 m   **9** 5.5169; 1.2%

CHAPTER 7

Section 2   page 159

1 $f(x)$ is strictly increasing, and $\mathrm{dom}(g) = \mathrm{range}(f) = \{\text{all } y\}$ **3** If $0 < x_1 < x_2$, then $1/x_2 < 1/x_1$, so $-1/x_1 < -1/x_2$, that is, $f(x_1) < f(x_2)$. Therefore $f(x)$ is strictly increasing, hence has an inverse, and $\mathrm{dom}(g) = \mathrm{range}(f) = \{y < 0\}$.   **5** $f(x)$ is strictly increasing. Since $3^2 = 9$ and $4^2 = 16$, $\mathrm{dom}(g) = \mathrm{range}(f) = \{9 < y < 16\}$.
7 Both x^5 and x^3 are strictly increasing; so is their sum, $f(x)$. It is clear from the graph of $f(x)$ and the facts that $f(x) \to \infty$ as $x \to \infty$ and $f(x) \to -\infty$ as $x \to -\infty$, that $\mathrm{dom}(g) = \mathrm{range}(f) = \{\text{all } y\}$.   **9** The function x^3 is strictly increasing; hence $f(x)$ is strictly decreasing, so has an inverse, and $\mathrm{dom}(g) = \mathrm{range}(f) = \{y > 0\}$.
11 $f(x)$ is strictly decreasing, so has an inverse, and $\mathrm{dom}(g) = \mathrm{range}(f) = \{y \geq 0\}$.
13 $x = 2\sqrt[3]{y}$ $\{\text{all } y\}$ **15** $x = -1/y$ $\{y < 0\}$ **17** $x = \sqrt{5/y}$ $\{y > 0\}$
19 $x = y^2 - 1$ $\{y \geq 0\}$ **21** $x = \sqrt{y^2 - 4}$ $\{y \geq 2\}$
23 $x = -\sqrt[4]{1/y - 10}$ $\{0 < y \leq \frac{1}{10}\}$ **25** $x = -\sqrt{-1/y}$ $\{y < 0\}$ **27** 0 **29** $\frac{4}{5}$

31   **33**   **35**   **37**

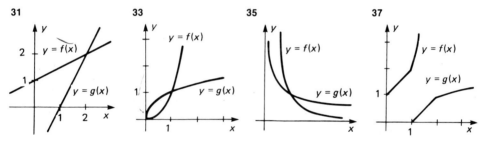

39 $x = -g(y)$ $\mathrm{dom}(-g) = \mathrm{range}(f)$

Section 3   page 164

1 $-\frac{1}{6}\pi$ **3** $-\frac{1}{4}\pi$ **5** $\frac{2}{3}\pi$ **7** $\frac{5}{6}\pi$ **9** 0.3 **11** $-\frac{1}{3}$
13 Set $\theta = \arcsin x$ so $-\frac{1}{2}\pi \leq \theta \leq \frac{1}{2}\pi$. Then $-\frac{1}{2}\pi \leq -\theta \leq \frac{1}{2}\pi$ and $\sin(-\theta) = -\sin\theta = -x$, hence $\arcsin(-x) = -\theta = -\arcsin x$.
15 Set $x = \cos\theta°$. Now $\theta° = (\pi\theta/180)$ rad, so $x = \cos(\pi\theta/180)$. But $0 \leq \pi\theta/180 \leq \pi$, hence $\pi\theta/180 = \arccos x$; that is, $\pi\theta = 180 \arccos x = 180 \arccos(\cos\theta°)$.
17 $-\frac{1}{2}\pi \leq \frac{1}{2}\pi - \theta \leq \frac{1}{2}\pi$ and $\sin(\frac{1}{2}\pi - \theta) = \cos\theta$, hence $\arcsin(\cos\theta) = \frac{1}{2}\pi - \theta$ **19** $55°$ **21** $\arccos\sqrt{1 - x^2} = -\arcsin x$
23 $\frac{2}{9}\sqrt{8}$ **25** $-\frac{16}{65}$

Section 4   page 166

1 0 **3** $-\frac{1}{6}\pi$ **5** 35 **7** -1
9 Set $\theta = \arctan x$ so $-\frac{1}{2}\pi < \theta < \frac{1}{2}\pi$. Then $-\frac{1}{2}\pi < -\theta < \frac{1}{2}\pi$ and $\tan(-\theta) = -\tan\theta = -x$. Hence $\arctan(-x) = -\theta = -\arctan x$.
11 $\cot(\arctan x) = 1/\tan(\arctan x) = 1/x$
13 $-\frac{1}{2}\pi < \frac{1}{2}\pi - \theta < \frac{1}{2}\pi$ and $\tan(\frac{1}{2}\pi - \theta) = \cot\theta$, hence $\arctan(\cot\theta) = \frac{1}{2}\pi - \theta$.
15 Set $\alpha = \arctan\frac{1}{2}$ and $\beta = \arctan\frac{1}{3}$ so $0 < \alpha < \frac{1}{4}\pi$ and $0 < \beta < \frac{1}{4}\pi$. We have $0 < \alpha + \beta < \frac{1}{2}\pi$ and $\tan(\alpha + \beta) = (\frac{1}{2} + \frac{1}{3})/(1 - \frac{1}{2}\cdot\frac{1}{3}) = 1$, hence $\alpha + \beta = \arctan 1 = \frac{1}{4}\pi$.
17 $\frac{3}{10}\sqrt{10}$ **19** $-\frac{5}{12}$ **21** $-\frac{2}{7}\sqrt{8} = -\frac{4}{7}\sqrt{2}$

23 $\cot \theta$ is strictly decreasing on $\{0 < \theta < \pi\}$ with range $\{$all $x\}$, hence it has an inverse, also strictly decreasing, with domain $\{$all $x\}$ and range $\{0 < \theta < \pi\}$.   **25**   Let $\theta = $ arc tan x, so $-\frac{1}{2}\pi < \theta < \frac{1}{2}\pi$ and hence $0 < \frac{1}{2}\pi - \theta < \pi$. Now $\cot(\frac{1}{2}\pi - \theta) = \tan \theta = x$, so $\frac{1}{2}\pi - \theta = $ arc cot x, that is, arc tan x + arc cot $x = \frac{1}{2}\pi$.

27   The function is strictly increasing, with range $\{|x| \geq 1\}$, hence has an inverse function with domain $\{|x| \geq 1\}$ and range $\{0 \leq \theta \leq \pi, \theta \neq \frac{1}{2}\pi\}$

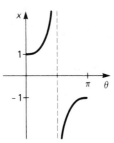

29   The function is strictly decreasing, with range $\{|x| \geq 1\}$, hence has an inverse function with domain $\{|x| \geq 1\}$ and range $\{-\frac{1}{2}\pi \leq \theta \leq \frac{1}{2}\pi, \theta \neq 0\}$

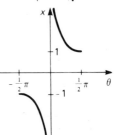

31   0.2527   **33**   1.2793   **35** -0.9582
37   6.91°   **39** $-89.81°$ **41**   90°
43

x	0.1	0.05	0.01
arc sin x	0.1001674	0.050020857	0.0100001667
$x + \frac{1}{6}x^3$	0.1001667	0.050020833	0.0100001667

Conclusion: arc sin $x \approx x + \frac{1}{6}x^3$ for $x \approx 0$
45 $\alpha + \beta = $ arc tan $\frac{1}{3}$ + arc tan $\frac{1}{2} = \frac{1}{4}\pi = \gamma$ by Ex. 15

Review Exercises   page 168

1 $x = \begin{cases} \frac{1}{3}y \\ \frac{1}{2}y \end{cases}$ for $\begin{cases} y \leq 0 \\ y \geq 0 \end{cases}$ **3** $-\frac{3}{14}\pi$
dom $= \{$all $y\}$ range $= \{$all $x\}$

5

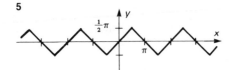

7 $g(y) = \pi - $ arc sin y

CHAPTER 8

Section 1   page 173

1 $\frac{1}{4}\pi$ $\frac{3}{4}\pi$ **3** ± 2.5185 **5** $\frac{1}{3}\pi$ $-\frac{2}{3}\pi$ **7** $\frac{1}{6}\pi$ $-\frac{5}{6}\pi$ $\frac{1}{3}\pi$ $-\frac{2}{3}\pi$
9 $\pm\frac{1}{12}\pi$ $\pm\frac{3}{4}\pi$ $\pm\frac{7}{12}\pi$ **11** $\frac{1}{6}\pi$ $\frac{5}{12}\pi$ $\frac{2}{3}\pi$ $\frac{11}{12}\pi$ $-\frac{1}{12}\pi$ $-\frac{1}{3}\pi$ $-\frac{7}{12}\pi$ $-\frac{5}{6}\pi$
13 ± 0.7782 ± 2.8726 ± 1.3162 **15**   0.4774 -2.6642 1.0934 -2.0482
17   210° 330°   **19**   30° 210°   **21** $\pm\frac{1}{4}\pi$ $\pm\frac{3}{4}\pi$ **23** ± 1.2310 **25** ± 1.9455
27 $-\frac{1}{4}\pi$ $\frac{3}{4}\pi$ 0.3218 -2.8198 **29** ± 1.2995 **31**   no solutions   **33**   0 π $\frac{1}{6}\pi$ $\frac{5}{6}\pi$
35   0 π **37**   0 π $\pm\frac{1}{2}\pi$ **39** $\frac{1}{4}\pi$ $-\frac{3}{4}\pi$ **41** $\pm\frac{2}{3}\pi$ **43**   0 $\pm\frac{2}{3}\pi$
45   all $\theta \neq n\pi, (\frac{1}{2} + n)\pi$ **47**   0 $\pm\frac{1}{2}\pi$ π **49**   0
51   0 $\pm\frac{1}{4}\pi$ $\pm\frac{1}{2}\pi$ $\pm\frac{3}{4}\pi$ π

Section 2   page 178
1 0.52256   **3** 0.52425   **5** 1.12862   **7** 0.5017   **9** 0.75487 767
11 0.85667 488

Section 3   page 182
1 1.5023   **3** 1.9346   **5** 0.6533   **7** 2.8064   **9** 6.5915   **11** 1.4973
13 0.4502

Review Exercises   page 183
1 $-\frac{7}{9}\pi$ $-\frac{1}{9}\pi$ $\frac{5}{9}\pi$ $-\frac{8}{9}\pi$ $-\frac{2}{9}\pi$ $\frac{4}{9}\pi$ **3** $\frac{2}{3}\pi n$ or $(2n + 1)\pi$
5 1.3987 -0.7966

CHAPTER 9

Section 1   page 189

1, 3, 5

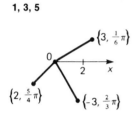

7

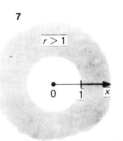

9

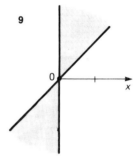

11 $(0,1)$ **13** $(\frac{1}{2}\sqrt{3}, -\frac{1}{2})$ **15** $(-\sqrt{2}, -\sqrt{2})$ **17** $\{\sqrt{2}, \frac{1}{4}\pi\}$ **19** $\{\sqrt{2}, \frac{3}{4}\pi\}$
21 $\{2, -\frac{1}{6}\pi\}$ **23** $\theta = \frac{1}{4}\pi$ **25** $r\cos(\theta - \frac{1}{4}\pi) = \frac{1}{2}\sqrt{2}$

Section 2   page 193
1 $r = 4\cos\theta$ **3** $r = 2\sin\theta$

5

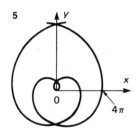

7

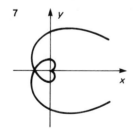

9

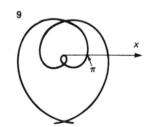

11

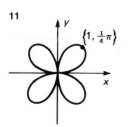

13

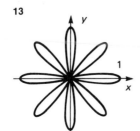

15

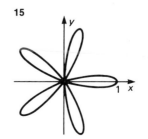

17
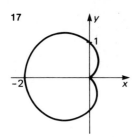

19
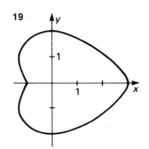

21
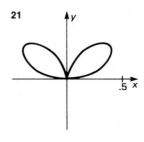

23
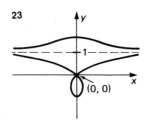

Section 3   page 199

1
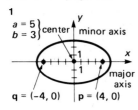

3
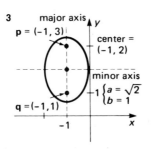

5
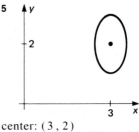

center: $(3, 2)$
foci: $(3, 2 \pm \frac{1}{2}\sqrt{2})$
vertices: $(3, 1), (3, 3),$
$\qquad (3 \pm \frac{1}{2}\sqrt{2}, 2)$
$a = 1, b = \frac{1}{2}\sqrt{2}$

7 $\frac{1}{81}(x - 1)^2 + \frac{1}{4}(y - 4)^2 = 1$ **9** $\frac{1}{25}(x - 5)^2 + \frac{1}{16}y^2 = 1$

11 $\frac{1}{16}(x - 1)^2 + \frac{1}{12}y^2 = 1$ **13** Describe the ellipse by $\mathbf{x} = (a \cos \theta, b \sin \theta)$. Then $x^2 + y^2 = (a^2 - b^2)\cos^2 \theta + b^2 \leq (a^2 - b^2) + b^2 = a^2$, with "=" only if $\cos \theta = \pm 1$.

15 From the figure, $(\frac{1}{3}x)^2 + (\frac{1}{6}y)^2 = 1$, so the answer is the first quadrant portion of the ellipse $\frac{1}{9}x^2 + \frac{1}{36}y^2 = 1$.

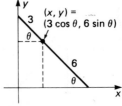

17 $\dfrac{(x + \frac{1}{2}a)^2}{(\frac{1}{2}a)^2} + \dfrac{y^2}{(\frac{1}{2}b)^2} = 1$; the ellipse with center $(-\frac{1}{2}a, 0)$, half the size of the original one, and parallel to it.

19

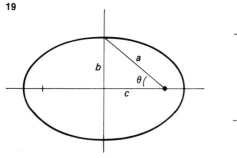

21 $e \approx 0.21$ **23** $x^2 + y^2 = \dfrac{(1 - t^2)^2 + (2t)^2}{(1 + t^2)^2} = \dfrac{(1 + t^2)^2}{(1 + t^2)^2} = 1$

Section 4   page 206

1

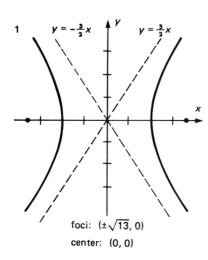

foci: $(\pm\sqrt{13}, 0)$

center: $(0, 0)$

3

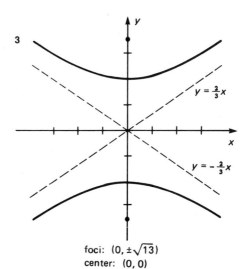

foci: $(0, \pm\sqrt{13})$

center: $(0, 0)$

5

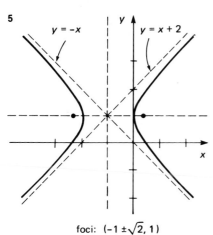

foci: $(-1 \pm \sqrt{2}, 1)$

center: $(-1, 1)$

7

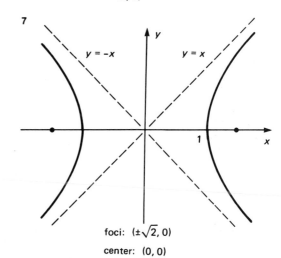

foci: $(\pm\sqrt{2}, 0)$

center: $(0, 0)$

9

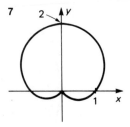

$y = 2x - 7$

$y = -2x + 5$

foci: $(3 \pm \sqrt{5}, -1)$
center: $(3, -1)$

11 $-\frac{1}{9}x^2 + \frac{1}{16}y^2 = 1$ **13** $\frac{1}{4}x^2 - \frac{1}{16}y^2 = 1$
15 $\frac{1}{3}(x - 1)^2 - \frac{1}{3}y^2 = 1$
17 Complete squares; the asymptotes are
$y - \frac{1}{2}b = \pm(x + \frac{1}{2}a)$, two perpendicular lines.
19

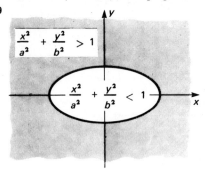

$$\frac{x^2}{a^2} + \frac{y^2}{b^2} > 1$$

$$\frac{x^2}{a^2} + \frac{y^2}{b^2} < 1$$

21 Let the circles be $\overline{xp} = r$ and $\overline{xq} = s$, with $\overline{pq} \geq r + s$. The desired locus is $\overline{xp} - r = \overline{xq} - s$; that is, $\overline{xp} - \overline{xq} = r - s$.

Section 5   page 213

1 $r = 2(\cos\theta + \sin\theta)$ **3** $r = 2\cos\theta + 2\sqrt{3}\sin\theta$ **5** $r = 10\cos\theta + 24\sin\theta$
7 $-\frac{1}{2}(\sqrt{2} - 1)\bar{x}^2 + \frac{1}{2}(\sqrt{2} + 1)\bar{y}^2 = 1$ **9** $\frac{1}{2}(\sqrt{2} + 1)\bar{x}^2 - \frac{1}{2}(\sqrt{2} - 1)\bar{y}^2 = 1$
11 ellipse $\alpha = \frac{1}{4}\pi$ $\frac{3}{2}\bar{x}^2 + \frac{1}{2}\bar{y}^2 = 1$ **13** hyperbola $\tan 2\alpha = \frac{1}{2}$
$\frac{1}{2}\bar{x}^2\sqrt{5} - \frac{1}{2}\bar{y}^2\sqrt{5} = 1$ **15** ellipse $\alpha = \frac{3}{8}\pi$ $(\frac{3}{2} + \frac{1}{2}\sqrt{2})\bar{x}^2 + (\frac{3}{2} - \frac{1}{2}\sqrt{2})\bar{y}^2 = 1$
17 hyperbola $\tan 2\alpha = -6$ $-(\frac{1}{2}\sqrt{37} - \frac{3}{2})\bar{x}^2 + (\frac{1}{2}\sqrt{37} + \frac{3}{2})\bar{y}^2 = 1$
19 $x^2 + y^2 = (\bar{x}^2\cos^2\alpha - 2\bar{x}\bar{y}\cos\alpha\sin\alpha + \bar{y}^2\sin^2\alpha)$
$\qquad + (\bar{x}^2\sin^2\alpha + 2\bar{x}\bar{y}\sin\alpha\cos\alpha + \bar{y}^2\cos^2\alpha)$
$\qquad = \bar{x}^2(\cos^2\alpha + \sin^2\alpha) + \bar{y}^2(\sin^2\alpha + \cos^2\alpha) = \bar{x}^2 + \bar{y}^2$.
This means that rotation does not change the length of a vector.
21 $\bar{a} + \bar{c} = (a\cos^2\alpha + b\cos\alpha\sin\alpha + c\sin^2\alpha)$
$\qquad + (a\sin^2\alpha - b\sin\alpha\cos\alpha + c\cos^2\alpha)$
$\qquad = a(\cos^2\alpha + \sin^2\alpha) + c(\sin^2\alpha + \cos^2\alpha) = a + c$.

Review Exercises   page 214

1 $\{\sqrt{2}, \frac{1}{4}\pi\}$ **3** $(1, -\sqrt{3})$ **5** $r = 6\cos\theta + 8\sin\theta$

7

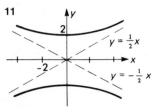

9

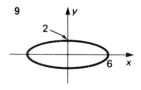

11

$y = \frac{1}{2}x$

$y = -\frac{1}{2}x$

13

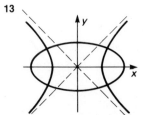

$(\frac{2}{5}\sqrt{10}, \pm\frac{1}{5}\sqrt{15})$
$(-\frac{2}{5}\sqrt{10}, \pm\frac{1}{5}\sqrt{15})$

CHAPTER 10

Section 1 page 219

1 $(0,3)$ **3** $(2,-2)$ **5** $(2,12)$ **7** $(3,0)$ **9** $(0,0)$ **11** $(1,1)$
13 $(10,4)$ **15** $(18,10)$

17
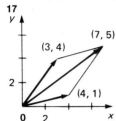

19
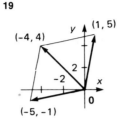

21 $(1,-10)$
23 $a = -2$ $b = 5$

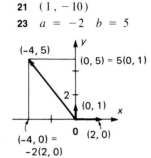

Section 2 page 225

1 13   **3** $\sqrt{61}$ **5** 6   **7** 0   **9** 3   **11** 31.35
13 $(2,3)\cdot(3,6) = 24$ $(2,3)\cdot(1,4) + (2,3)\cdot(2,2) = 14 + 10 = 24$
15 $(-3,4)\cdot(4,5) = 8$ $(-1,7)\cdot(4,5) - (2,3)\cdot(4,5) = 31 - 23 = 8$
17 $(12,15)\cdot(-1,2) = 18$ $3[(4,5)\cdot(-1,2)] = 3\cdot 6 = 18$
19 $\mathbf{v}\cdot\mathbf{w} = (v_1, v_2)\cdot(w_1, w_2) = v_1 w_1 + v_2 w_2 = w_1 v_1 + w_2 v_2 = (w_1, w_2)\cdot(v_1, v_2) = \mathbf{w}\cdot\mathbf{v}$
21 $(\mathbf{u} + \mathbf{v})\cdot\mathbf{w} = [(u_1, u_2) + (v_1, v_2)]\cdot(w_1, w_2) = (u_1 + v_1, u_2 + v_2)\cdot(w_1, w_2) = (u_1 + v_1)w_1 + (u_2 + v_2)w_2 = u_1 w_1 + v_1 w_1 + u_2 w_2 + v_2 w_2 = u_1 w_1 + u_2 w_2 + v_1 w_1 + v_2 w_2 = \mathbf{u}\cdot\mathbf{w} + \mathbf{v}\cdot\mathbf{w}$
23 $\|\mathbf{v} + \mathbf{w}\|^2 = (\mathbf{v} + \mathbf{w})\cdot(\mathbf{v} + \mathbf{w}) = \mathbf{v}\cdot(\mathbf{v} + \mathbf{w}) + \mathbf{w}\cdot(\mathbf{v} + \mathbf{w}) = \mathbf{v}\cdot\mathbf{v} + \mathbf{v}\cdot\mathbf{w} + \mathbf{w}\cdot\mathbf{v} + \mathbf{w}\cdot\mathbf{w} = \|\mathbf{v}\|^2 + 2\mathbf{v}\cdot\mathbf{w} + \|\mathbf{w}\|^2$
25 $\|\mathbf{v} + \mathbf{w}\|^2 - \|\mathbf{v} - \mathbf{w}\|^2 = (\|\mathbf{v}\|^2 + 2\mathbf{v}\cdot\mathbf{w} + \|\mathbf{w}\|^2) - (\|\mathbf{v}\|^2 - 2\mathbf{v}\cdot\mathbf{w} + \|\mathbf{w}\|^2) = 4\mathbf{v}\cdot\mathbf{w}$
27 $\frac{1}{2}\pi$ **29** arc cos $\frac{3}{5} \approx 53.13°$ **31** arc cos $\frac{3}{5} \approx 53.13°$
33 Take vertices $\mathbf{0}, \mathbf{v}, \mathbf{w}, \mathbf{v} + \mathbf{w}$, where $\|\mathbf{v}\| = \|\mathbf{w}\|$. Then one diagonal is $\mathbf{v} + \mathbf{w}$, the other is parallel to $\mathbf{w} - \mathbf{v}$. But

$$(\mathbf{v} + \mathbf{w})\cdot(\mathbf{w} - \mathbf{v}) = \mathbf{v}\cdot\mathbf{w} - \mathbf{v}\cdot\mathbf{v} + \mathbf{w}\cdot\mathbf{w} - \mathbf{w}\cdot\mathbf{v}$$
$$= -\|\mathbf{v}\|^2 + \|\mathbf{w}\|^2 = 0,$$

so the diagonals are perpendicular.
35 19.47°   **37** $\frac{1}{2}\mathrm{W}$ **39** 4.6 knots, 60° west of south

Section 3 page 231

1 $(\frac{1}{2}\sqrt{2}, -\frac{1}{2}\sqrt{2})\cdot\mathbf{x} = 0$ **3** $(\frac{3}{5}, -\frac{4}{5})\cdot\mathbf{x} = 0$ **5** $(\frac{3}{5}, \frac{4}{5})\cdot\mathbf{x} = \frac{2}{5}$ **7** $(\frac{12}{13}, \frac{5}{13})\cdot\mathbf{x} = \frac{1}{13}$
9 $(\frac{7}{25}, -\frac{24}{25})\cdot\mathbf{x} = 1$ **11** $(\frac{2}{13}\sqrt{13}, -\frac{3}{13}\sqrt{13})\cdot\mathbf{x} = \frac{4}{13}\sqrt{13}$
13 $(-\frac{2}{5}\sqrt{5}, \frac{1}{5}\sqrt{5})\cdot\mathbf{x} = \frac{1}{5}\sqrt{5}$ **15** $\dfrac{1}{\sqrt{a^2 + 1}}(-a, 1)\cdot\mathbf{x} = \dfrac{b}{\sqrt{a^2 + 1}}$
17 2   **19** $\frac{11}{5}$ **21** $\frac{13}{5}$ **23** $\frac{119}{13}$ **25** $\frac{1}{2}\sqrt{2}$ **27** $\frac{1}{4}\pi$ **29** arc cos $\frac{4}{5} \approx 0.6435$
31 arc cos $\frac{63}{65} \approx 0.2487$ **33** arc cos $\frac{3}{10}\sqrt{10} \approx 0.3218$ **35** $(\frac{2}{5}\sqrt{5}, \frac{1}{5}\sqrt{5})\cdot\mathbf{x} = \frac{2}{5}\sqrt{5}$
37 $(\frac{5}{34}\sqrt{34}, -\frac{3}{34}\sqrt{34})\cdot\mathbf{x} = \frac{7}{17}\sqrt{34}$

Review Exercises page 232

1 $(2,10)$ **3** $(3,-3)$ **5** $\frac{1}{2}$ 3   **7** $\frac{1}{3}\pi$
9 $(\frac{40}{41}, -\frac{9}{41})\cdot\mathbf{x} = 3$ **11** $\frac{39}{5}$

CHAPTER 11

Section 1   page 238

1 $9 + i$ **3** $-4 + 5i$ **5** $3 + 2i$ **7** $7 - i$ **9** $7 + 22i$ **11** $2i$
13 $\frac{3}{25} - \frac{4}{25}i$ **15** $1 + i$ **17** $\frac{22}{533} - \frac{7}{533}i$ **19** $-\frac{37}{145} - \frac{9}{145}i$ **21** $\frac{82}{85} + \frac{39}{85}i$
23 $3 - 3i$ **25** $8 - i$ **27** $-4 - 2i$ **29** 5 **31** $\sqrt{10}$ **33** $\sqrt{5} + \sqrt{17}$
35 $\frac{1}{3}\sqrt{5}$ **37** 1 **39** -1 **41** 1 **43** $(a + bi) + (c + di) =$
$(a + c) + (b + d)i = (c + a) + (d + b)i = (c + di) + (a + bi)$
45 $(a + bi)[(c + di) + (e + fi)] = (a + bi)[(c + e) + (d + f)i]$
$= [a(c + e) - b(d + f)] + [a(d + f) + b(c + e)]i$
$= [(ac - bd) + (ae - bf)] + [(ad + bc) + (af + be)]i$
$= [(ac - bd) + (ad + bc)i] + [(ae - bf) + (af + be)i]$
$= (a + bi)(c + di) + (a + bi)(e + fi).$
47 $\overline{a + bi} = \overline{a} - \overline{bi} = a + bi$ **49** $|a + bi| = \sqrt{a^2 + b^2} \ge 0$
51 $|a - bi| = \sqrt{a^2 + (-b)^2} = \sqrt{a^2 + b^2} = |a + bi|$
53 $|\alpha\beta|^2 = (\alpha\beta)(\overline{\alpha\beta}) = \alpha\beta\overline{\alpha}\overline{\beta} = (\alpha\overline{\alpha})(\beta\overline{\beta}) = |\alpha|^2|\beta|^2$, hence $|\alpha\beta| = |\alpha|\,|\beta|$.

Section 2   page 245

1 $3(\cos\frac{3}{2}\pi + i\sin\frac{3}{2}\pi)$ **3** $\sqrt{2}(\cos\frac{7}{4}\pi + i\sin\frac{7}{4}\pi)$ **5** $2(\cos\frac{7}{6}\pi + i\sin\frac{7}{6}\pi)$
7 $6(\cos\frac{1}{6}\pi + i\sin\frac{1}{6}\pi)$ **9** $2(\cos 0 + i\sin 0)$ **11** $4(\cos\frac{2}{3}\pi + i\sin\frac{2}{3}\pi)$
13 $\sqrt{3}, 3, 2\sqrt{3}, \frac{1}{3}\pi$ **15** $3\sqrt{2}$ **17** $\sqrt{65}$ **19** $2\sqrt{5}$
21 Set $\alpha = x + yi$ and $\beta = u + vi$, so
$\alpha\beta = (xu - yv) + (xv + yu)i$. Express $|\alpha|^2|\beta|^2 = |\alpha\beta|^2$ in terms of x, y, u, v.
The relation follows.
23

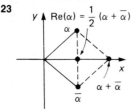

Section 3   page 248

1 $\sqrt{2}(\cos\frac{19}{12}\pi + i\sin\frac{19}{12}\pi)$ **3** $\frac{1}{4}\sqrt{2}(\cos\frac{17}{12}\pi + i\sin\frac{17}{12}\pi)$
5 $\frac{1}{2}(\sqrt{3} - 1), -\frac{1}{2}(1 + \sqrt{3}), \sqrt{2}, \frac{19}{12}\pi$ **7** $\frac{1}{2}\sqrt{2}$ **9** $\frac{2}{5}\sqrt{10}$ **11** $\frac{1}{29}\sqrt{290}$
13 $\pm(\frac{1}{2}\sqrt{2} - \frac{1}{2}i\sqrt{2})$ **15** $\pm\sqrt[4]{2}(\cos\frac{1}{8}\pi + i\sin\frac{1}{8}\pi)$ **17** $\pm\sqrt{2}(\cos\frac{11}{12}\pi + i\sin\frac{11}{12}\pi)$
19 $\pm\sqrt[4]{8}(\cos\frac{3}{8}\pi + i\sin\frac{3}{8}\pi)$
21

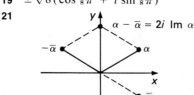

Section 4   page 253

1 $1 \pm 2i$ **3** $-\frac{1}{2} \pm \frac{1}{2}i\sqrt{23}$ **5** $\pm 2i, \pm i$ **7** $1, -\frac{1}{2} \pm \frac{1}{2}i\sqrt{3}$ **9** $\pm 1, \pm i$
11 $(x + 1)(x^2 - x + 1)$ **13** $z^5 + 1 = (z + 1)g(z)$ and $g(-1) = 5 \ne 0$, so
$z + 1$ is not a factor of $g(z)$. Hence $(z + 1)^2$ is not a factor of $z^5 + 1$.   **15** If $f(x)$ had

only irreducible quadratic factors, then $f(x)$ would have even degree. Hence $f(x)$ has a real linear factor $x - r$, so $f(r) = 0$.   **17** $a(x^2 - 4)(x^2 - 2x + 2)$ $a \neq 0$
19 $a(x^2 + b_1 x + c_1)(x^2 + b_2 x + c_2)$ $a \neq 0$, $b_1^2 - 4c_1 < 0$, $b_2^2 - 4c_2 < 0$
21 Set $\alpha = a + b\sqrt{2}$ $\beta = c + d\sqrt{2}$, a, b, c, d rational. Then
$\alpha \pm \beta = (a + b\sqrt{2}) \pm (c + d\sqrt{2}) = (a \pm c) + (b \pm d)\sqrt{2} \in$ **T**
$\alpha\beta = (a + b\sqrt{2})(c + d\sqrt{2}) = (ac + 2bd) + (ad + bc)\sqrt{2} \in$ **T**
23 $\alpha\alpha' = (a + b\sqrt{2})(a - b\sqrt{2}) = a^2 - 2b^2$. This is not zero, for if $a^2 - 2b^2 = 0$,
then $a^2 = 2b^2$, $\sqrt{2} = (a/b)^2$. But $\sqrt{2}$ is irrational. Now
$$(a + b\sqrt{2})\frac{a - b\sqrt{2}}{a^2 - 2b^2} = \frac{a^2 - 2b^2}{a^2 - 2b^2} = 1,$$
so $\alpha^{-1} = \dfrac{\alpha'}{\alpha\alpha'} = \dfrac{a}{a^2 - 2b^2} + \dfrac{-b}{a^2 - 2b^2}\sqrt{2} \in$ **T**

Section 5   page 259

1 $-32 - 32i$ **3** -64 **5** $-\frac{1}{2} + \frac{1}{2}i\sqrt{3}$ **7** $\cos 15° + i \sin 15°$
9 $2^{-11}(1 - i\sqrt{3})$ **11** $\cos \frac{11}{9}\pi + i \sin \frac{11}{9}\pi$ **13** $\pm 1, \pm i$
15 $\zeta_0 = 1, \zeta_1 = \frac{1}{2} + \frac{1}{2}i\sqrt{3}, \zeta_2 = -\frac{1}{2} + \frac{1}{2}i\sqrt{3}, \zeta_3 = -1, \zeta_4 = -\frac{1}{2} - \frac{1}{2}i\sqrt{3}$,
$\zeta_5 = \frac{1}{2} - \frac{1}{2}i\sqrt{3}$ **17** $2, -1 \pm i\sqrt{3}$ **19** $\frac{1}{2}\sqrt{10}(1 + i), \frac{1}{2}\sqrt{10}(-1 \pm i)$
21 $-i, \pm\frac{1}{2}\sqrt{3} + \frac{1}{2}i$ **23** $2i, \pm\sqrt{3} - i$
25 $\sqrt[6]{2}[\cos(\frac{1}{12}\pi + \frac{2}{3}\pi k) + i \sin(\frac{1}{12}\pi + \frac{2}{3}\pi k)]$ $k = 0, 1, 2$
27 $\cos(\frac{1}{9}\pi + \frac{2}{3}\pi k) + i \sin(\frac{1}{9}\pi + \frac{2}{3}\pi k)$ $k = 0, 1, 2$
29 $\beta_0 = \sqrt[4]{8}(\cos \frac{11}{24}\pi + i \sin \frac{11}{24}\pi), i\beta_0, -\beta_0, -i\beta_0$
31 $0 = \alpha^n - 1 = (\alpha - 1)(\alpha^{n-1} + \alpha^{n-2} + \cdots + \alpha + 1) = 0$ and $\alpha - 1 \neq 0$, so
$\alpha^{n-1} + \cdots + \alpha + 1 = 0$.   **33** $\alpha^5 = 1$ and $\alpha \neq 1$, hence
$\alpha^4 + \alpha^3 + \alpha^2 + \alpha + 1 = 0$, $\alpha^2 + \alpha + 1 + \alpha^{-1} + \alpha^{-2} = 0$. But
$\beta^2 = \alpha^2 + 2 + \alpha^{-2}$, so $(\beta^2 - 2) + \beta + 1 = 0$, $\beta^2 + \beta - 1 = 0$.
35 $\beta = \alpha + \alpha^{-1}$ $\alpha\beta = \alpha^2 + 1$ $\alpha^2 - \beta\alpha + 1 = 0$ **37** $(0, 0), (1, 1)$,
$(-\frac{1}{2} \pm \frac{1}{2}i\sqrt{3}, -\frac{1}{2} \mp \frac{1}{2}i\sqrt{3})$ **39** $\alpha^2 = 3 - 4i, \alpha^4 = -7 - 24i$. If z is any solution,
then $(z/\alpha)^4 = z^4/\alpha^4 = 1$, so z/α is a 4-th root of unity, hence equal to ± 1 or $\pm i$.
Therefore $z = \pm\alpha$ or $\pm i\alpha$.   **41** The 4-th roots of unity are $1, i, -1, -i$. Since
$1^1 = 1$ and $(-1)^2 = 1$, the roots 1 and -1 are not primitive. Since $i^1 \neq 1, i^2 \neq 1, i^3 \neq 1$,
it follows that i is primitive, and similarly $-i$ is primitive.   **43** The 6-th roots of unity
are listed in the solution of Ex. 15 above. Of these, $\zeta_0, \zeta_2, \zeta_3, \zeta_4$ are not primitive because
$$\zeta_0^1 = 1 \qquad \zeta_2^3 = 1 \qquad \zeta_3^2 = 1 \qquad \zeta_4^3 = 1.$$
Both $\zeta_1 = \frac{1}{2} + \frac{1}{2}i\sqrt{3}$ and $\zeta_5 = \frac{1}{2} - \frac{1}{2}i\sqrt{3}$ are primitive. Since $\zeta_k = \zeta_1^k$ for
$k = 0, \cdots, 5$, clearly ζ_1 is primitive. But $\zeta_5 = \zeta_1^{-1}$ is also primitive.
45 The 8-th roots of unity are
$$\zeta_k = \cos \frac{2\pi k}{8} + i \sin \frac{2\pi k}{8} \qquad k = 0, 1, \cdots, 7.$$
Of these, $\zeta_0, \zeta_2, \zeta_4, \zeta_6$ are the 4-th roots of unity, hence not primitive 8-th roots. The others,
$\zeta_1, \zeta_3, \zeta_5, \zeta_7$ are primitive. For
$$\zeta_3^3 = \zeta_1, \qquad \zeta_5^5 = \zeta_1, \qquad \zeta_7^7 = \zeta_1.$$
47 The 10-th roots of unity are $1, \zeta, \zeta^2, \cdots, \zeta^9$, where $\zeta = \cos \frac{2}{10}\pi + i \sin \frac{2}{10}\pi$. The
roots $1, \zeta^2, \zeta^4, \zeta^6, \zeta^8$ are 5-th roots of unity, hence not primitive, and $\zeta^5 = -1$ is also
not primitive. The others, $\zeta, \zeta^3, \zeta^7, \zeta^9$ are primitive. Note that
$$(\zeta^3)^7 = \zeta \qquad (\zeta^7)^3 = \zeta \qquad (\zeta^9)^9 = \zeta$$
49 The primitive 6-th roots of unity are $\frac{1}{2} \pm \frac{1}{2}i\sqrt{3}$ by Ex. 43. By the quadratic formula,
they are the roots of $z^2 - z + 1 = 0$.
51 Write $z^9 - 1 = (z^3 - 1)(z^6 + z^3 + 1)$. Each 9-th root of unity satisfies $z^9 - 1 = 0$.
Those which are cube roots of unity satisfy $z^3 - 1 = 0$; each of the other 6 roots satisfies
$z^6 + z^3 + 1 = 0$. (They are all primitive.)

Review Exercises   page 260

1 $-\frac{13}{10} - \frac{11}{10}i$ **3** $5 - i$ **5** i **7** $5(\cos\frac{3}{2}\pi + i\sin\frac{3}{2}\pi)$

9 -1, $\frac{1}{2} \pm \frac{1}{2}i\sqrt{3}$ **11** $2^{14}(\frac{1}{2} + \frac{1}{2}i\sqrt{3})$ **13**

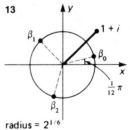

radius $= 2^{1/6}$

15 α^{-1} rotates clockwise on the circle $|z| = \frac{1}{2}$

17 Set $\omega = \dfrac{\gamma - \alpha}{\gamma - \beta}$. Then $|\omega| = \left|\dfrac{\gamma - \alpha}{\gamma - \beta}\right| = \dfrac{|\gamma - \alpha|}{|\gamma - \beta|} = 1$ and $\arg\omega = \pm 60° = \pm\frac{1}{3}\pi$,

hence

$$\omega = \cos\left(\pm\frac{1}{3}\pi\right) + i\sin\left(\pm\frac{1}{3}\pi\right) = \cos\left(\pm\frac{2\pi}{6}\right) + i\sin\left(\pm\frac{2\pi}{6}\right),$$

so ω is a 6-th root of unity (primitive, actually; see Ex. 43 of Section 5).

CHAPTER 12

Section 1   page 268

1 x^5y^4 **3** $a + \dfrac{1}{a}$ **5** $\dfrac{1}{b^{16}}$ **7** $\dfrac{-1}{16x^5y^2}$ **9** $\dfrac{1}{4^5x^{15}y^{10}z^5}$ **11** $\dfrac{2y^2}{x^2}$ **13** $\dfrac{b^2}{a^2}$

15 9 **17** $\frac{1}{3}$ **19** $3a^3$ **21** $5\sqrt{2}$ **23** $1/(2\sqrt{3}) = \frac{1}{6}\sqrt{3}$ **25** $3x\sqrt{2x}/(y + z)^2$

27 10 **29** $\frac{3}{4}$ **31** $\frac{1}{2}$ **33** $2\sqrt[4]{2}$ **35** $2ab^3$ **37** $2uv\sqrt[5]{2v}$ **39** x^2

41 $2y\sqrt{5xz}$ **43** $1/125x^6$ **45** u^3/v^9 **47** $x^{10}y^{15}/z^{20}$ **49** x^4/y^2 **51** $32x^2\sqrt{x}$

53 $2u - u^{1/6}$ **55** $\sqrt[6]{32}$ **57** 1 **59** $3/2\sqrt[3]{6a^2}$ **61** x^3 **63** $\sqrt[8]{a}$

65 $\sqrt[10]{3^5 \cdot 2^2}$ **67** $1/x^{1/5}y^{2/5}$ **69** $y^{9/20}/x^{3/2}$ **71** $1/x^{2/3}$ **73** 6.48×10^5 **75** 9.6×10^{-9}

77 6.4×10^{31} **79** 6.4×10^{-29} **81** 6.336×10^6 in.   **83** 4.8×10^{-18} grams

85 $275{,}854.74$ **87** 2.599273 **89** 1.7176×10^{-65} **91** $4.75682\ 85$

93 $1.79734\ 30$ **95** $1.07326\ 70 \times 10^8$ **97** 54.690036 **99** $1.25892\ 54$

101 2236.5045 **103** $0.04423\ 9617$ **105** $3.58338\ 94$ **107** $1.96157\ 06$

Section 2   page 273

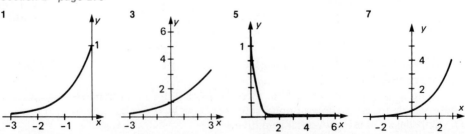

1   **3**   **5**   **7**

9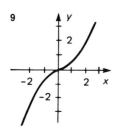

11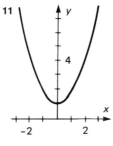

13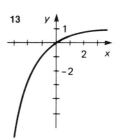

15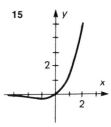

17 1 solution

19 2 solutions

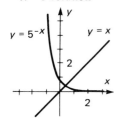

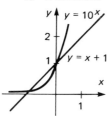

21 $f(x) = a^x$ by inspection **23** $2^{10/4.5} \approx 4.67$ times **25** (a) 155 °C (b) 25 °C (c) ≈ 38.46 °C **27** $(1.16)^{4+2/3} \approx 2.00$ **29** The 8% investment is worth \$1586.87, which is more than the 4% investment, worth only \$1539.45

Section 3 page 278

1

3

5

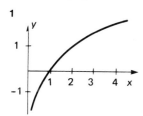

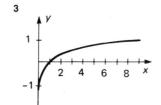

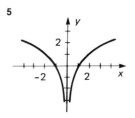

7 for $x = 10$, 10^2, 10^{10} **9** $\log_2 1024 = 10$ **11** $\log_8 \frac{1}{4} = -\frac{2}{3}$ **13** 7 **15** $\frac{4}{5}$ **17** $\frac{5}{2}$ **19** $\frac{1}{2}$ **21** 12 **23** -4 **25** 0 **27** -5 **29** 125 **31** 1 **33** 76 **35** $-\frac{1}{2}$ **37** $\pm\sqrt{19}$ **39** 7 **41** 2 **43** 0.7473 **45** -0.2400 **47** 1.5876 **49** -1.5681 **51** -0.8842 **53** $\log_a 48$ **55** $\log_a \frac{1}{3}$ **57** 3 **59** $\frac{1}{64}$ **61** -2 8 **63** Set $y_1 = \log_a x_1$ and $y_2 = \log_a x_2$. Then $x_1 = a^{y_1}$ and $x_2 = a^{y_2}$, so $x_1/x_2 = a^{y_1}/a^{y_2} = a^{y_1-y_2}$. Hence (2) follows. **65** $2 = 1 + 1$, but $\log_a 2 \neq 0 = \log_a 1 + \log_a 1$ **67** $\frac{1}{100} < x < \frac{1}{10}$

Section 4 page 283

1 0.377488 **3** 26.377488 **5** 1.144729 **7** 3.32193 **9** 5.65678 **11** 1.04822 **13** 1.10659 **15** 2303.74 **17** -0.990050 **19** 0.434294 **21** 2.01308×10^{10} **23** 0.224207 **25** 0.827087 **27** 0.333333 **29** 2.011155 **31** 4.83946×10^8 **33** 1 200.717 **35** 21.54435

Section 5 page 288

1 6.97 hr 9.97 hr **3** $N = 10^5 \times 10^{t/5}$ **5** 24 March 2004 **7** N triples in k units of time **9** 6.58×10^9 4.76×10^{10} 3.45×10^{11}

11 $k = 1.9806 \times 10^{11}$

	2000	2100	2200	ult.
	5.74×10^9	9.49×10^9	9.84×10^9	9.86×10^9

13 12.09 days **15** 0.693 sec **17** 2.19 days

Section 6 page 295

1 33 min **3** 2.59 min **5** 1.32×10^{10} times brighter **7** 6.3×10^5 times more intense
9 (a) $\log y = \log 3 + 2 \log x$ $v = 2u + \log 3$, a linear function (b) $y = cx^a$ $c > 0$
11 about 12,370 terms **13** $\approx 4.32 \times 10^{6001}$ **15** 1.57×10^{-4365} **17** $-0.86287\ 114$
19 $0.39901\ 298$ **21** $3.57715\ 21$ **23** $-0.82621\ 781$ **25** $-1.15413\ 266$
27 4.5364

Section 7 page 299

1

x	10	20	30	40	50	60	70	80	90	100
$2^x/x^{10}$	1.0×10^{-7}	1.0×10^{-7}	1.8×10^{-6}	1.0×10^{-4}	1.2×10^{-2}	1.9	4.2×10^2	1.1×10^5	3.6×10^7	1.3×10^{10}

3

x	1	2	3	4	5	6	7	8	9	10
$2^{-x}/x^{-2}$	0.5	1.0	1.1	1.0	.78	.56	.38	.25	.16	.098

5 167 **7** 10,000 **9** $2^x/x^{100} > 1$ for $x = 10^3$, $2^x/x^{100} > 10^{299,400}$ for $x = 10^6$
11 For $x = 10^{10}$, $(\log x)/x = 10^{-9}$. For $x = 10^{100}$, $(\log x)/x = 10^{-98}$
13 $x \log x \to 0-$ as $x \to 0+$ **15** $\log_2 x > \log_3 x$ for $x > 1$ **17** 1.0000053×10^{10}
on a Sharp EL-5806S

Review Exercises page 299

1 3^7 **3** $1.17073\ 53 \times 10^{83}$

5

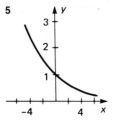

7 $x = \log_2(y - 1)$ dom $= \{y > 1\}$ range $= \{$all $x\}$

9

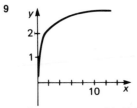

11

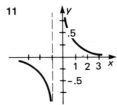

13 0 **15** $x \approx 21.434$ **17** to 4 places **19** 12 hr 1.2 min
21 3^x grows faster: $\dfrac{3^x}{2^{x+\log_2 x}} \to \infty$ as $x \to \infty$ **23** $1.39495\ 5$

INDEX

POLAR COORDINATES

$$\begin{cases} x = r \cos \theta \\ y = r \sin \theta \end{cases}$$

$$\begin{cases} x^2 + y^2 = r^2 \\ \\ \cos \theta = \dfrac{x}{r} \qquad \sin \theta = \dfrac{y}{r} \end{cases}$$

VECTORS

$$\mathbf{u} = (a, b) \qquad \mathbf{v} = (c, d)$$

$$\mathbf{u} + \mathbf{v} = (a + c, b + d)$$

Triangle inequality

$$\mathbf{u} \cdot \mathbf{v} = ac + bd = \|\mathbf{u}\| \cdot \|\mathbf{v}\| \cos \theta$$

$$\|\mathbf{u} + \mathbf{v}\| \leq \|\mathbf{u}\| + \|\mathbf{v}\|$$

$$\|\mathbf{u}\|^2 = a^2 + b^2$$

COMPLEX NUMBERS

$$i^2 = -1$$

$$(a + bi) + (c + di) = (a + c) + (b + d)i$$

$$(a + bi)(c + di) = (ac - bd) + (ad + bc)i$$

$$\overline{a + bi} = a - bi$$

$$|a + bi|^2 = a^2 + b^2$$

$$\overline{\alpha\beta} = \overline{\alpha}\,\overline{\beta}$$

$$|\alpha\beta| = |\alpha| \cdot |\beta|$$

Triangle inequality

$$|\alpha + \beta| \leq |\alpha| + |\beta|$$

Polar form $\quad \alpha = a + bi \neq 0$

$$\alpha = r(\cos \theta + i \sin \theta)$$

$$\begin{cases} a = r \cos \theta \\ \\ b = r \sin \theta \end{cases}$$

$$\begin{cases} r = \sqrt{a^2 + b^2} \\ \\ \cos \theta = a/r \qquad \sin \theta = b/r \end{cases}$$

De Moivre's Theorem

$$[r(\cos \theta + i \sin \theta)]^n = r^n (\cos n\theta + i \sin n\theta)$$

EXPONENTIALS AND LOGARITHMS

$$a^{x+y} = a^x a^y$$
$$(ab)^x = a^x b^x$$
$$(a^x)^y = a^{xy}$$
$$e \approx 2.71828\,183$$

$$\log_a (xy) = \log_a x + \log_a y$$
$$\log_a(x/y) = \log_a x - \log_a y$$
$$\log_a x^r = r \log_a x$$

Natural growth $\quad X = X_0 a^t \ (a > 1)$

Natural decay $\quad X = X_0 a^{-t} \ (a > 1)$